INTEGRATED PEST MANAGEMENT FOR

ALMONDS

SECOND EDITION

MW01285572

INTEGRATED PEST MANAGEMENT FOR

ALMONDS

SECOND EDITION

UNIVERSITY OF CALIFORNIA
STATEWIDE INTEGRATED PEST MANAGEMENT PROJECT

DIVISION OF AGRICULTURE AND NATURAL RESOURCES
PUBLICATION 3308
2002

PRECAUTIONS FOR USING PESTICIDES

Pesticides are poisonous and must be used with caution. READ THE LABEL BEFORE OPENING A PESTICIDE CONTAINER. Follow all label precautions and directions, including requirements for protective equipment. Use a pesticide only against pests specified on the label or in published University of California recommendations. Apply pesticides at the rates specified on the label or at lower rates if suggested in this publication. In California, all agricultural uses of pesticides must be reported. Contact your county agricultural commissioner for further details. Laws, regulations, and information concerning pesticides change frequently, so be sure the publication you are using is up-to-date.

Legal Responsibility. The user is legally responsible for any damage due to misuse of pesticides. Responsibility extends to effects caused by drift, runoff, or residues.

Transportation. Do not ship or carry pesticides together with food or feed in a way that allows contamination of the edible items. Never transport pesticides in a closed passenger vehicle or in a closed cab.

Storage. Keep pesticides in original containers until used. Store them in a locked cabinet, building, or fenced area where they are not accessible to children, unauthorized persons, pets, or livestock. DO NOT store pesticides with foods, feed, fertilizers, or other materials that may become contaminated by the pesticides.

Container Disposal. Dispose of empty containers carefully. Never reuse them. Make sure empty containers are not accessible to children or animals. Never dispose of containers where they may contaminate water supplies or natural waterways. Consult your county agricultural commissioner for correct procedures for handling and disposal of large quantities of empty containers.

Protection of Nonpest Animals and Plants. Many pesticides are toxic to useful or desirable animals, including honey bees, natural enemies, fish, domestic animals, and birds. Crops and other plants may also be damaged by misapplied pesticides. Take precautions to protect nonpest species from direct exposure to pesticides and from contamination due to drift, runoff, or residues. Certain rodenticides may pose a special hazard to animals that eat poisoned rodents.

Posting Treated Fields. For some materials, reentry intervals are established to protect field workers. Keep workers out of the field for the required time after application and, when required by regulations, post the treated areas with signs indicating the safe reentry date.

Preharvest Intervals. Some materials or rates cannot be used in certain crops within a specified time before harvest. Follow pesticide label instructions and allow the required time between application and harvest.

Permit Requirements. Many pesticides require a permit from the county agricultural commissioner before possession or use. When such materials are recommended in this publication, they are marked with an asterisk (*).

Processed Crops. Some processors will not accept a crop treated with certain chemicals. If your crop is going to a processor, be sure to check with the processor before applying a pesticide. Some chemicals may not be used if a portion of the crop will be used as animal feed.

Crop Injury. Certain chemicals may cause injury to crops (phytotoxicity) under certain conditions. Always consult the label for limitations. Before applying any pesticide, take into account the stage of plant development, the soil type and condition, the temperature, moisture, and wind. Injury may also result from the use of incompatible materials.

Personal Safety. Follow label directions carefully. Avoid splashing, spilling, leaks, spray drift, and contamination of clothing. NEVER eat, smoke, drink, or chew while using pesticides. Provide for emergency medical care IN ADVANCE as required by regulation.

ORDERING

For information about ordering this publication, contact:
University of California
Agriculture and Natural Resources
Communication Services
6701 San Pablo Avenue, 2nd floor
Oakland, California 94608-1239
Telephone 1-800-994-8849
(510) 642-2431
FAX (510) 643-5470
E-mail: danrcs@ucdavis.edu
Visit the ANR Communication Services website at
http://anrcatalog.ucdavis.edu
Publication 3308

Other books in this series include:
Integrated Pest Management for Walnuts, Publication 3270
Integrated Pest Management for Tomatoes, Publication 3274
Integrated Pest Management for Rice, Publication 3280
Integrated Pest Management for Citrus, Publication 3303
Integrated Pest Management for Cotton, Publication 3305
Integrated Pest Management for Cole Crops and Lettuce, Publication 3307
Integrated Pest Management for Alfalfa Hay, Publication 3312
Integrated Pest Management for Potatoes, Publication 3316
Pests of the Garden and Small Farm, Publication 3332
Integrated Pest Management for Small Grains, Publication 3333
Integrated Pest Management for Apples and Pears, Publication 3340
Integrated Pest Management for Strawberries, Publication 3351
Pests of Landscape Trees and Shrubs, Publication 3359
Integrated Pest Management for Stone Fruits, Publication 3389
Natural Enemies Handbook: The Illustrated Guide to Biological Pest Control, Publication 3386
Integrated Pest Management for Floriculture and Nurseries, Publication 3402
Integrated Pest Management in Practice, Publication 3418

ISBN 1-879906-52-X
Library of Congress Catalog Card No. 00-110969
© 2001 by the Regents of the University of California
Division of Agriculture and Natural Resources
First edition 1985
Second edition 2002
All rights reserved.
No part of this publication may be reproduced, stored in a retrieval system, or transmitted, in any form or by any means, electronic, mechanical, photocopying, recording, or otherwise, without the written permission of the publisher and the author.
Printed in Canada
3m-rev-12/01-LS/NS

To simplify information, trade names of products have been used. No endorsement of named or illustrated products is intended, nor is criticism implied of similar products that are not mentioned or illustrated.

The University of California prohibits discrimination against or harassment of any person employed by or seeking employment with the University on the basis of race, color, national origin, religion, sex, physical or mental disability, medical condition (cancer-related or genetic characteristics), ancestry, marital status, age, sexual orientation, citizenship, or status as a covered veteran (special disabled veteran, Vietnam-era veteran or any other veteran who served on active duty during a war or in a campaign or expedition for which a campaign badge has been authorized). University Policy is intended to be consistent with the provisions of applicable State and Federal laws. Inquiries regarding the University's nondiscrimination policies may be directed to the Affirmative Action/Staff Personnel Services Director, University of California, Agriculture and Natural Resources, 300 Lakeside Dr., 6th Floor, Oakland, CA 94612-3350; (510) 987-0096. **For information about ordering this publication, telephone 1-800-994-8849.**

This publication has been anonymously peer reviewed for technical accuracy by University of California scientists and other qualified professionals. This review process was managed by the ANR Associate Editor for Pest Management.

Contributors and Acknowledgments

Second edition revisions written by Larry L. Strand
First edition written by Barbara L. P. Ohlendorf
Photographs by Jack Kelly Clark
Mary Louise Flint, Technical Editor

Prepared by IPM Education and Publications, an office of the University of California Statewide IPM Project at Davis.

Technical Coordinators for Second Edition

Rex E. Marsh, Department of Wildlife, Fish, and Conservation Biology, University of California, Davis
Michael V. McKenry, Department of Nematology, University of California, Riverside; located at Kearney Agricultural Research Center, Parlier
Warren C. Micke, Department of Pomology, University of California, Davis
Timothy S. Prather, UC Statewide IPM Project, Kearney Agricultural Research Center, Parlier
Beth L. Teviotdale, Department of Plant Pathology, University of California, Davis; located at Kearney Agricultural Research Center, Parlier
Frank G. Zalom, UC Statewide IPM Project, University of California, Davis

Contributors to Second Edition

Entomology: Walter J. Bentley, Kent M. Daane, Lonnie C. Hendricks, Eric C. Mussen, Carolyn Pickel, Richard E. Rice, Frank G. Zalom

Nematology: Michael V. McKenry

Plant Pathology: James E. Adaskaveg, Greg T. Browne, Roger A. Duncan, Beth L. Teviotdale, Jerry K. Uyemoto

Pomology, Soil, and Water Relations: Patrick H. Brown, Joseph H. Connell, Warren C. Micke, Vito S. Polito, Wilbur O. Reil, Lawrence J. Schwankl, G. Steven Sibbett, Richard L. Snyder, Paul S. Verdegaal, Steven A. Weinbaum

Vegetation Management: Robert L. Bugg, Clyde L. Elmore, Bill B. Fischer, Kurt Hembree, Timothy S. Prather

Vertebrates: Richard A. Marovich, Rex E. Marsh, Terrell P. Salmon, Desley Whisson

Special Thanks

The following have generously provided information, offered suggestions, reviewed draft manuscripts, or helped obtain photographs: R. L. Coviello, S. H. Dreistadt, J. Edstrom, T. D. Eichlin, M. W. Freeman, C. E. Joshel, A. H. Purcell, K. A. Shackel, J. J. Stapleton, M. W. Stimmann, J. F. Strand, R. A. VanSteenwyk, and M. Viveros.

We would also like to acknowledge the important role of the contributors to the first edition of this manual, which was published in 1985: Martin M. Barnes, William W. Barnett, Walter J. Bentley, Richard M. Bostock, Joseph H. Connell, Robert K. Curtis, John E. Dibble, Clyde L. Elmore, W. Harley English, Bill B. Fischer, David A. Goldhamer, Marjorie A. Hoy, Dale E. Kester, Art H. Lange, Warren C. Micke, John M. Mircetich, George Nyland, Joseph M. Ogawa, Terry L. Prichard, Richard E. Rice, Terrell P. Salmon, G. Steven Sibbett, Richard L. Snyder, Beth L. Teviotdale, Mario Viveros, Craig V. Weakley, Frank G. Zalom.

Contents

Figure 1. Almond-growing areas of California.

Integrated Pest Management for Almonds

This manual is designed to help growers and pest control professionals apply the principles of integrated pest management (IPM) to almond crops in California.

> Integrated pest management is an ecosystem-based strategy that focuses on long-term prevention of pests or their damage through a combination of techniques such as biological control, habitat modification, modification of cultural practices, and use of resistant varieties. Pesticides are used only after monitoring indicates they are needed according to established guidelines, and treatments are made with the goal of removing only the target organism. Pest control materials are selected and applied in a manner that minimizes risks to human health, beneficial and nontarget organisms, and the environment.

An IPM program emphasizes anticipating and avoiding problems whenever possible, basing pest management decisions on established monitoring techniques and treatment guidelines, and using management methods that are least disruptive to the environment. In this manual, the term *pest* refers to insects, mites, pathogens, weeds, nematodes, and vertebrates that cause damage to almonds.

California is the primary place in North America where almonds are grown commercially. The majority of the state's 580,000 acres (235,000 ha) of almonds is in the Central Valley (Fig. 1). Almond orchards in California usually are irrigated by sprinkler, micro-sprinkler, flood, or drip systems. Major almond cultivars (varieties) are self-sterile. For this reason, more than one cultivar is planted in an orchard so the almond flowers can be cross-pollinated by honey bees as they transfer pollen from the flowers of one almond cultivar to the other.

Almond production is more successful in California's Central Valley than in other regions of the United States because the Mediterranean climate in the Central Valley is favorable to the crop's exacting requirements. Although almonds require winter chilling to break dormancy, the amount of chilling required is less than for most other deciduous fruit and nut crops. Almonds need considerable heat during the growing season to mature properly; almond trees usually produce best in the warm, dry interior valleys as long as irrigation is provided. Almonds bloom in February to early March, and from blossom time through April, the young nuts are susceptible to frost damage. Almond flowers are also susceptible to a variety of diseases during the bloom period, so rain should be minimal at this time and the days warm enough—55°F (13°C) or higher—for the bees to cross-pollinate the flowers. Rain at any time during the growing season can cause serious disease problems, and rain late in the season can interfere with harvest operations. This is why a Mediterranean-type climate with a rainy, mild winter followed by a warm, dry spring and summer is ideal for almond growing.

By mid to late March the trees have leafed out and small fruit can be seen. Fruit increase in size up to early May, when the embryo begins to appear. The embryo develops into the kernel during May and early June, and fruit maturation usually begins in July when hulls begin to split open. As the split widens, the hull separates from the shell. Hulls and kernels may dry either on the tree or on the ground after the nuts have been shaken from the trees. Almonds are ready to be shaken from the tree when 95 to 100% of the hulls have begun to split open. This generally occurs from early August to early October, depending on the cultivar and the location in the state where it is grown.

Disease management is very important during a rainy spring. Application of a selective insecticide in early spring is an effective alternative to dormant sprays of broad-spectrum insecticide for controlling peach twig borer, a major pest of almond. Other spring management activities include frost protection, crop pollination, surveying weeds, and monitoring vertebrate and insect populations.

The busiest management period in the orchard occurs during the warm part of the growing season. Activities include irrigation, weed management, sampling leaves for nutrient levels, and monitoring for insect, mite, disease, and vertebrate problems. As nuts begin to mature, the orchard floor must be prepared for harvest and steps must be taken to reduce crop loss to birds. Early harvest plays an important role in controlling damage by navel orangeworm, the most serious almond pest in California.

Nuts remaining on the tree after harvest should be removed because they provide overwintering sites for navel orangeworm and brown rot fungi. A good job of removing mummy nuts, combined with early harvest, can eliminate the need for insecticide sprays to control navel orangeworm. Other winter orchard management activities include pruning, monitoring for scale populations, applying dormant sprays and preemergence herbicides as needed, and surveying weeds.

This manual contains separate chapters on the five major categories of orchard pests: insects and mites, diseases, weeds, nematodes, and vertebrates. Introductory sections of these chapters explain where and when major pests occur and general monitoring guidelines. Each chapter gives detailed descriptions and photographs of individual pests, the damage they cause, and specific guidelines for enhancing natural control factors, designing a monitoring program, and using control actions effectively.

Introductory chapters on almond growth and development and general management practices provide the background upon which management guidelines in the pest chapters are based. This manual is intended to be used in conjunction with the *Almond Production Manual,* also published by the University of California Division of Agriculture and Natural Resources, listed in the suggested reading at the back of this book. The *Almond Production Manual* has more in-depth information on cultural methods, crop development, and specific crop requirements.

As knowledge about specific pests increases, and the interaction between these pests and the orchard environment is better understood, monitoring and control techniques will be modified and updated. Check regularly with your University of California Cooperative Extension farm advisor for new developments. Because pesticide registrations change frequently, few specific recommendations are given in this manual. For pesticide recommendations, new techniques, and other time-sensitive information related to pest management, see *UC IPM Pest Management Guidelines: Almond,* listed under "Pesticide Application and Safety" in the suggested reading.

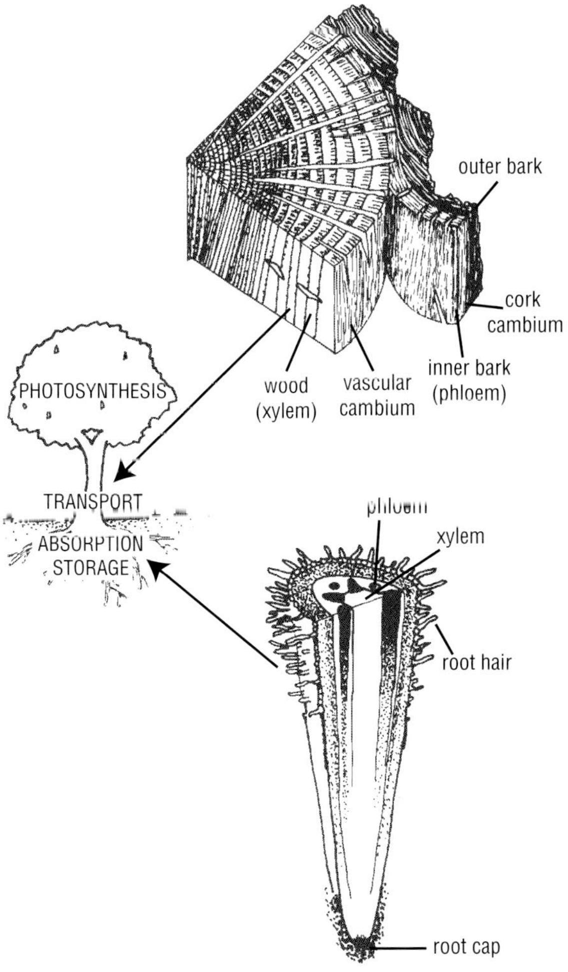

Figure 2. Basic tree structures and functions. Young roots absorb water and elements, which are transported in the xylem to the canopy, where leaves produce sugars in the process of photosynthesis. The tree uses sugars immediately for growth or transports them in the phloem mainly to the roots for storage.

The Almond Tree: Development and Growth Requirements

The objective of almond production is to harvest a large crop of high-quality kernels in the most cost-effective manner. The quantity or total yield of a crop is determined by the number of nuts produced and the size and weight of individual nuts. Each yield component is affected by different factors in the orchard environment, including pest management. The number of nuts produced on a tree, for example, depends on the number of flowers, the percentage of fruit set, and the proportion that matures. Because flowers are borne laterally on spurs and on shoots, adequate growth from terminal buds on spurs or growth of lateral shoots is essential to maintain high yields in consecutive years. Individual kernel size and weight depend on the cultivar, number of fruit per tree that reach maturity, and management practices.

Optimal almond production requires healthy trees. An orchard site must be properly prepared before planting to ensure that the trees get a good start. New trees require special irrigation, training, and pest control procedures during the nonbearing years while they are becoming established. Maintaining tree health once the orchard is established requires a basic understanding of the tree's seasonal cycle, the crop's specific requirements, and the impact that pest organisms and cultural and environmental factors have on tree and nut growth. Without this understanding, it is easy to overlook stress symptoms caused by the environment or to confuse them with pest damage. An important aspect of an IPM program is the evaluation of cultural practices and pest control methods for their impact on the total orchard system. In some cases, well-intended practices could compound tree stress and add to economic loss.

The Nonbearing Years

Basic tree structures and functions are illustrated in Figure 2. Almond trees are classified by the California Agricultural Statistics Service (CASS) as nonbearing until the fourth growing season (fourth leaf stage), or 3 years after planting, although some growers may harvest in the third growing season. During the nonbearing years, the major period of root growth occurs and the basic framework of the tree is developed. These years are critical to the future well-being and productivity of the tree.

The first year that the tree is in the ground is the most important for root development. Stress caused by disease, nematodes, weed competition, or insufficient irrigation greatly hinders root development and slows growth of the top, even in trees planted on resistant rootstocks. After the first few years, trees become more tolerant to stress.

The tree undergoes two to three growth flushes during its first year in the orchard. Like root development, however, the amount of top growth depends on weed competition, the availability of water, and other cultural and environmental factors. These growth flushes occur during spring and summer. By late summer, growth slows down.

After the first growing season, the tree is pruned (trained) to select the primary scaffold limbs, usually three. Shoot strikes (dead shoot tips) caused by the peach twig borer or oriental fruit moth can make it difficult to shape the tree at this time. Shoot strikes reduce the amount of terminal growth on branches and cause lateral buds to develop and grow. Excess lateral branch growth reduces the number of good branches to choose from when selecting the primary scaffolds.

Although tree growth during the second year also occurs in growth flushes, the major growing period may be reduced. Belowground growth of the roots takes place primarily in summer and fall. Another important training period follows the second growing season. At this time, primary branches are maintained and the secondary structure of the tree is selected, establishing its basic framework. Shoot strikes in the spring and early summer of the second year also create problems by reducing the amount of terminal growth and the number of good branches from which to choose the secondary structure.

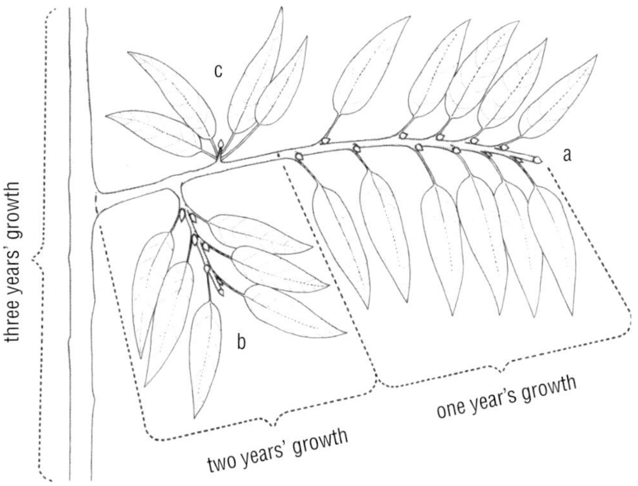

Figure 3. Schematic drawing of the three locations of vegetative growth: (a) From terminal bud of a long shoot (more than 25 nodes), lateral buds may be either vegetative or flower; if vegetative, they can produce shoot growth. (b) From terminal bud of elongated spur or short shoot (8 to 25 nodes), lateral buds mostly flower, sometimes vegetative. (c) From terminal bud of a spur (less than 8 nodes), lateral buds are flowers.

If the tree has adequate irrigation and other growing conditions are not limiting, flower buds are initiated in July on long shoots during the second or third growing season. Development within the flower bud proceeds during late summer and fall. These flower buds bloom the following season, and fruiting spurs then begin to develop on lateral shoots. If adequate shoot and spur development occurs during the second and third growing seasons, the tree usually becomes commercially productive by the fourth year. Spurs continue to develop laterally on these long shoots and bear fruit in subsequent growing seasons.

The Seasonal Cycle of Bearing Trees

In November, the almond tree enters a period of rest that lasts through December. During this time, the tree maintains a base level of water transport and starch consumption. In winter, the breakdown of starch increases sugar concentration in the cell sap, preventing the sap from freezing. To produce enough starch for this process, the tree must have been provided with sufficient water in the fall.

Developing vegetative and flower buds require adequate moisture and exposure to a certain amount of chilling in fall and winter to overcome the rest period. Once the chilling requirement has been met, bud growth begins when temperatures are favorable.

Flower buds usually begin expanding somewhat ahead of leaf and shoot buds, but this may vary with cultivar and season. Shoot growth begins from mid-February to mid-March—the time of flowering—and occurs at three locations on the twig (Fig. 3): from the terminal bud of a spur, from the terminal bud of a shoot, and from lateral buds of 1-year-old shoots. As leaves unfold and expand, photosynthetic production of carbohydrates starts, and growth is no longer entirely dependent on carbohydrates stored from the previous year.

The flowering stages for almonds are pictured in the Appendix at the back of this manual. Bloom time for a particular cultivar is important because almond flowers are not self-fertile and need cross-pollination. The amount of winter chilling they receive affects the degree to which the bloom periods for different cultivars overlap. In general, insufficient winter chilling results in a bloom period that is prolonged and straggly. Late-blooming cultivars usually show a delayed bloom, and they are affected more severely by insufficient winter chilling than are early blooming cultivars.

Pollination of almond flowers occurs when pollen grains are transferred from anthers of one cultivar's flower to the stigma of another cultivar's flower (Fig. 4). This work is done by insects; in commercial orchards, it is essentially all done by honey bees. It is critical for optimal crop set that enough bees are present in the orchard at the right time, because 50 to 60 pollen grains usually must be transferred for fertilization to be successful. To produce a harvestable nut, a pollen grain must germinate to form a pollen tube that grows down the style to

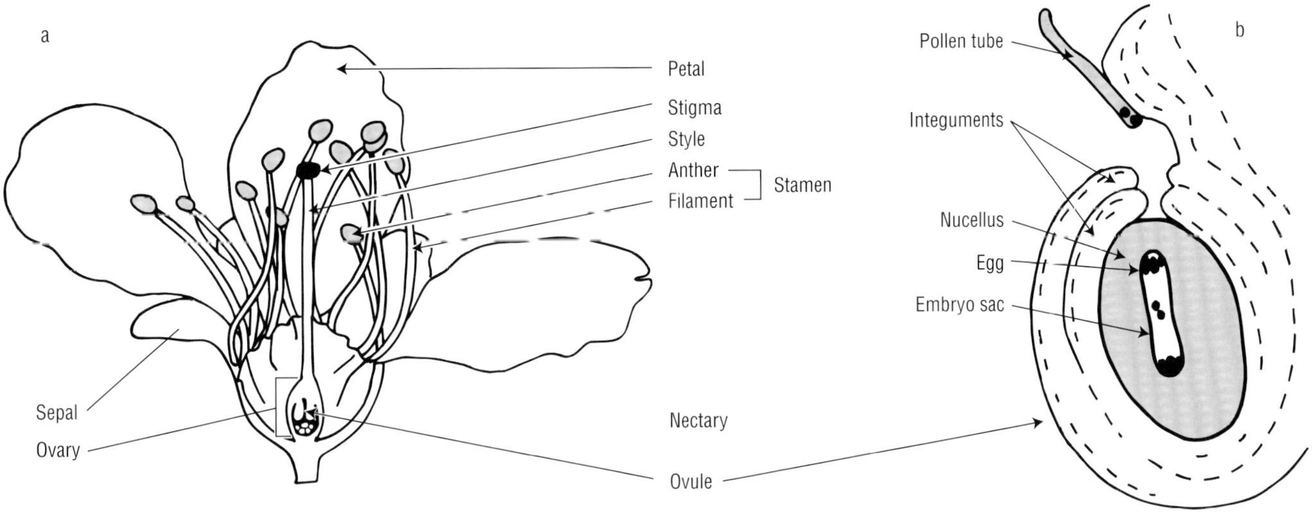

Figure 4. Almond flower and ovule: (a) Diagram of a Mission almond flower. (b) Enlarged diagram of a receptive ovule. The pistil of the flower contains two ovules. Typically, only one ovule becomes fertilized. The fertilized ovule develops into the kernel as the fruit matures. (Flower redrawn from *Insect Pollination of Cultivated Crop Plants*, S. E. McGregor, USDA Agricultural Handbook 496).

the ovary of the almond flower. Sperm are then released into the flower's embryo sac in the ovule to fertilize the egg.

Weather is important in the pollination and fertilization of the crop. Bees are not active when temperatures are below 55°F (13°C), in rainy weather, or when wind speed is at or above 12 miles per hour. Temperatures at which bees are active closely coincide with the temperature range that is best for growth of the pollen tube. If temperatures are too low, the pollen tube grows so slowly that the embryo sac may lose its viability before fertilization can take place. For more information, refer to *Almond Production Manual* and *Honey Bees in Almond Pollination*, both listed in the suggested reading.

A period of rapid hull and shell growth follows bloom and continues until early May. During this stage of growth, the potential size of the fruit (length and width) is established. The more fruit on a spur, the smaller the average nut size. Some of the small fruit will cease development and eventually drop. This loss is a natural thinning process that probably results from competition among too many fruit.

During fruit growth in April, the available energy of the tree is directed toward shoot and fruit growth. These two processes compete directly with each other for nutrients and water during this period, and both can be reduced by moisture stress or nutrient deficiencies.

As the fruit reaches full size in late April, the inner layer of the shell begins to harden, while the outer portion remains soft. This is a critical period with some cultivars because growth stress at this time can cause splits or cracks in the inner shell, while the outer shell, which is still soft, is capable of producing callus tissue. Cracks and splits may disrupt the conducting tissues in the nut and cause embryo abortion, gumming in the hull or kernel cavity, or corky layers. All of these conditions can lead to kernel loss or reduced quality.

The embryo or kernel begins to enlarge in early May. This process lasts until early June. Initially, the internal tissue of the embryo sac (nucellus) is clear and watery. Then a distinct translucent tissue, the endosperm, becomes evident at the tip of the nucellus. Within a few weeks, a tiny white embryo (kernel) develops within the endosperm (Fig. 5). Moisture stress or unseasonably cool, wet weather while the embryo is growing could result in an aborted or partially filled and shriveled kernel.

By early June, the kernel is formed, and dry weight accumulation begins in the kernel and continues until hull split. Because dry weight depends on the amount of carbohydrates produced by photosynthesis in leaves, good foliage and adequate moisture contribute to final kernel weight. Likewise, defoliation and moisture stress while the kernel is increasing in weight can adversely affect quality as well as final kernel weight and yield. Hardening of the shell is completed in June and July.

The approach of fruit maturity is signaled by the initiation of hull split—that is, separation of the hull along the suture and the separation (dehiscence) of the hull from the shell. The process of hull split is defined in six stages (Fig. 6). The first stage is the unsplit fruit (a). The second stage is divided into three parts and begins with the formation of a thin separation line at the suture depression that separates the two sides of the hull (b1). This suture line gradually deepens and extends from tip to base, but the hull cannot be squeezed open easily at the suture (b2). In the final part of Stage 2, the hull develops a separation layer. Although an opening may not be visible, the hull can be squeezed open by pressing both ends of the hull (b3). Soon the opening becomes visible (Stage 3, Fig. 6c), and the hull separates completely to expose the nut (Stage 4, Fig. 6d).

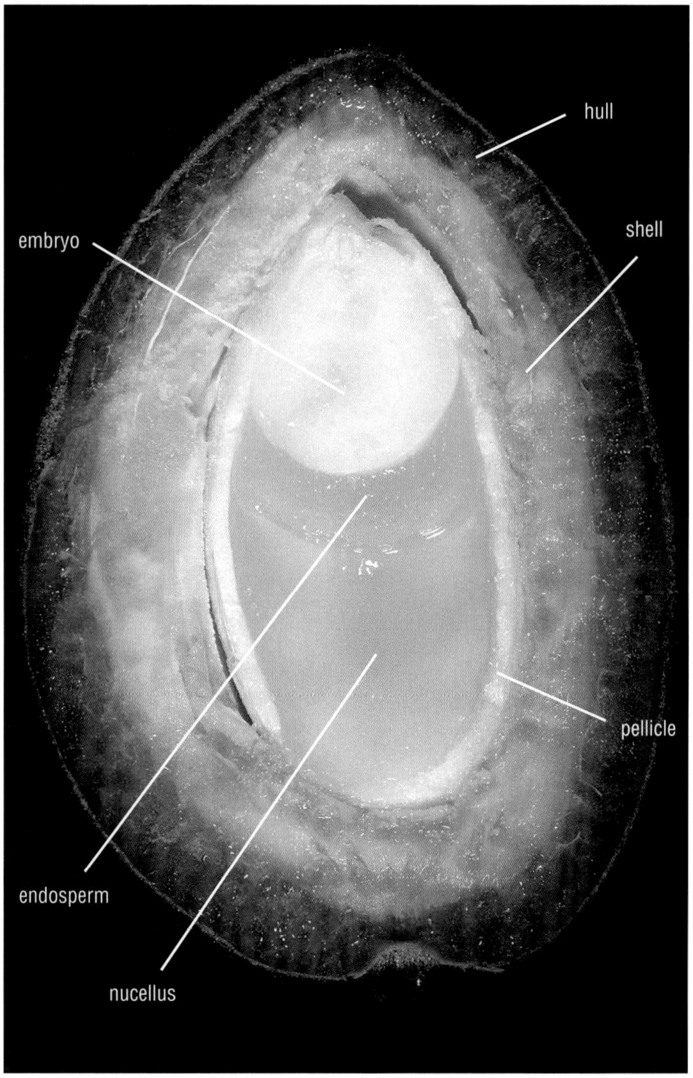

Figure 5. Cross-section of an immature nut showing developing embryo and endosperm.

At the same time that hull split is occurring, an abscission layer is being formed where the fruit attaches to the peduncle (fruit stem). Once this layer is completely developed, the fruit remains attached to the tree only by short fibers, which can be broken when the nuts are shaken or knocked.

As the abscission layer forms and the nut is exposed, the edge of the hull begins to dehydrate (Stage 5, Fig. 6e). With exposure and dehydration, the shell and the pellicle (skin of the kernel) change from white to brown. If the nuts are removed from the tree, dehydration and browning will continue on the ground until eventually the entire hull and nut are dried (Stage 6, Fig. 6f).

Hull split requires adequate tree moisture to proceed properly because the hull must be turgid. Moisture stress can change the sequence of ripening, triggering early abscission with or without hull split. Early-splitting hulls usually open improperly, or tighten on the shell, becoming "sticktights." On the other hand, hull split and abscission layer formation may be delayed in trees maintained under conditions that favor excessive vigor.

When hull split occurs is important in a pest management program because hull split exposes nuts to invasion by navel orangeworm, peach twig borer, and hull rot fungi. The longer the nuts remain on the tree after hull split, the more vulnerable they are. The exact time of hull split in an orchard is complicated, however, because ripening does not occur in all fruit on a tree at the same time. It begins in the upper and outer periphery of the tree and later extends to the lower and inner sections. The process of ripening throughout the tree may take 3 to 4 weeks.

Almonds should be harvested when 95 to 100% of the hulls have split (Stage 3 to 4) and adequate nut removal can be achieved. The start of almond harvest varies from year to year and district to district, but it usually begins in early to mid August. At this time, many of the fruit will have progressed past Stage 4, but some will still have hulls that are moist and green. However, these will dry rapidly on the ground when conditions are favorable. It is important to harvest at the earliest possible date to minimize navel orangeworm damage to nuts and to avoid complications caused by early rains. If the threat of a navel orangeworm infestation is severe, orchards can be harvested twice: once to remove the early-ripening nuts and a second time to remove the later-ripening ones.

While new fruit are maturing during summer, the tree is also initiating buds for next year's crop. As a result, stress during one growing season can have an impact on the next season's yields. Shoot bud initiation occurs mainly during April and the early part of May. Buds then enter a period of dormancy that lasts from June through mid-August, although they may be forced out of dormancy prematurely by severe water stress or defoliation followed by irrigation. In July to August, some of the buds in the leaf axils begin to differentiate into flower buds, while others remain as vegetative buds. About October, depending on the cultivar, buds enter a rest period that lasts through December; neither vegetative nor flower buds can be forced into growth during this period. A certain amount of chilling during this dormancy period is required for normal flowering.

During the growing season, sugar for the tree is synthesized in the leaves. By fall, the tree begins to store much of this

Figure 6. Stages of hull split: (a) Unsplit nut—no evidence of a separation at the suture. (b1) Initial separation—50% or more of a thin separation line visible. (b2) Deep V, unsplit—beginning of splitting with a deep "V" over at least 50% of the suture line, but the hull cannot be squeezed open at the suture. (b3) Deep V, split—a deep "V" in the suture, which is not yet visibly separated, but which can be squeezed open by pressing both ends of the hull. (c) Split, less than ⅜ inch (1 cm)—a visible opening of the suture less than ⅜ inch at midsuture. (d) Split, more than ⅜ inch—a visible opening of the suture greater than ⅜ inch at midsuture. (e) Initial drying—the edges of the hull are beginning to dry. (f) Completely dry. ▶

(a)

(b1)

(b2)

(b3)

(c)

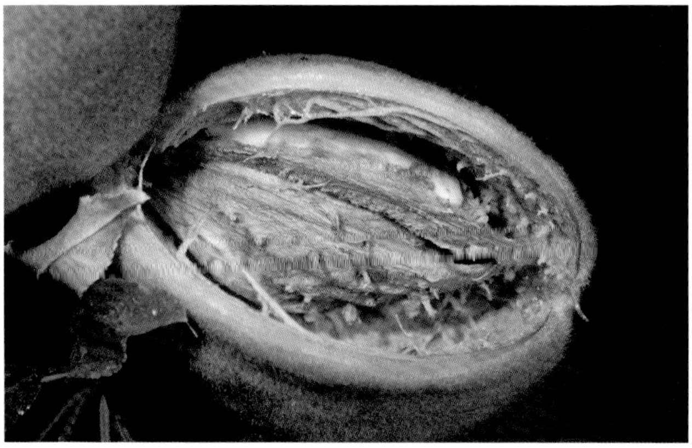
(d)

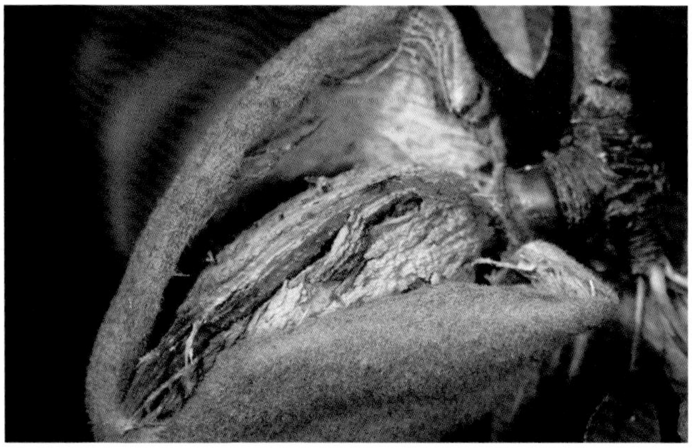
(e)

(f)

sugar as starch to be used for next year's growth flush. The tree also begins preparation for winter by decreasing its total moisture content and synthesizing proteins needed for the rest period.

Growth Requirements

Tree growth and fruit development require sufficient light, water, and nutrients, suitable soil, and adequate temperatures. Light is essential for photosynthesis, the process by which green plants manufacture sugars, their major food source (Fig. 7). Sugars are synthesized in the leaves from water taken from the soil, carbon dioxide taken from the air, and light energy from the sun. Sugar provides energy for tree growth, development, and maintenance and may be used immediately or stored as starch in leaves, stems, or roots.

Irrigation is necessary for profitable almond production in the semiarid climate of California. In the Central Valley, a mature almond orchard requires 30 to 60 inches (760–1,525 mm) per year of applied irrigation water, depending on rainfall, water use, and irrigation efficiency, for optimal yield of high-quality nuts. Only a small fraction of the water taken up by the tree is retained; most of it passes through the plant and out the leaf pores during transpiration. Transpiration is necessary to supply water to the leaves for photosynthesis, to carry nutrients to all parts of the tree, and to cool the plant. Climatic factors, such as day length and the amount of sunshine, humidity, heat, and wind, affect the rate of transpiration.

Trees need various nutrients for growth and nut production. California orchard soils may contain sufficient amounts of most essential nutrients, either because they are already present in the soil or are returned periodically through leaf decay. Relatively coarse-textured soils, such as sands and sandy loams, may require more frequent application of nutrients such as nitrogen (N) and zinc (Zn) to soil and foliage.

Almond trees grow best on deep, well-drained, nonstratified, coarse- to medium-textured soils. Under ideal soil conditions, roots may grow to depths of 9 feet (2.7 m) or more. Physical or chemical soil conditions that limit the rooting system or affect the health of roots have a direct influence on the trees' size, vigor, and crop potential.

Almond trees also grow best under warm temperatures but require a certain amount of winter chilling for uniform bloom and leaf production in the spring. Almonds are susceptible to extreme temperatures, which can cause frost damage to blossoms or young nuts in the spring or heat stress in summer.

Figure 7. In photosynthesis, green plants use energy from the sun to convert carbon dioxide from the air and water from the soil into sugars, their primary food source. In the process, oxygen is released into the air.

Managing Pests in Almonds

Integrated pest management (IPM) treats pests as part of a crop production system that includes not only the crop and its pests, but also the physical and biological environment in which the crop is grown. A good IPM program coordinates pest management activities with cultural operations to achieve economical and long-lasting solutions to pest problems while minimizing impact on the environment. The emphasis is on using monitoring techniques for optimal timing of control, developing a good prevention program, and selecting the best available control method based on knowledge of the pest's life history. Minimizing pest-caused crop losses, environmental damage, and risks to human health in the most cost-effective manner are the ultimate goals of IPM.

No single program works for all almond orchards. The best management program for a given location is determined by a number of factors including the almond cultivars being grown, soil characteristics, surroundings, local climate, and field history. The general discussion of management practices that follows will help you develop programs for your orchards.

An integrated pest management program has four major components:

- accurate identification of pests and beneficials
- field monitoring
- control action guidelines
- effective management methods

Figure 8 lists the cultural operations and pest management actions most likely to be used during the year. This summary also gives page numbers in this manual where details about a particular management practice can be found. Few growers will need to use all of these practices, and growers or their consultants will need to determine which are appropriate for each orchard.

Pest Identification

Correctly identifying pest species and stress symptoms is fundamental to choosing the most appropriate management strategy. Most pest management tools, including pesticides, are effective only against a select group of organisms; therefore, different practices may be needed even for closely related organisms. In some cases, symptoms caused by pest

MANAGEMENT ACTION	Page
WINTER (Nov–Jan)	
❑ Remove mummy nuts from trees and destroy	30, 64
❑ Prune trees	17
❑ Monitor prunings for scales and mite eggs	80, 94–95
❑ Apply dormant treatment as needed for peach twig borer, San Jose scale, and mite eggs OR apply dormant oil for scales andmite eggs followed by bloom treatment for peach twig borer	60, 76, 82, 94–95
❑ Monitor winter annual and perennial weed seedlings (Use record sheet, Figure 57)	168
❑ Monitor pocket gophers, control if needed	43
❑ Arrange to have beehives placed in orchard prior to bloom	25

MANAGEMENT ACTION	Page
SPRING (Feb–Apr)	
❑ Protect against frost as needed	18
❑ Mow ground cover prior to bloom and thereafter as needed	25
❑ Bloom and postbloom disease control	
1) Protect against brown rot as needed	128
2) Protect against anthracnose, scab, shot hole, and leaf blight as needed	131–134
3) Protect against green fruit rot, if needed	136
❑ Apply Bt or spinosad bloom sprays if no dormant spray were applied for peach twig borer and the pest is likely to cause damage	76
❑ Begin irrigation as needed	19–24
❑ Monitor for shot hole lesions and fruiting structures	130
❑ Monitor winter annual and perennial weeds (Use record sheet, Figure 56)	167
❑ Apply postemergence herbicides as needed	160–163
❑ Apply nitrogen fertilizer in April (first part of a split treatment)	24
❑ Begin monitoring insect and mite pests	55
1) Count mummy nuts remaining in trees	64
2) Monitor peach twig borer emergence from hibernacula	75–76
3) Set out egg traps for navel orangeworm and calculate degree-days	68–70
4) Set out pheromone traps for peach twig borer and calculate degree-days	76–78
5) Set out pheromone or sticky tape traps for San Jose scale if needed and calculate degree-days	80–83
6) Set out pheromone traps for oriental fruit moth, if needed	99
7) Monitor brown mites and European red mites, control if needed	94–95
8) Make an initial survey for ant colonies	97
9) Monitor for American plum borer, if needed	101
Monitor vertebrates (ground squirrels, pocket gophers), control if needed	39, 43

MANAGEMENT ACTION	Page
SUMMER (May–Jul)	
❑ Irrigate as needed	19–24
❑ Monitor summer annual weed seedlings and perennials (Use record sheet, Figure 57)	168
❑ Continue monitoring insects and mites	
1) Monitor webspinning and European red mites	88–94
2) Monitor ants after an irrigation	97–98
3) Monitor for American plum borer and other wood-boring insects if needed	100–103
❑ Have leaf analysis done (July)	24
❑ Monitor vertebrate activity (ground squirrels, birds, gophers), control if needed	39, 43, 51
❑ Monitor hull split for spray and earliest harvest date, and apply sprays for navel orangeworm if needed	5–7, 66–70
❑ Apply spray for San Jose scale and peach twig borer as needed, if not treated earlier in season	77, 82
❑ Prepare orchard floor for harvest	26
❑ Apply nitrogen fertilizer in July (complete application, or second part of a split treatment)	24

MANAGEMENT ACTION	Page
FALL (Aug–Oct)	
❑ Monitor webspinning mites	88–93
❑ Harvest early to avoid third navel orangeworm flight	66
❑ Pick up nuts as quickly as possible to minimize ant damage	97–98
❑ Have hulls analyzed for boron	24
❑ Monitor for shot hole lesions and fruiting structures	130
❑ Monitor vertebrates (ground squirrels, birds), control if needed	39, 51
❑ Apply potassium (soil) if needed	24
❑ Apply zinc spray at leaf fall if needed	24
❑ Irrigate if needed following harvest	19–24
❑ Apply preemergence herbicides	160–162
❑ Monitor for peach tree borer	100
❑ Remove weak trees, backhoe, and fumigate tree sites	116, 155
❑ Seed fall/winter cover crop if needed	181–185

Figure 8. Seasonal checklist of major management actions for an almond orchard and pages in this manual where activity is discussed in more detail.

organisms closely resemble those caused by nutrient deficiencies or soil problems. By making a habit of looking for arthropod and vertebrate pests, weeds, and disease symptoms, you will soon learn how to recognize them. You also need to recognize important beneficial organisms so you can assess the effectiveness of biological control.

The photographs, descriptions, and seasonal charts in this manual are intended to help identify pest problems and beneficial natural enemies. Other sources of information that will help in identifying pests are listed in the suggested reading at the back of this manual. You will need the assistance of experienced professionals to identify the causes of some problems. Do not hesitate to seek their help if you are not sure what is causing a problem. UCCE farm advisors (listed under "University of California Cooperative Extension Service" or "Cooperative Agricultural Extension, University of California" in the County Government listing of many local telephone directories), county agricultural commissioners, and pest control professionals can help and can direct you to other specialists when necessary.

Orchard Monitoring

Regular monitoring of each orchard gives you critical information for making pest management decisions. A good monitoring program involves surveys throughout the year to keep track of orchard conditions, cultural practices, and development of pest populations. You can use the information you gather to assess the orchard's performance and predict potential problems. Check regularly for pests, natural enemies, the maturity and health of the crop, soil moisture, and weather. To save time, plan to include pest monitoring whenever you are in the orchard for routine cultural practices.

Keep records of your monitoring results; they will help you forecast pest outbreaks and schedule cultural practices. Tables or graphs of pest counts will help you identify population patterns. Maps of pest damage or weed populations will help you identify localized problems and pest distribution. Records of monitoring results, such as total seasonal trap catches for insect pest species, are invaluable for long-term planning. They tell you what to expect, and which management techniques have been effective and which have not.

Monitoring Pests

The frequency of monitoring varies with the individual pest species and time of season. Monitoring programs for most species are needed at certain times of the year. Start monitoring well before populations begin to build and continue through the pest's damaging stages. Sample every orchard for insect and mite pests and their natural enemies at least once a week when pests are likely to be present and more frequently when insect and mite populations can be expected to flare up.

No weekly sampling programs have been developed to assess populations of weeds, diseases, nematodes, and vertebrates. It is important, however, to monitor their presence and watch for changes in their status. Prepare a record once each quarter (fall, winter, spring, and summer) for each orchard, noting the dominant weeds, diseases, vertebrates and any other significant stresses. Correct identification is essential. Sample for nematodes before planting the orchard and when you suspect they are causing losses.

Specific monitoring procedures are described in the discussions of individual pests.

Monitoring Weather

Weather greatly influences the development of the almond tree and its pests. Temperature controls the rate at which insects, mites, and diseases develop, and wetness from rainfall or fog is a primary factor favoring development of fruit and foliage diseases. A reliable source of weather information is critical to many pest management decisions. Keep track of daily high and low temperatures if you are using degree-day accumulations to schedule monitoring and management activities for pests such as peach twig borer and San Jose scale. Use evapotranspiration and rainfall data to schedule irrigations and calculate water requirements. Weather forecasts are essential for scheduling protective treatments for most blossom, foliage, and fruit diseases.

Many newspapers and radio stations provide local weather information. The National Weather Service broadcasts local and regional weather on NOAA Weather Radio, VHF channels 162.42, 162.50, and 162.55 MHz. Evapotranspiration information is available from the California Department of Water Resources' CIMIS program. Daily weather information for a number of locations throughout California is available from the University of California Statewide Integrated Pest Management Project (UC IPM). (See the suggested reading for ways to access CIMIS and UC IPM weather information; both of these programs are available via the World Wide Web.)

Significant local variations often occur in weather conditions, especially in temperature and rainfall. For the most accurate weather data, set up your own weather station in or near your orchard (Fig. 9). Weather instruments can range from simple devices such as a maximum-minimum thermometer to electronic devices that continuously monitor and record weather information for transfer to a computer. Even more sophisticated stations transmit the data to a remote computer. Some relatively simple devices are available that keep track of temperatures and accumulated

Figure 9. A weather station placed in or near your orchard will give you the most accurate weather information for making management decisions.

degree-days. Set up and maintain the weather instruments according to manufacturers' instructions, and calibrate them regularly to ensure accuracy. Keep records of all your observations.

Accumulating Degree-Days

The concept of degree-days (DD) is important in understanding crop and pest development. Currently, degree-days are used to forecast the development of navel orangeworm, peach twig borer, San Jose scale, and oriental fruit moth.

The growth rates of the tree and its invertebrate and microbial pests are closely related to temperature. In warmer years, pests such as peach twig borer appear sooner and generations develop more quickly than in colder years. Therefore, the calendar date alone is not a precise guide for carrying out crop and pest management activities. For more accurate predictions of crop and pest development, time and temperature have to be considered together. More accurate predictions of a pest's development can be made by measuring the amount of heat the pest is exposed to over time. Degree-day accumulations are used to measure this, and they are calculated using mathematical models developed for each pest.

Each species of plant and insect has specific lower and upper developmental thresholds. Degree-day models assume that no development occurs at temperatures below the lower threshold. Between the lower and upper thresholds, the development rate increases in a roughly linear fashion as temperature increases. For temperatures above the upper threshold, degree-day models may use a horizontal cutoff, which assumes that the development rate remains constant, or a vertical cutoff, which assumes that development stops above the upper threshold. A degree-day is defined as the area under the temperature-time curve, between the lower and upper developmental thresholds, equal to 1° X 1 day (24 degree-hours). To use degree-days to keep track of crop or pest development, you need to know how many degree-days are required to complete each growth stage or generation as well as the temperature thresholds for that specific organism.

The UC IPM World Wide Web site (http://www.ipm.ucdavis.edu) has programs that accumulate degree-days using data provided directly to UC IPM from remote weather stations. At the present time, degree-day calculations can be used to forecast the development of navel orangeworm, peach twig borer, oriental fruit moth, and San Jose scale.

You can estimate degree-days based on daily maximum and minimum temperatures using a computer program or by hand using a reference table. Electronic instruments that record temperatures hourly or more frequently often include software to compute degree-days. Degree-days estimated from daily maximum and minimum temperatures can differ widely from those computed from temperature data collected at more frequent intervals. Therefore, if you are calculating degree-days to schedule management activities for a pest, it is important that you use the same technique as was used to develop the predictive model for that pest.

Control Action Guidelines

Growers and pest control professionals use control action guidelines to help them decide when management actions are necessary to prevent losses to pests or to other stresses. Some guidelines are numerical thresholds based on specific sampling techniques, such as leaf analysis for nutrient levels or counts of insect numbers. Most guidelines require that you consider field history, stage of almond development, presence of pests or damage, weather conditions, and other observations. Control action guidelines may change as new cultivars and cultural practices are introduced and as new information on pests and management techniques becomes available.

Management Methods

The best IPM program prevents pest outbreaks and provides long-term, economical control. Preventive aspects of a good program include selecting a deep, well-drained orchard site with light- to medium-textured soil, properly preparing it before planting, choosing cultivars and rootstocks with some resistance to potential problems whenever possible, removing sources of new infection, choosing the appropriate irrigation and orchard floor management methods, and timing various cultural activities to discourage pest development and keep trees growing vigorously. Pesticides should be used only when careful orchard monitoring indicates they are needed to prevent economic loss, or if orchard history indicates preventive sprays are necessary.

The following sections briefly describe common cultural practices associated with almond production. For more information, consult the suggested reading.

Site Selection and Preparation

Because healthy, vigorous trees are best able to withstand external stresses, almond trees should be planted where they will be able to develop well above and below the ground. Many site problems, if they are detected before planting, can be overcome by careful preplant preparations, including soil modification, planting on berms or mounds, eradicating perennial weeds, fumigating, and choosing rootstocks or cultivars that offer some resistance to site-related problems.

A number of activities undertaken before you begin to prepare the planting site will help you plan your pest management program and will keep some potential problems from developing.

- Consult field records for cropping history, cultural practices, pesticide use, and problems with pests and soil conditions.
- Survey the field for weeds and plan appropriate management strategies. Infestations of perennials are easier to control at this time.
- Collect samples of irrigation water for analysis of nitrate and salinity.
- Collect soil samples for analysis of nematode populations.
- Check for soil compaction.
- Survey adjacent areas for pests that may move into the orchard, especially vertebrate pests such as rabbits, deer, and voles that can seriously damage newly planted orchards. Nearby abandoned or unmanaged orchards may be sources of pests that you will want to take precautions against, such as navel orangeworm, peach twig borer, and wood borers.

Several points need to be considered when choosing the location for your orchard, including the soil conditions, local weather history, availability of an adequate supply of good-quality irrigation water when you will need it, the cropping history and past pest problems of the site, and the availability of skilled labor to help manage and harvest your crop.

Almond trees grow on a wide variety of soils but produce best in deep, well-drained soils. Almonds usually do not do well on heavy, poorly drained, or stratified soils that restrict root growth unless micro-irrigation is used. Modify problem soils before planting by deep ripping soils that have a hardpan, or by slip plowing or backhoeing stratified soils that have no hardpan. If a clay pan is present, backhoe individual tree sites or slip plow the orchard.

Fine-textured, poorly drained soils provide ideal conditions for the development of Phytophthora crown and root rot, which requires soil saturation or standing water to reproduce and infect host plants. Avoid planting on these soils when possible, especially in areas of higher rainfall (Sacramento Valley). If problem areas must be planted, consider using resistant rootstocks, planting on berms or mounds, and carefully leveling the soil to prevent water from ponding in some parts of the orchard. Sites that have a history of periodic flooding should not be used for orchards.

Avoid sites that have a history of Armillaria root rot (oak root fungus). The pathogen can survive for many years in dead or living roots of fruit or nut trees, oak, aspen, and several fir and willow species. Once a tree is infected, little can be done to save it. If your orchard site has a history of Armillaria root rot, you will need to treat the soil before planting (see management guidelines for Armillaria root rot in the chapter "Diseases"), and plant your trees on Marianna 2624 rootstock. Even these precautions may not prove successful, however.

Preplant soil treatments may also be necessary in orchards where the soil contains root knot, root lesion, ring, or dagger nematodes. For additional information, see the chapter "Nematodes."

Once a site is selected, prepare the land in the fall before spring planting. Not only is proper land preparation important for optimal root growth by the trees, it is also important for optimal performance of fumigants and herbicides used in the orchard and to minimize potential phytotoxic effects of the chemicals on the trees. Fall is the best time for land preparation because land leveling, subsoiling, backhoeing, and slip plowing need to be done when the soil is dry. Land preparation in the fall also allows adequate time for the ground to resettle before planting. The land should be prepared early enough in the fall to allow time for fumigation and other activities. Any low areas with poor drainage need to be corrected. Irrigation systems should be designed at this time. It is best to install the irrigation system before or at the time of planting, although some systems can be installed after the trees are planted.

Soil Solarization

Soil solarization is a soil-heating procedure that involves covering prepared soil with special plastic for several weeks or months during the warmest, sunniest time of the year. This allows the sun to heat the moist soil to temperatures that are lethal to many weed seeds and some soilborne pathogens. It is a nonchemical technique that may be used to control a wide variety of weed species and Verticillium wilt. Some nematode pest populations may be reduced by soil solarization, but the technique does not control root knot nematodes. Weed control with solarization is discussed in the chapter "Vegetation Management."

You can solarize the soil before planting or lay down the plastic at planting and leave it in place for a year or more. Use clear plastic if you solarize before planting: it heats the soil to higher temperatures. If you begin solarizing at the time of planting, use black plastic; young trees are harmed by the higher soil temperatures reached under clear plastic. For solarization to be effective, soil moisture must be at field capacity. On coarser-textured soils, you may be able to preirrigate to fill the soil profile before laying plastic. On finer-textured soils, you will need to wait and apply water after placing the solarization plastic in order to avoid soil compaction from the application equipment. You can use drip lines or gated pipe placed underneath the plastic to bring soil moisture to field capacity after laying the plastic. It is difficult to use flood or furrow irrigation underneath solarization plastic, and these irrigation methods are not recommended where water must be applied after plastic is in place.

The preplant solarization treatment should last at least 6 weeks during the hottest and sunniest time of the year: July and August in the Central Valley. You will realize the greatest benefit if you solarize the soil all summer. Use plastic that is resistant to UV light; although it is more expensive, it will resist deterioration and make mechanical removal easier, especially if you are solarizing for an extended period. Working in a green manure cover crop before solarizing, especially a crop of a cereal grain or a mustard family plant, enhances the effect of solarization on soilborne pathogens.

If you lay solarizing plastic at planting, lay one sheet of black plastic on each side of the tree row, cut holes around the trees, seal the inner edges, and cover the outer edges with a layer of soil in the row middle. Drip irrigation must be used with this solarization technique. You can place drip lines on top of the plastic and run drip tubes to each tree, poked through the plastic. This makes it easier to check on the functioning of the drip tubes, since the plastic will remain in place for 1 to 3 years. The soil is heated to lower temperatures than under clear plastic, but it remains heated for a longer time. This procedure is very effective for controlling weeds in the tree rows, and it can reduce the trees' water requirement by 80% or more. If you use this type of solarization, cut the amount you irrigate by 80% and check the soil moisture regularly to be sure the root zone does not dry out. If soil moisture under black plastic is maintained at too high a level, tree roots may be severely injured. Trees tend to begin blooming sooner if the solarization plastic remains in place.

You can find more information about soil solarization in the publication *Soil Solarization: A Nonpesticidal Method for Controlling Diseases, Nematodes, and Weeds*, listed in the suggested reading.

Cultivar and Rootstock Selection

The potential for pest damage is greatly influenced by the choice of rootstocks and scion cultivars. At present, five rootstocks are available for use with almond trees. Each has advantages and disadvantages associated with its use (Fig. 10). Before planting an orchard, the soil properties and the history of the site should be thoroughly investigated so that the selection of a rootstock takes into consideration potential soil and pest problems along with horticultural concerns such as scion compatibility.

The five rootstocks used in almonds are almond, Lovell peach, Nemaguard peach, Marianna 2624 (plum), and peach-almond (PA) hybrid. Myrobalan plum and apricot rootstocks were used for almonds in the past, but both were unsatisfactory. Lovell peach and Nemaguard peach are the most commonly used rootstocks of the five available. Nemaguard peach is used primarily where root knot nematode is a problem in sandy soils. Lovell peach, on the other hand, is used mainly in well-drained loam to clay loam soil, in areas north of Sacramento, and in areas where root knot nematodes have not been serious. Marianna 2624 is tolerant of wet soil conditions and is resistant (but not immune) to oak root fungus and root knot nematodes. It is compatible with cultivars such as Mission, Padre, Ne Plus Ultra, Peerless, Thompson, Carmel, Merced, Price, and Jordanolo; apparently compatible with Aldrich, Fritz, LeGrand, Norman, Ripon, Ruby, and probably Wood Colony; but incompatible with Nonpareil, Solano, Milow, Mono, Dottie Won, Livingston, and some older cultivars such as Kapareil and Drake. The compatibility of Marianna 2624 rootstock with

Figure 10. Susceptibility of major rootstocks to pathogens, nematodes, and nutritional disturbances.

Susceptibility[1] to—	Almond	Lovell peach	Nemaguard	Marianna 2624	Peach-almond hybrid[2]
bacterial canker[3]	SR	SR	SR	VS	SR
Phytophthora crown and root rot	ES	MS	S	SR	VS
Armillaria root rot	S	S	S	SR	S
crown gall	VS	S	S	SR	S
Verticillium wilt	S	S	S	SR	S
root knot nematode	VS	S	R	R	SR
other nematodes[4]	S	S	S	S	S
nutritional disturbances	5	6	6	—	7
peachtree borer	S	VS	ES	S	—
pocket gophers and mice	VS	SR	SR	S	—

1. ES = extremely susceptible, VS = very susceptible, S = susceptible, MS = moderately susceptible, SR = somewhat resistant, R = resistant, — = no data.
2. Variable in response depending on the particular hybrid used. Not well documented.
3. Susceptibility of scions grown on this rootstock.
4. Other nematode pests of almonds include root lesion nematode (*Pratylenchus vulnus*), ring nematode (*Mesocriconema xenoplax*), and dagger nematode (*Xiphinema americanum*).
5. More resistant to excess boron and lime-induced chlorosis than other rootstocks listed.
6. More sensitive to calcareous soils and high boron or chloride than trees on almond rootstock.
7. More tolerant of lime, sodium, and chloride than Lovell peach, and somewhat resistant to excess boron.

some of the newer cultivars is not clear at this time; check with your nurseryman. Finally, trees on Marianna 2624 are smaller than those on peach rootstock and may be shorter-lived. Because of those limitations, Marianna 2624 often is used in spot situations where other rootstocks would not survive, such as on heavier soils or where Armillaria root rot has occurred. If an entire orchard is to be planted on Marianna 2624, space the trees 10 to 20% closer together than you would for other rootstocks.

Trees planted on almond or on peach-almond hybrid rootstocks are suited to a limited range of soil types, and the range for peach-almond hybrid rootstocks may vary depending on the kind of hybrid.

While once important, almond rootstock seldom is used now. When it is used, trees on almond rootstock should be planted in deep, well-drained soils where root knot nematode is not a problem. Because of the deep-rooting characteristic of this rootstock and its susceptibility to Phytophthora crown and root rot, trees on almond rootstock must be planted on land where the water table is deep and drainage is good. Almond rootstock provides more drought tolerance than peach or plum rootstocks.

Peach-almond hybrid rootstocks produce large, vigorous, deep-rooted trees. These trees are somewhat more tolerant of drought conditions during late summer and fall and of calcareous soils than trees on peach rootstock, and their deep root systems provide excellent anchorage. However, trees on this rootstock may become so vigorous when grown in deep, fertile soil that framework limbs tend to split during the development period, and the trees' larger size could become a problem for harvest and general orchard management. In addition, trees on this rootstock should not be planted where conditions are favorable for crown rot. For these reasons, trees on peach-almond hybrid rootstock seem best suited for weaker soils, coarse-textured soils, soils high in lime, locations where severe moisture stress occurs at harvest, or windy areas where excellent anchorage is important. They require good drainage because they cannot tolerate wet conditions. Padre sometimes has shown incompatibility symptoms on peach-almond rootstock. Young trees on peach-almond hybrid rootstock are more sensitive to problems related to cold storage.

Many almond scion cultivars have been introduced during the past 25 years, and additional cultivars will be forthcoming in the future. The major scion cultivars currently growing in California include Nonpareil, Carmel, Butte, Mission, Price, Monterey, Fritz, Peerless, Sonora, Ne Plus Ultra, Ruby, and Padre. When planning an orchard, consider the disease and insect susceptibilities of the different scion cultivars (Fig. 11) along with horticultural considerations such as bloom time, time and ease of harvest, nut quality, and marketability. If the orchard is located in an area where a particular disease is known to be damaging, select cultivars, when available, that have some resistance to that disease. Frequently, the growth characteristics of the tree or properties of the nut can influence the potential for pest damage. Navel orangeworm and peach twig borer, for example, are most damaging to cultivars with nuts that have poor shell seals or soft shells and to cultivars that have an extended hull split period. Finally, the cultivars and rootstocks chosen for an orchard (as well as the soil they are planted in) may influence the selection of herbicides that are used in the orchard. For example, Mission trees are more sensitive to simazine damage than other cultivars, particularly if they are growing in sandy soils that are low in organic matter.

Planting and Managing a New Orchard

When you receive young trees from the nursery, keep in mind the following points.

- Do not allow the roots to dry out or to be exposed to freezing temperatures.
- If you cannot plant the trees immediately, heel them in damp soil or damp sawdust, but do not cover the tree trunks above the depth at which they grew in the nursery.
- Do not let the heeling-in area become waterlogged or the trees may develop bundle rot (see Phytophthora crown and root rot in the chapter "Diseases.").
- Keep the roots moist but not wet, and avoid large air pockets around the roots.
- Make sure the soil you use to heel-in the trees is free of disease organisms. If you fumigated your orchard site, use soil that has been fumigated and is free of herbicides. Take the same precautions with the soil placed around the roots when you plant the trees.
- Plant the trees as soon as soil conditions permit.

Tree hole sites are generally marked and dug with an auger. In sandy to loam soils it is often nearly as fast, and in some cases better, to dig the holes with a shovel. Be sure not to dig the holes any deeper than is necessary to cover the roots. Trees planted in holes that are too deep may settle in the soil, favoring development of crown rot. Because trees will settle deeper into the soil after irrigation, plant them high, with the uppermost root just under the soil surface. After settling, the trees should stand no deeper than they did in the nursery.

If the soil is too moist, the auger may cause glazing or sealing of the sides of the hole. Or, if the hole was dug a day or so before planting, a crust may form on the sides and bottom of the hole. If either situation has occurred, be sure to break down the sides of the hole before planting to prevent confinement of the tree roots or poor water penetration. Trees should be watered the same day they are planted, unless the soil moisture is high. "Tanking" with two to three gallons (7.5–11.5 l) of water per tree is usually sufficient. Dry soil, air pockets, and poor soil contact resulting from insufficient water can kill newly planted trees. Even a rain immediately after planting, unless it's quite heavy, is not sufficient to settle the soil around the root system. However, trees can

Susceptibility to	Butte	Carmel	Fritz	Merced	Mission	Monterey	Ne Plus Ultra	Nonpareil	Padre	Peerless	Price	Ruby	Sonora	Thompson
almond leaf scorch	R	R	R	SR	SR	—	S	VS	SR	VS	R	R	VS	R
Alternaria leaf spot	VS	VS	VS	S	S	SR	—	VS	SR	—	S	S	VS	—
anthracnose	S	S	VS	VS	S	VS	VS	SR	S	S	VS	—	—	VS
brown rot	VS	VS	S	VS	R	S	VS	SR	S	S	VS	VS	VS	VS
ceratocystis	S	VS	S	S	VS	S	S	VS	—	S	S	S	VS	VS
foamy canker	VS	VS	—	—	S	—	S	S	—	S	S	—	—	—
hull rot	S	SR	SR	SR	SR	SR	SR	VS	SR	SR	SR	SR	VS	SR
leaf blight	S	S	VS	VS	S	—	VS	S	—	VS	—	—	—	S
scab	S	VS	S	VS	S	S	VS	S	—	S	VS	VS	VS	S
shot hole	S	S	S	VS	S	—	VS	VS	S	VS	S	S	S	VS
Verticillium wilt	S	VS	S	SR	S	S	S	S	—	S	S	S	VS	S
yellow bud mosaic	R	R	R	R	VS	—	SR	SR	—	R	R	R	—	R
spider mites	VS	VS	VS	VS	SR	—	VS	VS	SR	S	S	S	VS	S
navel orangeworm/ peach twig borer	SR	SR	S	VS	R	SR	S	VS	R	R	S	SR	S	VS

Key: VS = very susceptible S = susceptible SR = somewhat resistant R = resistant — = no information

Figure 11. Susceptibility of major scion cultivars to diseases and other pests.

also be lost from excess moisture caused by tanking already moist soils or excessive rain occurring after tanking. If you use water to settle the soil around a tree, do not firm the soil further by stepping on it. This compacts the soil and may interfere with root development and water penetration around the base of the tree.

At planting time, trim off any roots that are smashed, broken, or too long to fit in a reasonable-sized planting hole without being bent. After trimming, it's a good idea to dip the remaining root system in a solution of commercially available biological control agent for crown gall, especially if the site has a history of the disease.

After planting the trees, head them back to 32 to 36 inches (80–90 cm) and place a tree protector or a quart milk carton around the tree. If you don't plan to use a tree protector or milk carton, plant the trees so that the bud union is on the southwest side to help reduce sunburn on the stock. You can also protect the tree trunks from sunburn by painting the trunks with white interior latex paint, especially if the trees are planted late. It is usually best to use both trunk protectors and white paint. Where prevailing winds are a problem, plant trees at a slight angle into the wind.

Planting on Berms or Mounds. On heavier soils, trees survive better when planted on berms or mounds. Planting on berms or mounds improves drainage away from the crown of the tree and helps reduce the risk of crown rot. Berms reduce the chances of tree damage due to certain herbicides and also reduce the amount of weeds around tree trunks because berms remain drier during irrigation.

Prepare the berm before planting on ground that first has been leveled to provide a slight slope. The berm should be about 2 to 3 feet (60–90 cm) wide and 6 to 8 inches (15–20 cm) higher after settling than the surrounding soil, unless irrigation practices dictate an even higher berm. On uneven ground, where berms are not feasible, plant the trees on mounds. Make the mound about 3 feet (90 cm) square and about 6 inches (15 cm) high. Plant the tree on top of the berm or mound as discussed previously, making sure trees do not settle too deep.

Irrigating a New Orchard. If the soil has good moisture at planting time, work the soil around the roots by hand. If the soil is dry at planting, apply 2 to 3 gallons (7.5–11.5 l) of water to each tree at that time. Do not irrigate again until after the trees have grown about 4 to 6 inches (10–15 cm).

Be sure to supply adequate water to trees during the first year. Use frequent, light irrigations, particularly on shallow or sandy soils. Almond trees planted in a deep soil that has been wetted to field capacity by rains or irrigation may need less frequent irrigations the first year, provided the soil is kept absolutely weed free to eliminate competition for water. Under such circumstances, trees are able to get part of the water they need as the root system expands into new moist soil. Lack of readily available moisture can seriously reduce the growth of newly planted trees. If trees remain dry for even a short time during the summer, they may be weakened, sunburned, and/or attacked by wood borers. Dry trees may not begin growing again even if subsequently irrigated.

During the remaining nonbearing years, adequate soil moisture is the most important consideration in obtaining good tree growth. When the soil has not been wetted adequately by winter rains, you may want to give a partial irrigation to the orchard before the start of the growing season. In areas of normally high rainfall, be careful that an early irrigation aggravated by subsequent rain does not induce root and crown rot. Control weeds to reduce competition for soil moisture and nutrients.

If you irrigate with surface or sprinkler systems, plan your last irrigation for early August in the Sacramento Valley or late August in the northern San Joaquin Valley. If you irrigate first- and second-year trees much later than late August in the Sacramento Valley or mid-September in the northern San Joaquin Valley, the trees may be killed by Phytophthora crown and root rot. Substantial early fall rains can cause similar damage. (Fall irrigations in the southern San Joaquin Valley, where rainfall is considerably lower than the northern areas, do not generally cause crown rot problems.) It is safer to keep a young orchard weed-free in the fall and to allow the trees to use moisture remaining in the soil from earlier irrigation than to risk a Phytophthora infection by irrigating too late.

Pruning (Training) a New Orchard. Prune the tops of newly planted trees by topping the trunk with one cut 32 to 36 inches (80–90 cm) above ground. The primary scaffolds, which develop within 6 to 10 inches (15–25 cm) of this cut, will form the tree crotch. Because mechanical shakers are commonly used for harvest, you must provide sufficient unbranched trunk space for shakers to attach between the soil and the lowest scaffold. If lateral branches are present, they can be removed; if they are in desirable locations, they can be headed back, leaving 1 to 2 inches (2.5–5 cm) of growth with one or two lateral buds.

After the first growing season, usually three primary scaffold limbs are selected by pruning. These limbs should be spaced evenly around the tree and several inches apart vertically. Because almond trees normally branch readily, the primary scaffold limbs can be left unheaded or tipped at 36 to 42 inches (91–107 cm) from the crotch. Select limbs so that the main scaffold does not form a pocket that will collect rain or irrigation water.

The secondary structure is established following the second growing season. Secondary limbs usually are not headed unless the tree needs balancing or additional branching is desired. Two secondary limbs usually are allowed to develop on each primary. Secondary scaffold limbs should not be allowed to develop less than 30 inches (75 cm) and preferably at least 36 inches (90 cm) from the crotch to allow room for shaking scaffold limbs if the trunk becomes too large for shaking.

Fertilizing a New Orchard. During the first growing season, trees frequently benefit from light applications of nitrogen fertilizer. Apply 1 to 2 ounces (28–56 g) of actual nitrogen per tree after growth is started in either a single application (late May) or split application (May and early July). You can place the fertilizer in the irrigation furrows, broadcast it around the tree if using sprinkler or flood irrigation, or place it under emitters if using drip or micro-sprinklers. However, always keep it at least 18 inches (46 cm) away from the trunk. Do not put fertilizer in the holes at planting time. Have your irrigation water tested for nitrate, and take this into account when calculating nitrogen applications.

Intercropping. In young orchards, an annual crop is sometimes grown between the tree rows for 1 to 2 years to obtain additional income. These crops must be chosen with great care or the income may be negligible and the almond trees may be damaged. The welfare of the trees must always come first.

If the orchard is intercropped, do not use crops that require the soil to be dried out during the summer for harvesting. Examples of such crops are grains and oat hay. Drying can seriously stunt almond trees. In addition, if the drying crop catches fire, the orchard trees can be seriously damaged or destroyed. Many broadleaf crops, including cotton, lettuce, and alfalfa, are good hosts for root knot nematodes. If nematode populations build up on these hosts, they can invade and feed on the roots of Nemaguard rootstock, causing significant damage even though they cannot reproduce on the resistant rootstock. Mites also can be serious in intercropped orchards, especially if beans are planted as the intercrop. The pathogen that causes Verticillium wilt, which can damage young almond trees, usually builds up in the soil when tomatoes, cotton, or related crops are grown. Therefore, these crops may not be good intercrops. However, cotton has been used successfully in new orchards planted on old cotton ground, but care must be taken not to damage trees with cotton defoliant. Union mild etch, a viruslike disease that affects trees on Marianna 2624 rootstock, occurs more frequently in orchards that are intercropped and overirrigated.

Any herbicide or other pesticide you use on the intercrop may damage the almond trees. Moreover, because of the presence of the intercrop and because of residue problems on it, you may not be able to use the most suitable pest control material on the almond trees. Finally, care must be taken that the equipment used for the intercrop does not damage the trees.

Pest Management in Nonbearing Orchards. Once the orchard is planted, protect young trees from sunburn and wood borer damage by wrapping the trees with a protective tree wrap, painting the trunk of the trees with white interior latex paint, or both. Because the crown of the tree is extremely susceptible to sunburn and wood borer damage, dig down slightly below the soil line and paint exposed surfaces. After painting, place the soil back around the trees, making sure the painted surface extends below the soil line.

Monitor your trees in spring for the presence of disease and pest problems. Although it usually is not necessary to control diseases such as brown rot or shot hole the first year, the trees may need to be protected from peach twig borer. Treat trees from February to early March with Bt or spinosad, timing treatments based on monitoring of peach twig borer emergence in your area, and monitor frequently in the spring for peach twig borer or oriental fruit moth shoot strikes. Both insects damage young trees by destroying terminals and causing buds below the terminal to grow. Excess lateral growth often makes pruning and shaping the tree extremely difficult. If shoot strikes are common, treat the trees according to the guidelines given for controlling peach twig borer in the spring. Use your monitoring results to develop disease and pest control strategies for the following year.

If you are planning to use nontillage or to disc the orchards only in one direction and use strip weed control down the row, herbicide applications are generally safe after the first growing season. When using a contact (postemergence) herbicide, make sure it doesn't come into contact with the tender bark on young trees. Preemergence herbicides should be applied in the fall following the first growing season before winter weeds and grasses begin to grow, and when rain or sprinkler irrigation is available to move the herbicides into the ground. Make sure the herbicides are safe to use around young trees.

Apply dormant oil to first- through third-leaf trees to control San Jose scale and the eggs of brown mite and European red mite. By keeping scale from reaching high levels, you will avoid the need for more harmful pesticides later in the season or in subsequent seasons.

Handle second- and third-leaf trees the same as newly planted trees, except that in prolonged wet springs, a fungicide for control of brown rot and shot hole may be needed if disease symptoms occurred in the orchard the previous year. If moderate shoot damage was experienced the previous season, continue to monitor for peach twig borer and oriental fruit moth in mid-April.

Watch for wood-boring insects, which can quickly cause severe damage to young trees. In certain areas of the state, American plum borer is a serious problem on 2- and 3-year-old trees. If American plum borer becomes a problem, follow the guidelines for this pest in the chapter "Insects and Mites."

Replanting in an Established (Bearing) Orchard

When replanting in an established orchard, treat the stump as soon as possible after the tree is removed with an approved formulation of translocated herbicide. This treatment will kill the roots and help reduce the population of pests such as root knot nematodes that may be surviving on the living roots. Use a backhoe to remove the old tree stump and as many of the large roots as possible. The backhoe will also cut the roots of surrounding trees that would compete with the replant. Then refill the hole, add any necessary soil amendments, apply any necessary soil treatments, settle the soil, and make the planting hole in the usual manner at planting time. Carry out this operation in fall before planting to allow time for the soil to settle.

If root knot nematodes are a severe problem in the orchard, use a rootstock that is resistant to this nematode, even if you fumigate the soil. Similarly, use a rootstock that is tolerant of oak root fungus (Armillaria root rot) if this disease is the reason for replanting. (See "Cultivar and Rootstock Selection" earlier in this chapter.)

Replants do not do well if they are heavily shaded by surrounding trees. If shading is a problem, cut back the tops of surrounding trees.

While it is important to supply water and nutrients to any newly planted trees, it is essential in the case of a replant because of competition from the older surrounding trees. Give replants extra irrigations, at least during the first 2 years, but do not allow the soil to become too wet or the trees may develop crown and root rot. Also apply small amounts of a nitrogen fertilizer once or twice during the summer to help overcome competition for nutrients. Apply nitrogen fertilizers at least 18 inches (46 cm) from young trees. Do not allow weeds to grow around a replant; they can become severe competitors for water and nutrients.

Frost Protection

The dormant almond tree is quite tolerant of temperatures below freezing. Because almonds bloom earlier than most other deciduous fruit trees, however, the blossoms and young nuts have a greater likelihood of being damaged by frost.

Sprinkling under the tree during bloom protects almond blossoms from frost damage, if temperatures do not drop more than a few degrees below freezing, but it may also increase disease problems due to continual wetting of blossoms and fruit in the lower parts of the tree. Also, if frost occurs several nights in a row and if soil drainage in the orchard is restricted, root drowning and rotting could occur.

The presence or absence of a cover crop can influence the frost hazard potential of an orchard. Bare, dark, moist, firm soil stores more heat during the day for release at night than soil with a cover crop or soil that has been cultivated and left loose. A closely mowed cover crop, however, is almost as warm as bare, moist, dark soil. For more detailed information on frost protection in almond orchards, see the *Almond Production Manual* listed in the suggested reading.

Water Management

Adequate quantities of water are essential for growth and crop production in almonds. Allowing an almond tree to become stressed for water early in the season will slow shoot growth. This is especially serious in young trees where maximum shoot growth is needed to hasten orchard development. Early-season moisture stress will also lower yields by reducing nut enlargement. Severe water stress during the summer may cause leaf drop and sticktight nuts. Too much water, on the other hand, damages roots by depriving them of oxygen and creates conditions that favor infection by soilborne disease organisms. In particular, Phytophthora crown and root rot, one of the most destructive diseases of almonds, is favored by saturated soils.

Irrigation Methods. Several methods can be used to irrigate almond orchards, including surface (flood, furrow), sprinkler, and micro-irrigation (drip, micro-sprinklers). Each system can supply water adequately if managed properly. Major considerations in choosing a system are soil texture and depth, slope of the land, and the cost, source, and availability of water. Installation and operation costs are also important, as well as frost hazards in the orchard.

Surface Irrigation. Irrigation water can be most rapidly applied on level or leveled-to-grade soil by using furrow or flood irrigation if a good head of water is available. Rapid application of a large amount of water can be helpful in extremely hot weather if the need arises for immediate irrigation. Also, because gravity is used to distribute water across the orchard, energy inputs with surface methods are relatively small. However, because it takes some time for water to reach the far end of the field, more water will usually infiltrate the soil at the head end of the orchard. Also, because soil infiltration rates can vary greatly throughout the orchard, the uniformity of infiltrated water can be low with surface irrigation. Finally, the temporary levees used for surface irrigation require considerable labor and may make it difficult to smooth the soil for mechanical harvesting. If the soil-smoothing operations are not well timed, clods left on the surface may interfere with harvesting. Under nontillage, however, soil smoothing is not a problem in flood-irrigated orchards where berms are present in the tree row.

Use of surface irrigation may affect certain pest problems. For example, if water comes from canals, it may be contaminated with weed seeds and pathogens, which can spread through the orchard as the water moves across the soil surface. Ants tend to be less troublesome in flood-irrigated orchards. Flood and furrow irrigation may discourage ground squirrels and pocket gophers from digging burrows in the orchard; however, where trees are on berms, flood irrigation may encourage burrow-building in tree rows, where it can be even more damaging.

Sprinkler Irrigation. Properly designed solid-set sprinkler systems apply water uniformly over the soil surface, and the orchard floor requires minimal soil-smoothing operations to produce a smooth surface for mechanical harvesting. In addition, sprinklers require less labor than do most other methods of irrigation. (Considerable energy is required, however, to distribute the applied water.) A well-designed sprinkler irrigation system can be used for frost protection and the application and incorporation of some chemicals. Because sprinklers usually apply water at a rate lower than the soil infiltration rate, the uniformity of water infiltration is primarily determined by the system design. Thus, with sprinkler irrigation, water infiltration is generally more uniform throughout the orchard than it is with surface irrigation methods. Sprinkler systems with low rates of application can be used on land where runoff or slow penetration would be a problem if surface irrigation methods were used.

The use of a sprinkler irrigation system, however, can increase disease problems if water is applied so that branches, leaves, flowers, or developing nuts are wetted.

Micro-irrigation. Micro-irrigation, or localized irrigation, which includes drip (trickle) and micro-sprinklers, is another irrigation method used for almonds. With micro-irrigation, soil moisture used by the tree is replaced by applying water to a relatively small percentage of the orchard floor on a frequent (often daily) basis. Trees irrigated by this method have a reduced root zone compared to trees irrigated by other methods. Also, water volume stored in the soil surrounding the roots of micro-irrigated trees is lower than with other methods. As a result, the soil water reservoir must be refilled frequently.

Advantages of this system are more uniform application of water, often resulting in more uniform tree growth and productivity, potential water savings in the early years of the orchard, and usually minor interference with most cultural operations. In mature orchards, water savings under micro-irrigation may occur because water application is

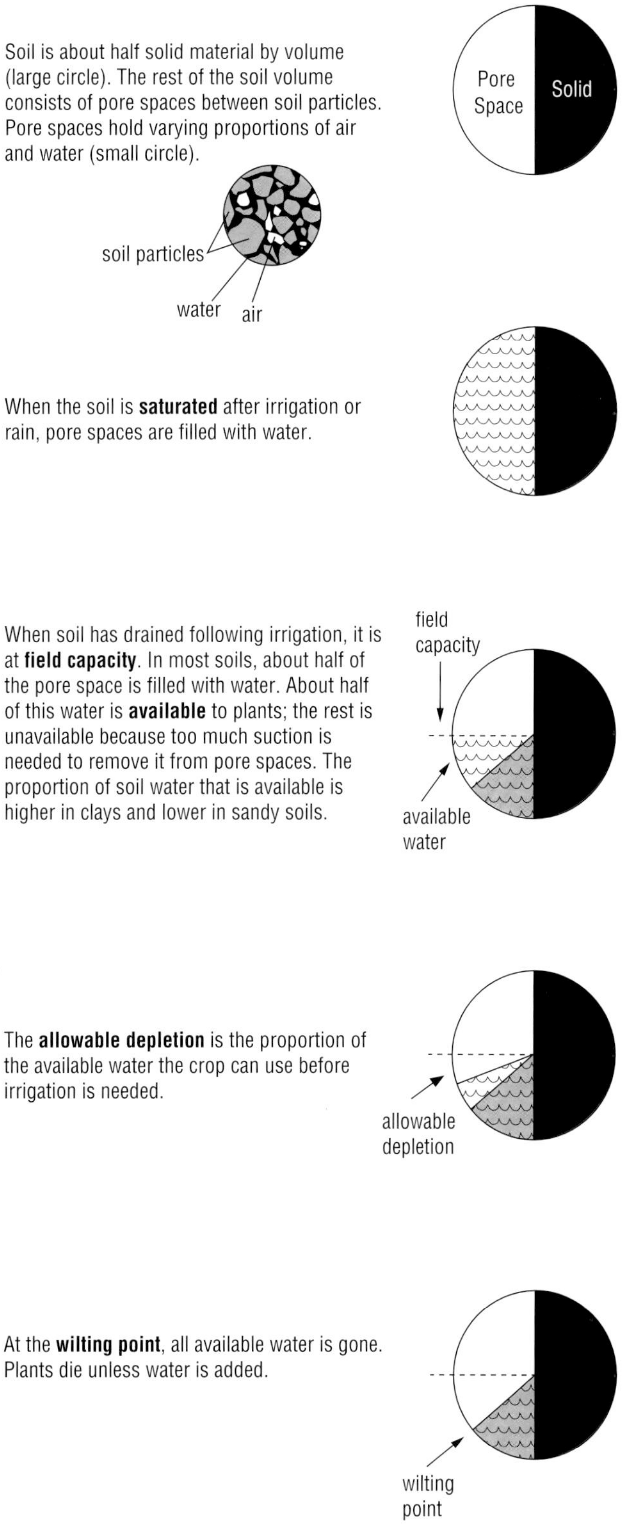
Soil is about half solid material by volume (large circle). The rest of the soil volume consists of pore spaces between soil particles. Pore spaces hold varying proportions of air and water (small circle).
Pore Space
Solid
soil particles
water
air
When the soil is **saturated** after irrigation or rain, pore spaces are filled with water.
When soil has drained following irrigation, it is at **field capacity**. In most soils, about half of the pore space is filled with water. About half of this water is **available** to plants; the rest is unavailable because too much suction is needed to remove it from pore spaces. The proportion of soil water that is available is higher in clays and lower in sandy soils.
field capacity
available water
The **allowable depletion** is the proportion of the available water the crop can use before irrigation is needed.
allowable depletion
At the **wilting point**, all available water is gone. Plants die unless water is added.
wilting point

Figure 12. The soil reservoir.

more efficient than with other methods. Micro-irrigation usually works well on soils with a low infiltration rate and is well suited for automatic operation with timers.

The disadvantages of micro-irrigation are higher costs than with surface irrigation and higher maintenance, emitter clogging, and lack of reserve moisture in the soil. Clogging, a major difficulty, usually can be reduced by filtering or by chemical injection, depending on whether the clogging results from physical particulates, chemical precipitates, algae, or slimes.

Drip irrigation must be engineered to meet the maximum expected water use demand, or trees can become stressed under hot conditions and production may be reduced more rapidly than with other irrigation methods. Drip irrigation hinders weed control because it keeps the area around the emitters moist, facilitating rapid breakdown of preemergence herbicides while promoting germination of weed seeds. Recent research indicates that micro-sprinklers, which wet a larger surface area than drip emitters but do so on a less frequent basis, allow the soil surface to dry and thus increase herbicide effectiveness. Subsurface drip systems may reduce or eliminate the need for preemergence and postemergence herbicides because the upper layer of soil, where weed seeds germinate, is kept dry.

Irrigation Scheduling. For orchards grown on well-drained soils, a water budget is the best tool for determining when to irrigate the orchard and how much water to apply. To use this method, you need to know the depth of the root zone, the water-holding capacity of the soil in the root zone, the rate of water consumption by the crop, and the irrigation efficiency of the method you plan to use. Another principal method of irrigation scheduling involves making soil moisture measurements and attempting to relate them to the well-being of the trees.

Although water budgets can be used with all irrigation methods, they are easiest to use with localized irrigation because soil water-holding capacity and root zone depth are of minor importance in localized irrigation. Also, irrigation efficiency for localized irrigation is simpler to compute than for other irrigation methods. Water budgets are not recommended for use in scheduling irrigations on soils with high water tables or in areas where reliable evapotranspiration (ET) data are not readily available.

The first step in creating a water budget is to determine how much water the soil in the tree's root zone can hold—that is, the size of the soil water reservoir. Begin by finding out how much water is available when the soil is at *field capacity:* the water content of the soil 2 to 3 days after a heavy rain or irrigation (Fig. 12). About half of the soil volume consists of solid particles. The rest of the soil volume is pore spaces holding varying proportions of air and water. The portion of the soil moisture that can be extracted by plant roots is the *available moisture*. Like field capacity, the available moisture content of a soil depends largely on soil texture. For example, the amount of available moisture is greater in clay soils than sands, which have lower total pore space. Table 1 shows the approximate water-holding capacity of various soils. You will need to determine the texture of the soil throughout the root zone of the tree and the total root zone depth. Use a soil auger to examine the soil profile. One way to determine soil texture is to use the feel method described in Table 2. When available, soil maps prepared by the U.S. Natural Resources Conservation Service can be used to obtain information on soil texture. To determine the rooting depth, visually examine the soil core for the presence of tree roots.

Don't wait until all of the available water in the tree's root zone has been extracted before irrigating because water stress begins before complete depletion. The proportion of available moisture that can be used before irrigation is needed *(allowable depletion)* depends on the plant species, its growth stage, soil texture, and atmospheric conditions. Although no single recommendation can be made for all orchard situations, generally when 50% of the available moisture has been used, it is time to irrigate.

Once you determine the available moisture and the allowable depletion level of your soil, the total water-holding capacity can easily be calculated by estimating the rooting depth of the tree. Figure 13 shows an example for a profile containing two soil types. The total water-holding capacity is calculated by determining the available water in each layer of the root zone. Multiply the total available water-holding capacity by the allowable depletion level to determine the amount of water use permitted between irrigations. When soil moisture loss reaches this predetermined depletion level, you need to irrigate to refill the soil water reservoir.

You can estimate the orchard's water consumption by keeping records of crop evapotranspiration (ETc), which is the combined total of water taken up by the crop and water that evaporates from the soil surface. ETc tables based on long-term average conditions are available for mature almond orchards with or without a ground cover. An example is given in Table 3. ETc rates 10 to 25% above or below average may occur for short periods in certain years, particularly during spring and fall. Historical data for a normal year provides a better approximation for midsummer when the weather does not vary much from year to year. ETc can be calculated from daily or weekly reference evapotranspiration (ETo), using a crop coefficient (Kc) (see the *Almond Production Manual*, listed in the suggested reading). Current ETo

Table 1. Available Water for Various Soil Types.

	AVAILABLE MOISTURE	
Type of soil	Range (in/ft)	Average (in/ft)
very coarse to coarse-textured sand	0.5–1.00	0.75
moderately coarse-textured—sandy loams and fine-textured sandy loams	1.00–1.50	1.25
medium-texture—very fine sandy loams to silty clay loams	1.25–1.75	1.50
fine and very fine texture—silty clay to clay	1.50–2.50	2.00
peats and mucks	2.00–3.00	2.50

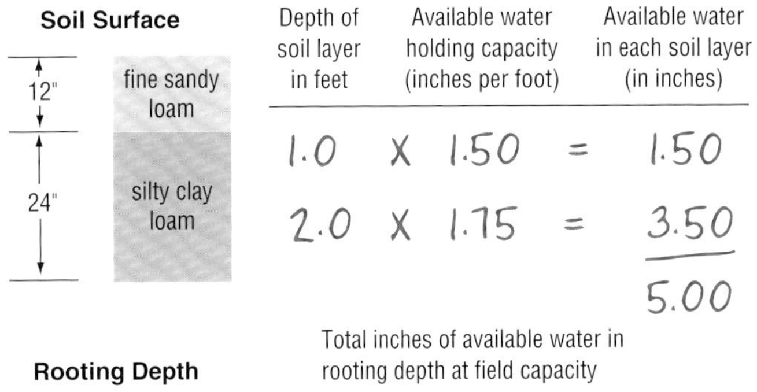

Figure 13. To calculate how much water the soil can hold at field capacity, first examine the soil profile to determine the thickness of different soil textures present in the root zone. Multiply the thickness of each layer, in feet, by its water-holding capacity and add the results for all layers to get the total water-holding capacity of the root zone.

Table 2. Guide for Determining Soil Texture and Water Content by Feel and Appearance.

Coarse-textured soils	Inches of water needed[1]	Medium-textured soils	Inches of water needed[1]	Fine-textured soils	Inches of water needed[1]
Soil looks and feels moist, forms a cast ball, and stains hand.	0.0	Soil dark, feels smooth, and will form a ball; when squeezed it ribbons out between fingers and leaves wet outline on hand.	0.0	Soil dark, may feel sticky, stains hand; ribbons easily when squeezed and forms a good ball.	0.0
Soil dark, stains hand slightly; forms a weak ball when squeezed.	0.3	Soil dark, feels slick, stains hand; works easily and forms ball or cast.	0.5	Soil dark, feels slick, stains hand; ribbons easily and forms a good ball.	0.7
Soil forms a fragile cast when squeezed.	0.6	Soil crumbly but may form a weak cast when squeezed.	1.0	Soil crumbly but pliable; forms cast or ball, will ribbon; stains hand slightly.	1.4
Soil dry, loose, crumbly.	1.0	Soil crumbly, powdery; barely keeps shape when squeezed.	1.5	Soil hard, firm, cracked; too stiff to work or ribbon.	2.0

1. Numbers in each column are inches of water needed to restore 1 foot of soil depth to field capacity when soil is in the condition indicated.

Table 3. Estimated Monthly Evapotranspiration from Almond Orchards in the Sacramento and San Joaquin Valleys, Based on Long-Term Weather Data.

ESTIMATED MONTHLY EVAPOTRANSPIRATION (INCHES)

Location	Jan	Feb	Mar	Apr	May	Jun	Jul	Aug	Sep	Oct	Nov	Dec	Total
Sacramento Valley													
With ground cover[1]	0.9	1.0	3.3	4.9	7.0	8.7	9.5	8.0	5.9	2.1	1.2	0.8	53.3
Clean cultivated	0.0	0.6	2.1	3.4	5.2	6.9	7.6	6.5	4.4	1.4	0.0	0.0	38.1
San Joaquin Valley													
With ground cover[1]	0.8	1.4	3.2	5.1	7.5	9.0	9.4	7.9	5.6	3.5	1.2	0.6	55.2
Clean cultivated	0.0	0.0	1.9	3.4	5.4	6.9	7.5	6.3	4.4	2.5	0.0	0.0	38.3

1. Either a green, actively growing cover crop or weeds. If the ground cover dies or is senescent, consider the orchard to be clean cultivated with respect to water use.

data are often provided in local newspapers, on the radio, or by irrigation districts. By adding up weekly (or daily, if available) ETc figures until they approach the calculated allowable depletion, you can estimate when the next irrigation is needed (Fig. 14).

To calculate the actual amount of water that needs to be applied, you must determine what portion of the applied water is actually available for use by the trees. Certain unavoidable losses occur during irrigation, mostly from deep percolation and runoff. Well-managed surface irrigation is normally 60 to 80% efficient, sprinkler irrigation is around 75% efficient, and localized irrigation is about 85% efficient. Private irrigation consultants can be hired to assess your irrigation efficiency, or you can learn how to estimate the effectiveness yourself by contacting the U.S. Natural Resources Conservation Service or your local UCCE farm advisor. Divide the net amount of irrigation water needed at each irrigation by the calculated efficiency. The Natural Resources Conservation Service also can provide methods for estimating the efficiency of rainfall.

At the end of the week following an irrigation, record the ET amount from the newspaper for the days following irrigation. In midsummer when the need to irrigate may be weekly or biweekly, use long-term weekly average ET information (divide monthly estimates by 4 to get weekly averages) to project the next irrigation date. When the next published weekly ET data becomes available, use it to confirm your projected irrigation date. If you have already irrigated, based on a projected irrigation date, and you underestimated the depletion level, make the appropriate adjustment when planning the next irrigation. For example, if you discover when the weekly ET rate is published that you miscalculated the amount of water extraction from long-term data and only replaced 90% of the soil water reservoir, when you start your next water budget, begin with a 10% deficit.

The water consumption of a young orchard is less than that of a mature orchard. The ETc for a young orchard can be calculated as a percentage of mature orchard ETc using the following equation:

$$P = 2G$$

where P is the percentage of a mature orchard's ETc and G is the percentage of the orchard floor that is shaded by young trees at noon. The young orchard's water consumption becomes equal to that of a mature orchard when ground shading reaches 50% (i.e., P = 100 when G = 50). You can minimize ET losses in young orchards by applying water in one furrow on either side of the tree row. Such localized irrigation has little or no benefit in a mature orchard.

You may not be able to apply at one time the full amount calculated by the water budget method. The amount you can apply at one time depends on the intake rate of the soil. To minimize problems with root diseases, do not apply more water than the soil can take up in 48 hours. For best protection of rootstocks susceptible to *Phytophthora*, apply only as much water as can be taken up in 24 hours. Adjust the allowable depletion to match the soil's ability to take up water, and irrigate more often.

Soil moisture measurements, the second major approach to irrigation scheduling, can be used effectively to confirm your calculations and verify that the soil profile has been filled. Check the soil moisture after an irrigation and periodically between irrigations with soil tubes, soil augers, gypsum blocks, tensiometers, or neutron probes. You can use a soil tube or soil auger to take soil samples from the root zone; check the moisture content using the soil-feel technique described in Table 2.

Tensiometers, which are installed at different soil depths, tell how tightly water is held in the soil. Quantitative evaluation of tensiometer readings is difficult because they depend on soil type, depth of instrument placement, and irrigation method. Readings can be used qualitatively, however, to check whether the water has reached a certain depth after an irrigation (when suction decreases) and also to check how dry the soil gets between irrigations (when suction increases). Tensiometers are particularly useful where micro-irrigation is used because they can indicate whether the system is maintaining adequate moisture in the root zone.

Tensiometers work best in sandy soils. Clay and clay loam soils contain substantial available water that is beyond the measurement range of a tensiometer. In these heavier soils, using gypsum blocks to measure soil moisture requires less

Orchard/Crop San Joaquin Valley bare soil/Almonds (mature)

Soil fine sandy loam, silty clay loam

Available Water 1.50 - 1.75 in/ft

Rooting Depth 3 ft

Total Available Stored Water ≈ 5.0 in.

Allowable Depletion 50%

Allowable Level of Water Depletion 2.5 in.

Irrigation Efficiency 75% (sprinklers)

Gross Irrigation Requirement 2.5 in. ÷ 0.75 = 3.3 in.

Date	ETc (inches)	Cumulative ETc between irrigations	Remarks
3/1 - 3/7	0.48	0.48	Assume soil at field capacity until leaf-out
3/8 - 3/14	0.52	1.0	
3/15 - 3/21	0.56	1.21	Rain 0.5 in. 3/17. Subtract 0.35 (70% of 0.5)
3/22 - 3/28	0.65	1.86	Irrigate soon. (Approaching allowable depletion)
3/29 - 4/4	0.79	0	Irrigated 4/3
4/5 - 4/11	0.85	0.85	
4/12 - 4/18	0.92	1.77	
4/19 - 4/25	1.25	0.75	Irrigated 4/21
4/26 - 5/2	1.35	2.10	Irrigate soon.
5/3 - 5/9	1.40	1.10	Irrigated 5/4
5/10 - 5/16	1.42	2.52	Irrigate 5/16
5/17 - 5/23	1.67	1.67	
5/24 - 5/30	1.90	1.41	Irrigated 5/25

Figure 14. Sample form for scheduling irrigation using crop evapotranspiration (ETc) data. Start accumulating ETc at weekly intervals as soon as the trees bloom. If substantial rainfall occurs, adjust ETc by subtracting effective rainfall (about 70% of total) from accumulated ETc. The first irrigation is needed during the week of March 29, as the cumulative ETc approaches the allowable depletion of 2.5 inches (6.4 cm). Schedule the second irrigation for about April 21, when another 2.5 inches are about to be depleted. Apply 3.3 inches (8.4 cm), the gross irrigation requirement, at each irrigation.

maintenance and may be more appropriate. This technique measures soil moisture by measuring electrical conductance through gypsum blocks buried at different soil depths.

Neutron probes use a detection tube set at a specific depth and a radiation source to measure water content of the soil. These devices are significantly more expensive than tensiometers or gypsum blocks, and the operator must be licensed to handle radioactive material. However, once calibrated, neutron probes give direct readings of the percentage of water content and can also store moisture readings.

More detailed information on tensiometers, gypsum blocks, and neutron probes can be found in the *Almond Production Manual,* listed in the suggested reading.

Fertilization

Trees weakened by nutrient deficiencies produce less and are more susceptible to pest damage than healthy trees. Most California orchard soils contain sufficient amounts of most essential nutrients except nitrogen (N), which has to be replenished on a yearly basis. If you maintain a nonlegume ground cover, you will need to apply additional nitrogen for use by the cover plants. Legume cover crops may contribute significantly to the amount of nitrogen available to the trees. Potassium (K) also needs to be replenished on a routine basis because almond trees consume high amounts of this nutrient. Potassium deficiencies are common on sandy soils in the Central Valley.

Other nutrients that may be deficient in almond orchards are zinc (Zn), copper (Cu), iron (Fe), and boron (B). Zinc deficiencies are fairly common in sandy soils or young, vigorously growing orchards and should be added to prevent growth reduction and yield loss. Foliar applications usually are the most effective. Copper and iron deficiencies are more limited in occurrence. Severe boron deficiencies occur in a few areas in the northern San Joaquin Valley and in the Sacramento Valley. More moderate boron deficiencies are widespread and can reduce yield while causing few or no vegetative symptoms. Magnesium, manganese, and phosphorus deficiencies are rare in almonds.

The tree's ability to take up nutrients depends on the type of fertilizer, soil moisture, soil type, and temperature. Make sure you apply the appropriate fertilizer at the right time and in the right place. For information on application methods, rates of application, materials, and timing to correct various nutrient deficiencies, see the *Almond Production Manual,* listed in the suggested reading.

Testing Nutrient Levels. Leaf tissue should be analyzed on a regular basis either yearly or every other year to determine the levels of nutrients except boron, which is measured in samples of hulls taken at harvest. Also, sample nutrient levels if you suspect a problem is the result of a nutrient deficiency or toxicity. Some nutrient deficiencies produce characteristic leaf symptoms (see "Environmentally-Caused Diseases" in the chapter "Diseases"). Confirm suspected deficiencies by having a laboratory analyze leaf samples from trees exhibiting symptoms.

On bearing trees, sample mature leaves on nonfruiting spurs every year in late June or July to detect deficiencies or possible toxicities. Take leaf samples from 6 feet (1.8 m) above the ground, selecting trees at random from the area being sampled. On young, nonbearing trees, sample basal shoot leaves. For boron analysis, hull samples can be taken from the tree after hull split or at harvest. Use the same sampling strategy as for leaf samples. Table 4 lists the critical levels of nutrients that are most often deficient or present at toxic levels. See *Soil and Plant Tissue Testing* listed under "Soil, Water, Weather, and Nutrients" in the suggested reading for more details on how to test for nutrient levels.

If a leaf or hull analysis reveals toxic levels of elements such as sodium, boron, or chlorine, have the soil or water analyzed to determine where these toxic levels are coming from.

Ground Covers

Ground covers are frequently maintained in almond orchards, especially those growing in heavy, fine-textured soils. A ground cover that includes grasses often improves water penetration and aeration of the soil, two important factors for a healthy root system. In addition, a well-managed ground cover provides cooling and reduces dust, which often contributes to mite outbreaks. With a perennial cover, the cooling effect of the ground cover can be advantageous in areas with hot summers but a disadvantage in areas with frost hazards. The frost hazard can be minimized by mowing the cover crop before bloom. A ground cover may be important in providing food and shelter for many predators and parasites of insects and mites. However, the ground cover also may host certain disease-causing pathogens and nematodes and may also provide food or shelter for vertebrate pests.

Table 4. Critical Nutrient Levels in Almond Leaves[1] or Hulls.

Nutrient	Deficient below	Adequate	Adequate over	Toxic above
nitrogen	2.0%	2.2–2.5%	—	—
potassium	1.0%	—	1.4%	—
zinc	15 ppm	—	—	—
copper	—	—	4 ppm	—
boron[2]	80 ppm	80–150 ppm	—	300 ppm
sodium	—	—	—	0.25%
chlorine	—	—	—	0.3%
manganese	—	—	20 ppm	—
magnesium	—	—	0.25%	—
calcium	—	—	2.0%	—
phosphorus	—	0.1–0.3%	—	—

1. Determined by leaf analysis performed in July; percent or parts per million (ppm) based on dry weight.
2. Boron levels determined by analyzing hull samples taken at harvest; parts per million (ppm) based on dry weight.

— = no data

A ground cover may consist of volunteer annual plants or a planted annual or perennial crop. Low-growing winter annuals that dry up in the summer heat and decompose by harvest make good ground cover plants. If the volunteer ground cover is not adequate, however, it may be necessary to plant an annual fall-seeded crop. Winter annuals such as mustard, wild radish, and fiddleneck, which bloom at the same time as almonds, are not desirable cover plants because they may be preferred by bees. To avoid competition by flowering weeds, mow the ground cover just before almond bloom.

To maintain a ground cover properly, be sure to mow frequently enough to avoid trash buildup at harvest. Keep plant growth away from the tree trunks, where it provides conditions favorable for crown rot infections. Depending on the mix of plants in the ground cover, it may use more water and nutrients than a clean-cultivated orchard; the cost of adjusting irrigation and fertilization programs to account for this extra use must be considered. More information on cover crops can be found in the chapter "Vegetation Management," and in *Almond Production Manual*, *BIOS for Almonds*, and *Cover Cropping in Vineyards* listed in the suggested reading.

Pollination

All major almond cultivars grown in California require cross-pollination from compatible cultivars (pollenizers) because their own pollen is not able to fertilize their flowers. The standard practice is to plant alternating single rows of two cross-compatible cultivars so that each acts as a pollenizer for the other. This single-row arrangement maximizes yield over the life of the orchard, may make harvest easier, and helps avoid mixing of the nuts during harvest. Other practices for achieving cross-pollination such as alternating trees within a row, grafting pollenizer limbs, and using bouquets of pollenizer flowers are discussed in the *Almond Production Manual*, listed in the suggested reading.

Managing Bees. Many wild insects forage for nectar and pollen in almond blossoms, but honey bees are the primary pollinators. As a general rule, professional beekeepers are hired to place hives in or adjacent to the orchard. Two hives per acre (5 per ha) are usually sufficient. The desired minimum-strength hive has at least six frames of bees and a queen actively laying eggs. With an active queen and at least one frame of "open brood" (larvae), the bees will be foraging more actively for pollen, which they use primarily to produce food for the queen and larvae. You can hire inspectors to verify and certify hive strength (check with your county agricultural commissioner). Follow these guidelines for placing and managing hives.

- Place hives in the orchards when the first blossoms open. Bees immediately scout for food sources and will work the first acceptable blossom they find until it is no longer attractive.
- Two hives per acre (5 per ha) of orchard usually are sufficient. Commercial beekeepers often cluster the bees in groups of 6 to 12 hives to facilitate management.
- Place hives around the perimeter of orchard blocks smaller than 40 acres (16 ha). For a long orchard, place more of the hives near the middle of the long sides.
- For larger orchards, place hives along drive rows in the middle of the orchard blocks.
- Place hives in the warmest, driest locations possible, in weed-free locations away from tall grass, and preferably where morning sun will warm the hives. The sooner hives warm up in the morning, the sooner the bees will begin to forage. Bees do not fly when the air temperature is below 55°F (13°C), when the wind velocity is at or above 12 miles per hour (19 km/h), or when it is raining. You can put black plastic or tar paper under hives to increase warming.
- Control weeds and keep cover crops mowed as needed to eliminate the presence of competing blooms in the orchard.
- Hive inserts containing pollenizer pollen, bouquets of pollenizer flowers, pollen traps, and bee attractants have been shown to improve pollination only under abnormal circumstances, and are in most cases unnecessary.

Africanized Bees. At the time of this writing, Africanized bees are present in the southern San Joaquin Valley, and they are expected to expand their range northward. In parts of the world where these bees have become established, beekeepers have adopted new practices that allow them to continue managing their hives, both for honey production and crop pollination.

Management techniques for Africanized honey bees include:

- Wear adequate protective clothing when working with the hives.
- Move the hives at night to reduce agitation and to reduce the likelihood that colonies will abscond.
- When foraging conditions are poor, provide the bees with syrup to reduce the likelihood that colonies will abscond. If you move a colony to a new location where food is less available, the bees usually leave.
- Keep hives replenished with European honey bee queens that were mated in areas where Africanized honey bees do not occur.

Harvesting

Harvest time varies according to district, season, and cultivar, but it usually begins in August and extends into October. Table 5 lists early, early-to-mid, mid, late, and very late harvesting categories of almond cultivars. If trees are shaken

Table 5. Harvest Times of Different Almond Cultivars.

Early	Early-mid	Mid	Late	Very late
Jeffries	Harvey	Aldrich	Butte	Drake
Kapareil	Milow	Carrion	Carmel	Fritz
Nonpareil	Mono	Jordanolo	LeGrand	Mission
	Peerless	Ne Plus Ultra	Livingston	Monterey
	Price	Ripon	Merced	Planada
	Rosetta	Thompson	Padre	Ruby
	Sauret #1	Tokyo	Sauret #2	
	Solano	Yosemite		
	Sonora			

too early, nuts will not easily fall from the tree, and the tree will require extensive shaking to remove them if they can be removed at all. Extensive shaking can result in tree injury, which may lead to Ceratocystis canker. If nuts of some cultivars are allowed to hang on the tree too long, they may tend to stick and will not shake easily. Nuts allowed to remain on the trees may sustain a higher infestation of navel orangeworm because they are exposed for a longer period to ovipositing moths. The incidence of hull rot may increase the longer nuts remain on the tree. Nuts may also be difficult to remove from trees stressed by lack of moisture or excess mite damage, and the nuts that are removed may be difficult to hull.

Inspect your trees to determine when to start harvesting. When 95 to 100% of the almonds have hulls that are visibly split (Stage 3 or more; see Fig. 6), shake a few trees to determine if nut removal is satisfactory. If too many nuts remain in the tree, try again in a few days. Split almonds will still be green at harvest but will dry rapidly on the ground without any loss in nut quality as long as weather conditions are satisfactory for drying. These nuts are also relatively easy to hull. Nuts that are removed from the tree before splitting, however, will usually dry in such a way that hull removal is difficult or impossible.

The orchard floor must be properly prepared before harvest to allow for rapid and efficient nut pickup after the trees are shaken. The ground should be smooth, dry, and free of weeds, ants, trash, and other debris. Orchards that are regularly cultivated should be disced to break up clods and then rolled to make the surface smooth and firm. Some growers apply their last preharvest irrigation after smoothing the soil. Orchards where the soil is not regularly cultivated, but where the weeds are mowed or managed with herbicides, require fewer preharvest operations. Weeds growing between the tree rows should be flail-mowed close to the ground so that they will sufficiently decompose before harvest. With either ground management system, the last preharvest irrigation should be 1 to 4 weeks before harvest, depending on soil texture and depth. Irrigation may be closer to harvest where micro-irrigation is used.

Harvest at the earliest possible date to avoid worm damage to nuts and to minimize damage from hull rot and early rains. If danger of a navel orangeworm infestation is severe, it may be cost-effective to harvest before maximum nut removal is possible. Nuts left behind can be harvested along with later-maturing pollenizing cultivars. However, if too many nuts are left, mixed nuts could be a problem with later-harvested cultivars.

After nuts are shaken onto the ground, the hulls are allowed to dry before the nuts are swept into windrows and picked up. During this period they are susceptible to damage by ants; the longer they remain on the ground, the greater the potential for damage. If an economically damaging number of ants are present in orchards, they should be controlled before harvest (see the chapter "Insects and Mites" for further details).

Unless nuts can be hulled rapidly and delivered to the handler, consider on-farm fumigation immediately after harvest to stop the amount of damage to the crop by navel orangeworm and peach twig borer.

Pruning

Nonbearing almond trees are pruned in order to shape and train the tree for structural strength while developing maximum fruiting area and fruitfulness. Bearing trees are pruned to renew the fruiting wood for consistently high yields from year to year. In addition, diseased, weakened, and broken limbs, as well as interfering branches and unneeded water sprouts, are removed.

Almond trees bear a majority of their fruit laterally on spurs that live about 5 years. Depending on vigor and cultivar, some crop is formed on long shoots. Prune about one-fifth of the fruiting wood each year to stimulate adequate shoot growth. Usually the removal of three to five 1- to 3-inch (2.5- to 7.5-cm) older limbs is sufficient to maintain vigor and renew older fruiting wood. The actual amount of pruning, however, depends on cultivar, fertilization, tree vigor, and the size of the previous crop. When large limbs weaken, die, or break, prune those limbs flush at the base.

Sanitation

Sanitation is an important management tool in almond orchards because of the long life of the trees. Pests such as perennial weeds, nematodes and soilborne pathogens are difficult to control after trees are planted. Prevent introducing or spreading infestations of these and other pests by following the precautions listed below. The most important of all sanitation practices is the removal of mummy nuts from the orchard following harvest, which may keep navel orangeworm from reaching damaging levels.

- Shake mummy nuts from the trees and sweep them into the row middles, where they may deteriorate in the cover vegetation or be destroyed by flailing. Try to achieve a level of 2 or fewer mummy nuts per tree by

February. This deprives the navel orangeworm of overwintering sites. Consider poling remaining nuts from trees after harvest as this may yield some return to help offset sanitation costs.

- Clean all equipment after using it in areas infested with soilborne pests or troublesome weeds and before you move it into other orchards.
- Keep tree crowns free of weeds to reduce the threat of crown rot and vole infestations.
- Spot-treat infestations of perennial weeds as soon as they are discovered in the orchard to prevent them from spreading throughout the orchard.
- Don't irrigate with tailwater from fields infested with nematodes or other soilborne pathogens, and don't use water that may carry harmful herbicide runoff.
- Destroy stands of problem weeds along the borders of the orchard to reduce the buildup of voles and prevent the production of weed seeds that will infest the orchard. However, cover crops along orchard borders may help reduce dust and provide habitat for beneficial predators and parasites.
- Remove roosting, nesting, or resting areas, when possible, that might attract large numbers of crop-destroying birds.
- Remove stumps, brush piles, and debris to limit refuge for ground squirrels and prevent population buildups of wood-boring insect pests.

Biological Control

Pest populations in the almond orchard are frequently controlled by other organisms inhabiting the orchard. These biological control agents include predators, parasites, pathogenic microorganisms, and competitors. They may be native or introduced from other areas. In almond orchards, biological control is most important in controlling mites, San Jose scale, and navel orangeworm. Natural enemies are also important in the control of other scales and certain caterpillars.

Biological control may be disrupted by chemical treatments for key pests, resulting in outbreaks of mites or scales. You can reduce such secondary outbreaks by carefully selecting types and rates of insecticides and timing applications so that they are least disruptive to the natural enemies or by using cultural practices that substitute for pesticide application.

Biological control of diseases and weeds is very limited at present. One almond disease, crown gall, can be controlled on young trees at planting time by dipping the trees in a solution of a nonpathogenic strain of the crown gall organism. The nonpathogenic strain prevents the pathogenic strain from infecting the roots by outcompeting it. Biological control of puncturevine by a stem and seed weevil may occur in nontilled orchards. Where tillage occurs, however, control is disrupted. Weevil populations may build up in puncturevine growing in nontilled areas adjacent to the orchard, but not sufficiently to control the weed in a tilled orchard.

Pesticides

Properly used, pesticides provide effective, economical control for many of the pests that affect almonds. Within an IPM program, pesticides are valuable tools because they can reduce pest populations drastically within a short time. In many cases, they are the only control tools available. Whenever possible, use pesticides only when monitoring indicates they are needed to prevent a crop loss. When available and effective, select pesticides that present less risk to human health, the environment, and nontarget beneficial organisms.

Using Pesticides. Careful use of pesticides makes them more effective and reduces the likelihood of pesticide resistance, pest resurgence and secondary outbreaks, crop injury, and hazards to humans and the environment. In many cases, several different chemicals or formulations are available for use to control a single pest. In some cases one chemical formulation can be used to control more than one pest problem at a time. Some materials can be applied together, while other pesticide combinations are incompatible and must be avoided. The best choice for a given situation depends on the degree of control needed, effects on other pests and beneficials, economic considerations, and legal restrictions. Use the following guidelines when selecting and applying pesticides.

- Make sure you have identified correctly the target pest or problem.
- Choose the least-disruptive material whenever possible to minimize problems with secondary pest outbreaks and pest resurgence. For example, some synthetic organic materials used to control navel orangeworm and peach twig borer are more disruptive to natural enemies of mites than are other synthetic organic materials.
- Consider the history of pesticide resistance within the target pest population.
- Treat for insects and mites only when monitoring indicates that damaging populations are present or anticipated.
- Apply pesticides when the target pest is most vulnerable and when natural enemies are least likely to be affected.
- Apply postemergence herbicides when weeds are at the growth stage recommended on the label.
- Use techniques and equipment that apply pesticides most efficiently. Calibrate application equipment properly, use correct rates, and when possible, place the material selectively. By increasing the efficiency of pesticide treatments, you will reduce costs, reduce the rate

of development of pesticide resistance, and reduce the likelihood of secondary outbreaks and pest resurgence.

- Before choosing a pesticide, check the latest *UC IPM Pest Management Guidelines: Almond*.
- Consult your UCCE farm advisor for the lowest effective dosage and best application method for the pest problem in your orchard. Recommended rates and frequency of application given on pesticide labels may be higher than necessary for good control when properly timed and applied. Recommended rates for herbicides usually have to be adjusted for soil type, climatic conditions, and irrigation method.
- Be sure that all personnel handling or applying pesticides are familiar with necessary safety precautions and know what to do in case of an emergency.
- Always read the label carefully before using any pesticide. Follow directions and observe all safety precautions.

Because label restrictions may change and availabilities of pesticides are constantly changing, specific pesticide recommendations are not made in this manual. Information on pesticide recommendations can be found in the *UC IPM Pest Management Guidelines: Almond*, listed under "Pesticide Application and Safety" in the suggested reading. Keep up to date with label changes. Check with your county agricultural commissioner, UCCE farm advisor, or pest control professionals if you have any questions about label changes or label directions.

Ground Application. Nearly all ground spraying for insects, mites, and diseases in commercial almond orchards is done with air carrier (air blast) sprayers (Fig. 15). Most air carrier sprayers pump a pesticide-water mixture to a series of nozzles that direct the pesticide mixture into a blast of air from a fan that blows the spray into the tree. The air volume produced by air blast sprayers ranges from small (4,000 cubic feet per minute) to large (more than 100,000 cubic feet per minute), and the air velocity ranges from low (80 miles per hour) to high (180 miles per hour). Both high air velocity–low air volume and low air velocity–high air volume sprayers are available.

As the sprayer moves through the orchard, the air carrying the spray must mix with the still air in the interior of the tree. If the sprayer moves too quickly or the sprayer's air output is too low, you will not achieve complete mixing with the interior air, and coverage and control will be poor. The ideal sprayer distributes the spray mixture uniformly over all tree surfaces. Low velocity–high volume sprayers are more effective for large or tall trees and for trees with dense canopies.

The nozzles of an air blast sprayer are usually mounted in a semicircle in or adjacent to the air stream. Most sprayers have one bank (manifold) of nozzles on each side; some have two banks. Nozzles should be arranged to deliver two-thirds of the spray volume (in gallons per minute) from the top half of the sprayer manifold. Use Figure 16 or the manufacturer's nozzle chart to arrange nozzles on the manifold. You will also need to know the ratio of air discharge, which should have the same top to bottom ratio, but this is not always adjustable. This arrangement prevents the sprayer from overspraying the bottom part of the tree and underspraying the top, where most of the persistent pest problems occur. For certain pests, when you are applying a pesticide that may be harmful to beneficials, you may be able to close the bottom nozzles completely to ensure survival of the beneficials in the lower part of the canopy.

Figure 15. Air carrier (air blast) sprayers are used most commonly for ground applications of pesticides in almond orchards.

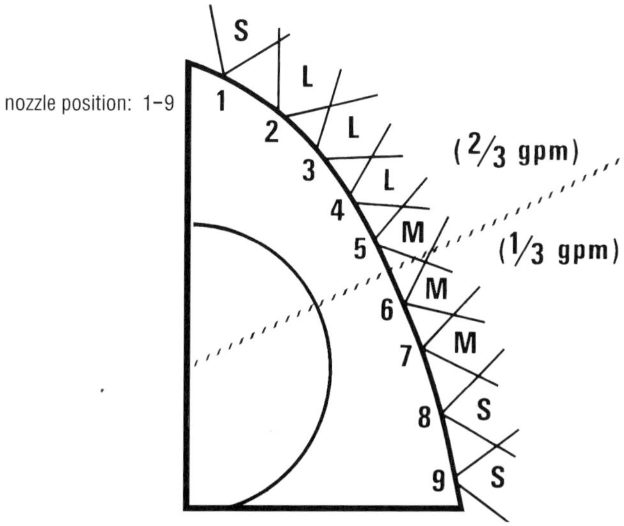

Figure 16. Nozzle arrangement on one manifold of an air carrier sprayer. Arrange nozzles so that two-thirds of the spray discharge, in gallons per minute, comes from the top half of the manifold. Relative nozzle sizes: L = large, M = medium, S = small.

Check nozzles frequently for wear. When nozzles wear out, they cannot reliably emit the proper droplet size or the correct rate of discharge. Wettable powder formulations are more abrasive than liquid formulations and tend to wear nozzles out faster. Hollow-cone and nonplugging types of nozzles are used for dilute sprays. These nozzles are made from common steel, brass, nylon, stainless steel, or ceramics. For more abrasive formulations, use only nozzles made from stainless steel, ceramic, or tungsten carbide, because they are more resistant to the abrasive wear. Low-volume air-shear nozzles operate at low pressures and do not have serious wear problems.

Most of the air blast spray rig is made up of the spray tank. Spray tanks usually are made of coated or stainless steel; in some cases they are made from chemical-resistant plastics or fiberglass. Stainless steel or fiberglass tanks are preferable. Strainers inside the tank keep foreign material from clogging the nozzles, and agitators keep the pesticide chemicals uniformly mixed with the water during application. Most sprayers have mechanical agitators; some use hydraulic agitation. Wettable powder formulations require mechanical agitation.

Spray Coverage. Complete coverage of all parts of the tree that may harbor the target pest is essential for maximum control. The most important causes of poor coverage are improper selection and arrangement of nozzles, too fast a ground speed, inadequate pump pressure, and incorrect gallonage for the pest. Poor coverage can also be caused by failure to make adjustments when moving between orchards having different-sized trees and incorrect use of sticker-spreaders or other spray additives. Always be sure to read the pesticide label and follow directions for proper application.

Do not move ground spray rigs through the orchard more quickly than 2.5 miles per hour (4.2 km/h) when applying dilute sprays or 2.0 miles per hour (3.4 km/h) when applying concentrate sprays. Trees will not receive adequate coverage if the ground speed is faster. Speeds even slower than this (2.0 and 1.5 mph [2.5 km/h] respectively) are recommended for dense canopies, for large trees, and for treatment of heavy scale infestations. To maintain proper ground speed, use a tractor with an accurate speedometer or time your travel over a set distance by counting the number of trees you pass in a minute. Figure 17 shows an example of a worksheet for calculating ground speed from tree spacings.

1. Calculate how many feet you travel in 1 minute (this example is for a desired speed of 2 mph).

$$\text{ft/min} = \frac{\text{mph x ft/mile}}{\text{min/hr}} \quad \frac{2.0 \times 5280}{60} = 176 \text{ ft}$$

2. Calculate how many tree spacings you should pass in 1 minute (this example uses 22-ft tree spacings).

$$\frac{176 \text{ ft}}{22 \text{ ft/tree}} = 8 \text{ tree spaces}$$

If it takes you more or less than 1 minute to pass 8 tree spaces adjust your ground speed.

Figure 17. Worksheet for calculating ground speed.

Gallonage. Pesticides are applied either as concentrate sprays (low-volume or no-drip sprays) or as dilute sprays (high-volume or runoff sprays). Concentrate sprays are applied in smaller quantities of water per acre, with the spray equipment adjusted to produce more, smaller droplets that are more uniform in size. Concentrate sprays usually are applied at rates of 25 to 100 gallons per acre, discharged at 2 to 5 gallons per minute from each manifold. High-volume sprays are applied at 300 to 800 gallons per acre and 12 to 40 gallons per minute from each side. The exact amount of spray you use in a given orchard depends on the target pest, tree density and size, the equipment used, economics, and past experience.

Pesticide efficacy is dependent on dose. Using the same amount of pesticide with increased water volume decreases the dose of the pesticide and may be less toxic to the target pest. As gallonage is changed, make certain that sufficient pesticide is used within label restrictions to ensure that an effective dose is delivered. A few pesticides should always be applied as concentrate sprays. For example, Bt is most effective when applied as a low-volume concentrate spray.

For some pests, concentrate sprays are as effective as dilute sprays because the smaller droplets allow a smaller volume to cover the same area as a high-volume spray (Fig. 18), although the concentrate requires greater precision during application. The advantages of concentrate spraying are several: the equipment is smaller, may be less costly, and is cheaper to maintain; the concentrate uses 70 to 90% less water, so tanks need to be refilled less frequently; and up to 25 or 30% less chemical may be required because the spray does not run off the tree as do dilute sprays. When applied from a properly calibrated and adjusted sprayer, the small droplets of a concentrate spray stick to the tree surface on contact.

Certain applications are more effective when applied as high gallonage dilute sprays. For example, spray oils applied during the dormant season for scales and mite eggs are most effective when sufficient oil is used in a dilute application.

You can modify an older spray rig designed for dilute applications to enable it to accommodate concentrates by adjusting the pump pressure, the spray manifold, and the number, size, and placement of nozzles. Newer-model spray rigs designed for dilute application are more easily adapted to concentrate spraying because they produce pressures of 100 pounds per square inch at the manifold and air velocities of 100 to 160 miles per hour, and they are also provided with a

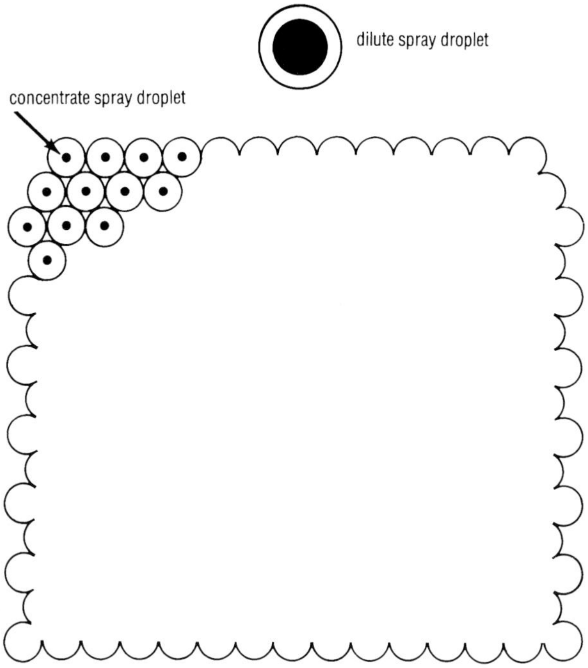

Figure 18. This diagram shows how coverage depends on the droplet size of the spray. In the diagram, each droplet is surrounded by a 100-micron zone in which each spray droplet is effective. The spray from a concentrate sprayer covers a much larger area (below) than the same amount of spray from a dilute sprayer (single droplet above). A droplet 300 microns in diameter from a dilute sprayer has the same volume as 216 droplets of 50 microns in diameter discharged from a concentrate sprayer. (300 ÷ 50 = 6; 6 X 6 X 6 = 216)

a. List the known criteria of your operation:

1. Row spacing (trunk to trunk, ft) 22
2. Gallons of spray per acre (gpa) 100
3. Desired ground speed (mph) 2.0

b. To find the gallons per minute (gpm) for the nozzle manifold (one side), use the following equation:

$$\text{gpm} = \frac{\text{gpa x mph x row spacing}}{1{,}000}$$

$$\text{gpm} = \frac{100 \times 2.0 \times 22}{1{,}000} = 4.4$$

4.4 gallons per minute need to be discharged from one side (8.8 gpm from both sides) to deliver 100 gpa at 2.0 mph.

Two-thirds (2.9 gpm) should be discharged from the top half of the manifold.

One-third (1.5 gpm) should be discharged from the bottom half of the manifold.

Figure 19. Worksheet for calibrating air carrier sprayers.

second manifold arrangement with smaller, hollow-cone, disc-type nozzles for concentrate sprays. Other spray rigs are available that are designed specifically for concentrate application. Some use the same nozzles, pressure, and air velocities as dilute application models. Others use a low-pressure system and special nozzles designed so the air discharge shears the spray into smaller droplets. Whatever nozzle system you use, it should produce a relatively narrow range of droplet sizes.

Figure 19 shows how to calculate the number of gallons per minute needed from each nozzle manifold. This formula can be used for either dilute or concentrate sprays. Publications with more information on the calibration and use of orchard sprayers can be found under "Pesticide Application and Safety" in the suggested reading.

Aerial Application. Aircraft applications of pesticides to orchards are usually not as satisfactory as ground applications. Coverage is not as thorough, especially after trees are fully leafed out. Dormant treatments to control San Jose scale or mite eggs cannot be applied effectively by air. Aerial applications are useful for bloomtime applications of Bt at very low gallonage and for early-season fungicide applications when the orchard is too wet for ground application. Aerial application may be the only alternative when a large area needs to be covered in a short period of time, for example, when a hull split spray is needed for navel orangeworm. Because the first almonds to split are in the upper, outer portions of the tree canopy, a properly applied aerial application can provide effective control against the navel orangeworm at this time. If you are spraying for mites at the same time, however, aerial sprays may not give adequate mite control, particularly if mite predators are not present in adequate numbers or tree canopies are dense. To achieve the best coverage with an aerial application, apply concentrate sprays by helicopter at a rate of no fewer than 25 gallons per acre and an air speed of less than 30 miles per hour. Droplet size must be large enough to prevent pesticide drift, yet small enough to stick to the trees without bouncing off.

Pest Resurgence and Secondary Outbreaks. Pesticides that kill or disrupt natural enemies may cause pest resurgence or secondary pest outbreaks. In almond orchards, these problems are most likely to occur with mites and some scale insects, which are normally kept under control by natural enemies but are able to build up rapidly when natural enemy populations are disrupted.

Pest resurgence occurs when a nonselective pesticide kills both the target pest and its natural enemies. The natural enemy populations take longer to recover because they require the pest for food and are usually more susceptible to the pesticide. Without the restraint of natural control, the few pests that survive treatment or invade the orchard from an untreated area multiply quickly, sometimes increasing to populations higher than those that prompted treatment in the first place. To reduce pest resurgence, use selective materials that are more toxic to the target pest than to natural enemies whenever possible. You can also use lower rates of selective materials to reduce pest populations well below damaging levels and still leave some pests as a food source for the natural enemies. Relative toxicities to natural enemies and honey bees for pesticides used in almonds are listed in Table 6.

Pesticide applications sometimes cause nontarget pests to increase to damaging levels. These increases, called *secondary outbreaks*, occur when pesticides destroy natural enemies that have kept the nontarget pest under control. Mites often cause problems in almonds after broad-spectrum insecticides have been used to control such pests as navel orangeworm, peach twig borer, and San Jose scale.

Pesticide Resistance. The repeated use of one type of pesticide to control a pest may result in the buildup of resistance in the pest population. In some cases, a pest population that develops resistance to one pesticide may also gain resistance to certain other pesticides, a phenomenon called cross-resistance.

Pesticide resistance is most likely to develop in pest populations that reproduce rapidly (e.g., mites) or diseases that

Table 6. Relative Toxicities of Pesticides Used in Almonds to Natural Enemies and Honey Bees.

Pesticide[1]	Predatory mites[2]	General predators[3]	Parasites[3]	Honey bee adults	Honey bee larvae
abamectin (Agri-Mek)	M	L	L	H	—
azinphosmethyl (Guthion)	L/M	H	H	H	H
Bacillus thuringiensis	L	L	L	L	L
benomyl (Benlate)	L/H	L	L	L	L
captan	L	L	L	L	L
carbaryl (Sevin)	L/H	H	H	H	H
chlorpyrifos (Lorsban)	M	H	H	H	H
clofentizine (Apollo)	L	L	L	—	—
diazinon	L/M	H	H	H	H
esfenvalerate (Asana)	H	M	H	H	H
fenbutatin-oxide (Vendex)	L	L	L	L	L
fixed copper	—	—	—	L	—
iprodione (Rovral)	L	L	L	L	—
liquid lime sulfur	H	—	—	L	—
maneb/mancozeb	L	—	—	L	—
mefenoxam (Ridomil Gold)	L	—	—	L	—
methidathion (Supracide)	H	H	H	H	H
myclobutanil (Rally)	L	—	—	L	—
naled (Dibrom)	M	H	H	H	—
oils	L	L	L	L	—
permethrin (Pounce)	H	H	H	H	H
phosmet (Imidan)	H	H	H	H	H
propargite (Omite)	L/M[4]	L	L	L	L
propiconazole (Break)	—	—	—	—	—
spinosad (Success)	L	L[5]	L	H[6]	L
sulfur	L/H	L	H	L	L
thiophanate methyl (Topsin)	L	L	L	L	—
ziram	L	—	—	L	L

H = high M = moderate L = low — = no data

1. Check current label restrictions. Registrations may change.
2. Toxicities to western predatory mite, *Galendromus occidentalis*. Where differences have been measured, these are listed as pesticide-resistant strain/native strain.
3. Toxicities are averages of reported effects and should be used only as a general guide.
4. Use the lowest label rates for best management of western predatory mite/spider mite ratio.
5. Toxic to sixspotted thrips.
6. Toxic if spray contacts bees, but there is no residual toxicity. Do not spray when bees are present in the orchard or allow spray drift to contact hives.

produce large quantities of spores (e.g., brown rot), and when multiple applications of the same pesticide or same class of pesticide are used to control a given pest during the season. To minimize the likelihood that resistance will develop, be sure to alternate the types of chemicals you use to control a pest and use pesticides only when needed.

Pesticide Hazards. Pesticides used in almonds may affect bees, wildlife, livestock, and humans. You must take precautions to avoid the hazards associated with pesticide application.

Hazards to Bees. Many of the pesticides used in almond orchards are harmful to bees (see Table 6). Because bees may forage 2 to 3 miles (3–5 km) from their hives, community-wide efforts are needed to protect bees from pesticide damage. Before you apply certain pesticides, you are required to contact the county agricultural commissioner to get the names and telephone numbers of registered beekeepers, who must be notified of the application. Check with your county agricultural commissioner to find out about legal restrictions before you apply any pesticides in the spring. Take the following precautions when using pesticides in the spring to minimize harm to bees:

- When applying a delayed-dormant insecticide spray, observe the minimum interval between application and placement of hives in the orchard.
- During bloom, use pesticides—even those considered nontoxic to adult bees—only when they are absolutely necessary. Pesticides, including some fungicides, can damage the honey bee brood when carried back to the hive by foraging bees.
- Notify beekeepers when you are planning to spray.
- Do not spray over hives or allow pesticides to get into hive entrances.
- Take precautions to avoid the drift of dormant sprays onto blooming winter annuals if there are bee hives within 2 miles (3 km) of the orchard.

Hazards to Wildlife and Water Quality. Pesticides applied to almonds may harm wildlife if runoff or spray drift contaminates bodies of water or nearby natural areas. Nontarget wildlife may be affected by toxic baits used to control vertebrate pests. Dormant sprays may cause runoff problems when winter rains wash residues into streams, rivers, or other natural bodies of water. The California Department of Pesticide Regulation (DPR) is developing special regulations and guidelines for dormant-season applications of pesticides in orchard crops. Stay informed about pesticide regulations that are designed to protect water supplies and wildlife in your area. Take special precautions, select less-harmful materials, or avoid using pesticides where runoff or spray drift is likely to contaminate nearby bodies of water or sensitive wildlife habitat.

In areas where endangered species occur, special guidelines apply to the types of pesticides you can use and the means of application. For each county in California, bulletins are available that list the endangered species that occur, map their location, and specify for each pesticide the guidelines that should be followed when using them in the vicinity of endangered species habitat. These bulletins can be obtained from the DPR World Wide Web site (www.cdpr.ca.gov/docs/es/index.htm) and from your county agricultural commissioner. Following is a list of guidelines designed to prevent adverse effects of fungicides, herbicides, or insecticides when used near sensitive habitat. Guidelines for the use of rodenticides are discussed in the chapter "Vertebrates."

- Avoid pesticide use in occupied habitat. Herbicides may be used in organized habitat recovery programs or for selective control of invasive exotic plants.
- Irrigate efficiently to prevent excessive runoff.
- Schedule irrigations and pesticide applications to maximize the amount of time between the pesticide application and the first subsequent irrigation.
- Allow at least 24 hours between pesticide application and an irrigation that may result in surface runoff into natural bodies of water.
- Time pesticide applications to allow sprays to dry before rain or sprinkler irrigation.
- Do not make an aerial pesticide application while irrigation water is in the field unless surface runoff is contained for at least 72 hours after the application.
- When air is calm or moving away from nearby habitat, begin spray or dust applications on the side of the orchard nearest the habitat and move away. When air is moving toward nearby habitat, do not make an aerial spray or dust application within 200 yards (180 m) of the habitat or a ground application within 40 yards (36 m) of the habitat. The probability of drift into nearby habitat (and the size of untreated buffer strip needed) can be reduced by intervening physical barriers such as hedgerows, windbreaks, and riparian corridors.
- Do not apply pesticide within 30 yards (27 m) uphill of habitat unless a suitable method is used to contain or divert runoff.
- Provide a buffer strip of at least 20 feet (6 m) of vegetation that is not treated with pesticides along rivers, streams, wetlands, vernal pools, stock ponds, and the downhill sides of fields from which runoff may occur. Several rows of legumes or other cover crops are recommended next to off-target water sites.
- Prepare the land to contain as much runoff as possible within the orchard site.

- Mix pesticides in areas not prone to runoff, such as concrete pads, flat disked soil, or graveled mixing pads, or use methods that contain spills and rinses.
- Triple-rinse and properly empty all pesticide containers at the time of use.

Hazards to Livestock. Almond hulls may be used as livestock feed. Be sure to follow all label directions regarding pesticides that may be used and proper preharvest intervals when almond hulls will be used for livestock feed.

Hazards to Humans. Some of the pesticides applied to almonds are hazardous to humans. Applicators are most at risk, but orchard crews and others who enter orchards may also be exposed. Pesticides may drift to nearby areas where they can cause problems. Read and follow all label directions regarding the handling and application of pesticides, reentry and preharvest intervals, and disposal of pesticide containers. Special precautions are needed when you are applying pesticides near residential areas and schools. Be sure that all workers who handle or apply pesticides are trained in proper safety procedures and wear the correct protective clothing. Keep all application equipment in good working order, with all necessary safety features for the materials being used. For detailed information on pesticide selection, handling, application, storage, and disposal, see *The Safe and Effective Use of Pesticides* and other publications listed under "Pesticide Application and Safety" in the suggested reading.

Squirrels can damage branches by gnawing on the bark and occasionally girdling a limb.

Vertebrates

Almond orchards provide food and shelter for a variety of vertebrates that can cause significant damage. Species that are considered pests include rodents (squirrels, pocket gophers, and voles), rabbits, deer, and birds. Rodents, rabbits, and deer are potentially more serious pests than birds because they can cause long-term damage by killing or seriously stunting the growth of trees. Ground squirrels may feed on nuts, girdle and kill young trees, and sometimes cause problems by gnawing on drip lines. Pocket gophers may also girdle trees or seriously debilitate or kill them by feeding on the roots. The burrowing of ground squirrels and pocket gophers can interfere with orchard management activities such as mowing and irrigating. In almond orchards located adjacent to wooded areas, tree squirrels may take large amounts of nuts. Deer, voles (meadow mice), rabbits, and coyotes may occasionally damage trees or irrigation lines in some growing districts. In some orchards birds may be the most serious vertebrate pests. Crows, scrub jays, and magpies can cause substantial losses by consuming or carrying off large numbers of nuts. House finches and, to a lesser extent, crowned sparrows may destroy flower buds during the dormant season or when buds begin to swell.

The vertebrate pest problems in a given orchard are determined in large part by the orchard's location. Rodents are potential pests in all orchards, but they are more likely to invade orchards next to rangeland or unmanaged areas, where their populations may build up unchecked. Orchards adjacent to unmanaged areas are more likely to be damaged by species such as rabbits, deer, and birds that live in these areas and feed in the orchards. Orchard management activities also have some effect on vertebrate pest problems. Flood irrigation discourages ground squirrels and, to a lesser extent, voles and pocket gophers. Management of orchard floor vegetation to reduce preferred food sources or shelter may reduce some vertebrate problems.

Managing Vertebrate Pests

Successful management programs for all vertebrate pests involve four basic steps:

- identifying the damaging species
- assessing management options
- implementing appropriate control actions
- monitoring to detect reinfestation of the orchard

Table 7. Control Methods for Vertebrate Pests of Almonds.

PEST	CONTROL METHOD Habitat Modification	Trapping	Baiting	Fencing	Tree protectors	Repellents	Frightening	Shooting	Fumigating
ground squirrels	•	•	•					•	•
pocket gophers	•	•	•						•
voles	•		•		•				
rabbits	•	•[1]	•	•	•	•		•	
deer				•		•	•	•[2]	
birds	•	•[1]					•	•[1]	

1. Useful for some species.
2. During hunting season or with a permit.

Observation and Identification. If you are going to choose an appropriate control action, it is critical that you correctly identify the species causing the damage. In the case of birds, you need to use direct observation to distinguish between nonharmful species that frequent orchards and birds that actually cause damage. Direct observation as well as signs such as tracks, feces, and burrows may be used to identify other species causing damage. The descriptions, line drawings, and photographs in the following sections will help you identify vertebrate pests that are causing problems in your orchards. Several publications listed under "Vertebrates" in the suggested reading have more information on vertebrate identification and biology.

Control Actions. For most vertebrate pests, more than one control method is available to reduce damaging populations. Table 7 lists control options you can use against important vertebrate pests in almond orchards. Details on how to use these controls are given in the pest sections that follow. It is a good idea to consult your county agricultural commissioner before you use any of these controls to find out which procedures work best in your location and what the restrictions are on these techniques. The timing of control actions is often critical and is determined in large part by the life cycle of the target pest. Become familiar with the biology of the vertebrate pests affecting your orchards and the available control options so you will be able to plan the most cost-effective management strategies.

You can take a number of steps when preparing and planting the orchard to prevent or reduce potential problems with certain vertebrates. Properly installed fencing protects young trees against deer and rabbits. Deep plowing and discing destroys or disperses resident vole populations, and it also destroys much of the burrow systems of pocket gophers (and to a lesser extent ground squirrels), reducing the risk of reinvasion. A vegetation-free zone around the orchard greatly reduces the risk of invasion by voles. Take steps to eliminate pocket gopher and ground squirrel populations before you plant a new orchard; it is much easier to control them prior to planting than afterward. Trapping, baiting, shooting, or burrow fumigation can be used, depending on the pest involved and the situation. Once trees are planted, tree guards protect against damage by rabbits and, to some degree, by voles.

Once the orchard is in place, you should develop and implement some type of management program to address any vertebrate pest problems that may arise.

Habitat Modification. Changes to the environment in and around the orchard may affect vertebrate pest problems and their management. Brush piles near the orchard may provide shelter and resting places for ground squirrels, rabbits, and birds. By removing brush piles, you can sometimes reduce bird activity in the orchard and make it easier to observe and take control actions against other vertebrate pests. Thick ground cover and mulches provide habitats that are preferred by voles. By reducing ground cover around trees and around the edges of orchards, you can greatly reduce problems with voles. Certain ground covers are attractive to pocket gophers, voles, and even rabbits, so you can practice specific management of orchard floor vegetation in conjunction with other controls to reduce vole and gopher problems, especially vole infestations. Flood irrigation and regular cultivation of the orchard floor also discourage voles as well as pocket gophers and ground squirrels, especially when trees are not planted on berms. The presence of berms tends to increase rodent problems.

Biological Control. Vertebrate populations are affected most by the availability of food and cover, while diseases and predators play a relatively minor role. A number of predators such as hawks, owls, foxes, coyotes, and snakes feed on some of the vertebrates species that can become orchard pests. However, natural enemies seldom keep vertebrate pests from reaching damaging levels. Take precautions to avoid harming predators when you use toxic baits or traps. Although you can install nesting boxes for barn owls and perches for other raptors, there is no evidence that this has a measurable impact on pest numbers.

Table 8. Guidelines on the Use of Toxic Baits and Burrow Fumigants within the Ranges of Endangered Species. Where more than one endangered species occurs, the most restrictive limitations apply. If in doubt about any portion of these guidelines, contact your county agricultural commissioner's office for additional information or clarification.

Endangered species	Control	Target pest	Guidelines
blunt nosed leopard lizard **giant garter snake** **kangaroo rats** **San Joaquin kit fox**	burrow fumigants	ground squirrels pocket gophers	Use must be supervised by someone trained to distinguish active burrows or dens of target species from those of nontarget species. Only active burrows of target species are treated. Contact your county agricultural commissioner for information on training.
giant garter snake	burrow fumigants	ground squirrels pocket gophers	In addition to above restrictions, use is prohibited from October 1 through April 30 except in areas under active cultivation[1] or on the inner bank (water side) of water supply channels.
all	acute poison baits	pocket gophers	No restrictions on bait placed directly in burrows by hand or mechanical bait applicator.
kangaroo rats	anticoagulant and acute poison baits (bait station)	ground squirrels	Use of toxic baits is prohibited unless • used in bait stations specially designed to allow access by pest species but not by kangaroo rats (see Fig. 20) OR • bait stations are elevated to prevent access by kangaroo rats (Fig. 21) and designed to prevent spillage OR • bait is placed in stations only during daylight hours and removed, or station entrances closed by dusk
	anticoagulant and acute poison baits (broadcast)	ground squirrels voles	Broadcast treatments can be made only in areas under active cultivation that are separated from native vegetation by at least 10 yards (9.1 m) of untreated crop or cultivated ground.[1]
San Joaquin kit fox	all baiting programs listed below for these target pests	ground squirrels voles jackrabbits	Carcass removal should be part of any baiting program that may result in carcasses remaining above ground. Begin monitoring for carcasses 3 days after baiting started and until at least 5 days after baiting stopped. Handle carcasses carefully to avoid contact with parasites such as fleas and ticks. Bury carcasses deep enough or otherwise dispose of them so that they are inaccessible to wildlife. Prebaiting: Prebaiting with untreated grains such as oats or barley is recommended to make baiting more effective and shorten the time required for baiting. Do not prebait with milo or cracked corn, which are highly attractive to birds.
	anticoagulant baits (bait station)	ground squirrels	Formulation: Active ingredient not to exceed 0.005% in bait formulations used in bait stations. Bait station design: Openings not to exceed 3 in (7.5 cm) in diameter, designed to control spillage, staked to prevent tipping, not filled beyond capacity and never with more than 10 pounds (4.5 kg) of bait (see Fig. 20, 21). Bait station monitoring: Must be monitored for signs of spillage, tampering, moisture, and depletion; monitored at least weekly after bait feeding begins; kept replenished; bait removed immediately after feeding ceases. If subsequent baiting needed, wait at least 2 weeks; this minimizes exposure of nontarget species without jeopardizing good pest control.
		jackrabbits	Self-dispensing bait stations can be used only if bait acceptance is first determined, stations are monitored and carcasses removed as described above, baiting ceases when feeding stops, and stations are used only where rabbits are active. Pelletized baits are prohibited.
	anticoagulant baits (broadcast)	ground squirrels	Spot baiting: Scatter handfuls of bait (about 10 handfuls per pound or 22 handfuls per kilo) evenly over 40 to 50 square feet (about 4–5 square m) near active burrows or runways. Repeat every other day until feeding ceases. Mechanical bait spreader: apply at rate of 10 pounds per swath acre (11.2 kg per ha) through infested area. Make second application 2 or 3 days later. Pelletized baits may not be broadcast or used for spot treatments within kit fox ranges.
San Joaquin kit fox	anticoagulant baits (broadcast)	ground squirrels voles	Broadcast or spot baiting: Active ingredient no greater than 0.01%; apply only after prebaiting with untreated bait and determining that untreated bait is taken; do not pile bait or place directly in burrows; do not reapply when significant quantities of previously applied bait remain; survey for carcasses as described above. Pelletized baits may not be broadcast or used for spot treatments within kit fox ranges.

Table 8. (continued)

Endangered species	Control	Target pest	Guidelines
San Joaquin kit fox (continued)	acute poison baits (bait station)	ground squirrels	Pelletized baits: May be used for ground squirrels only in bait stations designed to exclude the kit fox as described above.
	acute poison baits (broadcast)	ground squirrels voles	Grain baits that are not pelletized may be broadcast for ground squirrel or vole control. Follow recommendations for prebaiting and carcass removal.

1. Areas under active cultivation are defined as areas that have been tilled within the previous year or that are irrigated by furrow, flood, or overlapping sprinklers.

Information compiled from U.S. Environmental Protection Agency draft bulletins Protecting Endangered Species Interim Measures for Use of Pesticides, February, 2000. Bulletins for each county and restrictions pertaining to specific locations can be found on the World Wide Web at www.cdpr.ca.gov/docs/es/index.htm and in bulletins available from agricultural commissioners' offices. Consult the latest bulletins for the most up-to-date information. ALWAYS FOLLOW LABEL DIRECTIONS.

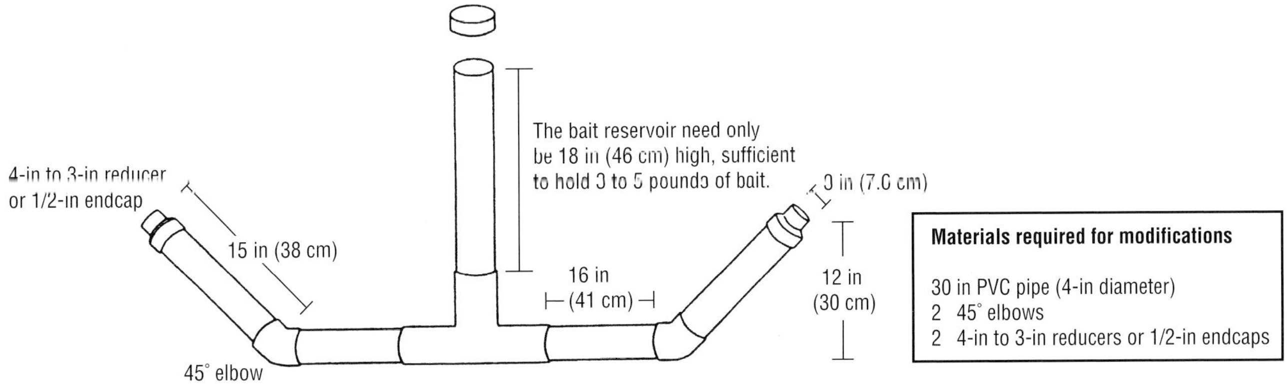

Figure 20. Ground squirrel bait station design for use within the range of endangered kangaroo rats and San Joaquin kit fox.

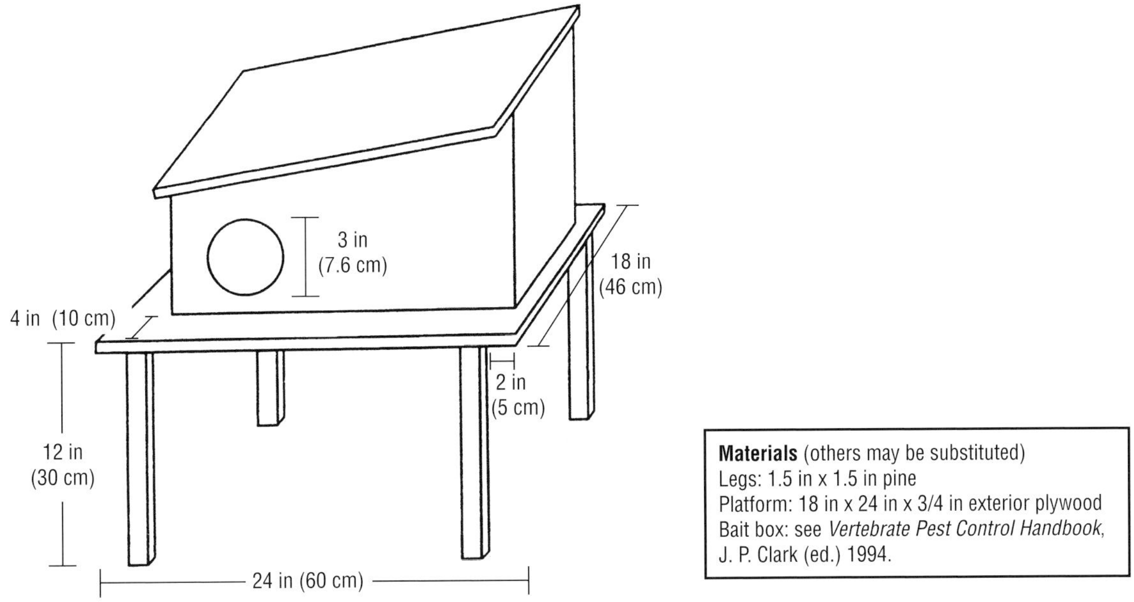

Figure 21. Platform design for elevating ground squirrel bait station to prevent access by kangaroo rats. This station must be constructed and anchored to the ground rigidly enough so that it cannot be tipped easily.

Endangered Species Guidelines. In some areas of the Central Valley, almond orchards are located within the range of federally- and state-protected endangered species. The species likely to be of concern are the San Joaquin kit fox, several species of rare kangaroo rats, and, where burrow fumigants are used, the blunt-nosed leopard lizard. See the composite map showing ranges of these species (Fig. 22). Special guidelines apply to the use of toxic baits and fumigants for vertebrate pest control in these areas; these guidelines are described briefly in Table 8. Your county agricultural commissioner has the latest detailed maps that show the ranges of endangered species and the latest information on restrictions that apply to pest control activities in those areas. You also can get more information on endangered species regulations from the DPR World Wide Web site (http://www.cdpr.ca.gov/docs/es/index.htm).

Monitoring. Follow the recommendations in the pest sections below on when and how to monitor for specific vertebrate pests. After you take a control action, establish a routine monitoring program to assess the effectiveness of the control and to detect any reinvasion. Keep detailed records of the procedures you use and their effects on vertebrate activity. These will help you plan future control strategies.

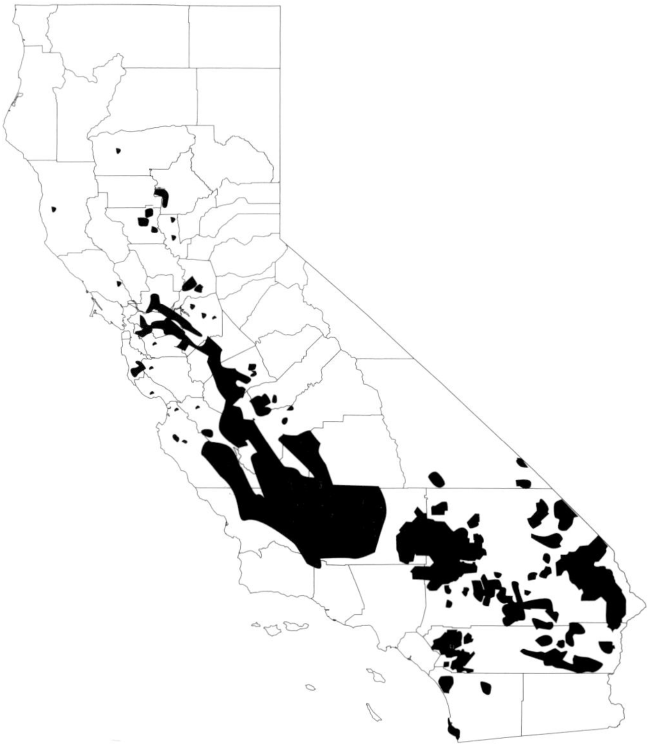

Figure 22. Distribution of endangered vertebrate species in California. This is a composite of current maps showing approximate locations of endangered species that may affect pest management options in or near almond orchards. Check with the county agricultural commissioner or the California Department of Pesticide Regulation web site (www.cdpr.ca.gov/docs/es/index.htm) for the latest information regarding restrictions that may apply to a specific location.

Ground Squirrels

Spermophilus beecheyi

Ground squirrels are likely to be the most damaging vertebrate pests in almond orchards. Damage may be especially high in orchards adjacent to uncultivated areas where squirrels are not controlled. Flood-irrigated orchards tend to have fewer squirrel problems because flooding discourages squirrels from digging burrows in the orchard, although they may simply locate their burrows on higher ground at the orchard border. If trees are planted on berms, flood irrigation may encourage squirrels to dig burrows in the berms and close to the base of trees, which can increase the severity of the damage they cause.

Ground squirrels occasionally gnaw on the bark and can girdle the trunk or scaffold, and they rarely feed on blossoms and buds. Their greatest damage is to developing and mature nuts. Ground squirrels readily climb trees and strip branches of large numbers of nuts. They also damage plastic irrigation lines by gnawing. Their burrows may disrupt irrigation, can cause erosion, and often interfere with harvest operations. In some areas, ground squirrels may present a potential health problem because they carry ectoparasites that may transmit diseases such as bubonic plague to humans.

The adult California ground squirrel has a head and body 9 to 11 inches (23–28 cm) long and a somewhat bushy tail that is about as long as the body. The fur is mottled dark and light brown or gray. Ground squirrels live in colonies that may grow very large if left uncontrolled. They are active only during the day and are usually most active in morning and late afternoon.

Each ground squirrel burrow system can have several openings with scattered soil in front. Individual ground squirrel burrows may be 5 to 30 feet (1.5–9 m) long, 2.5 to 4 feet (75 cm–1.2 m) below the surface, and about 4 to 6 inches (10–15 cm) wide. Burrows provide a place to retreat, sleep, hibernate, rear the young, and store food. Ground squirrels often dig their burrows along ditches, fence rows, and on other uncultivated land. When uncontrolled, they frequently move into the orchard and dig their burrows beneath trees.

Most ground squirrels hibernate during the winter, emerging around January in warmer locations such as the southern San Joaquin Valley and in February or March in cooler locations of the northern Sacramento Valley and foothills. In spring the squirrels feed on green vegetation, including new growth on trees. They switch to seeds and nuts in late spring and early summer as vegetation dries up. Females have one

California ground squirrels have mottled dark and light brown or gray fur and a long tail that is somewhat bushy. One subspecies (shown here) found in northern California, *Spermophilus beecheyi douglasii*, has a dark patch on its back.

Ground squirrels often dig burrows at the base of trees when they invade orchards. Their burrows have large, conspicuous openings that are not plugged.

litter, averaging 8 young, in the spring. The young squirrels emerge from the burrow when about 6 weeks old. Adults often go into a temporary dormant state (*aestivation* or *estivation*) during the hottest part of the summer. Young ground squirrels do not aestivate their first summer, and some may not hibernate during their first winter. Figure 23 illustrates the periods of activity for the California ground squirrel.

Management Guidelines

The type of control action needed for ground squirrels depends primarily on their activity patterns and feeding preferences during the time of year when control is to be undertaken. The choice of control action is also influenced by the location of the infestation and the number of squirrels present. Watch for signs of squirrel activity within the orchard, especially the appearance of burrows, during routine orchard activities. Check the perimeter of the orchard at least once a month during the times of year when squirrels are active. Midmorning usually is the best time of day for observing squirrel activity. Keep records of when squirrels emerge from hibernation, the approximate number of squirrels you see, and the location and number of burrows.

To keep populations from increasing, begin to apply controls as soon as you see squirrels or burrowing activity within the orchard. Select the control method best suited for the time of year (see Fig. 23). If you have access to nearby squirrel infestations, try baiting, trapping, or burrow fumigation. From a biological point of view, the most effective time of year to control ground squirrels is in early spring when adults have emerged but before they have reproduced. For best control use burrow fumigation about 3 weeks after the first squirrels emerge from hibernation. Because squirrels are

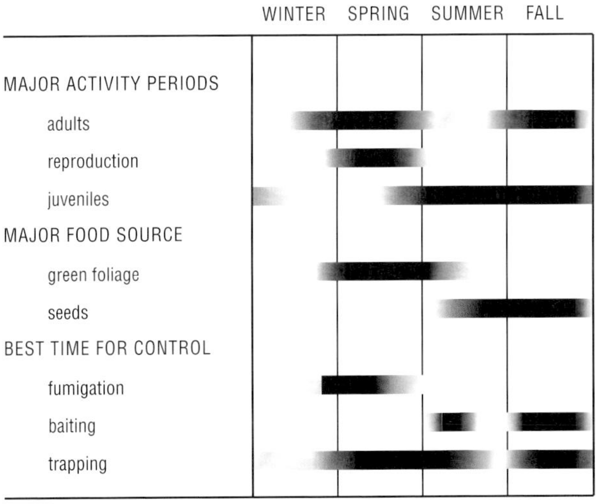

Figure 23. Activity periods and preferred food sources for the California ground squirrel. Activity periods vary somewhat from one growing area to another, depending on local climate. To choose the most effective control action for ground squirrels and the proper timing, you need to know when they are active and what their preferred food sources are.

feeding nearly exclusively on green vegetation early in the season, poisoned grain baits are generally not effective until late spring and early summer. In late spring or summer where squirrels are moving from adjacent lands into the orchard to feed, baiting or trapping along the perimeter offers the most effective control if access to the neighboring property is not possible.

Removal of brush piles, stumps, and debris in and around the orchard may to some extent help limit the buildup of squirrel populations, and will make it easier for you to monitor

Ground squirrel burrows may be located around the orchard's edges, such as along roadsides.

Figure 24. Conibear traps are placed in the entrance to ground squirrel burrows. Secure traps to a stake.

(a)

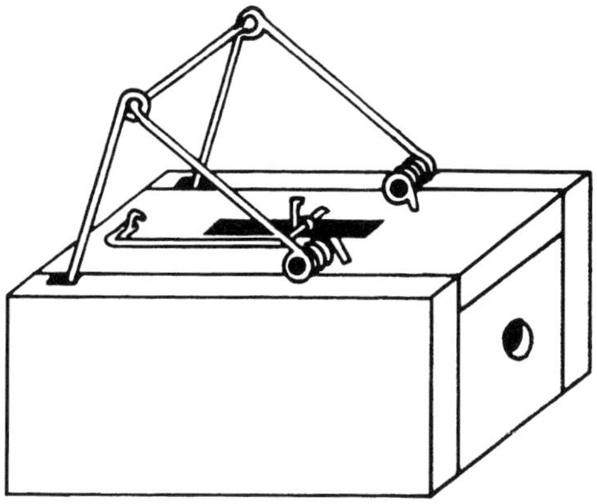

(b)

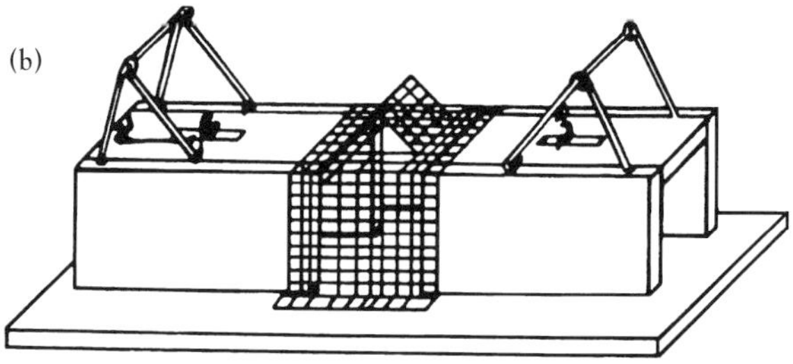

Figure 25. Single box-type gopher traps (a) can be used for ground squirrels. They can be used in pairs by removing the backs, connecting the two traps with wire mesh, and attaching them to a board (b).

squirrel activity. Ground squirrels quickly occupy abandoned burrow systems. After controlling squirrel infestations outside your orchard, use thorough cultivation or deep plowing to destroy burrow entrances and help slow the rate of reinvasion.

Traps. Trapping is an effective tool for controlling small populations of squirrels, and it can be used any time of year when squirrels are active. Trapping is especially effective from midspring through the fall. The most commonly used ground squirrel traps are kill traps such as Conibear traps or modified gopher box traps.

Conibear traps are most commonly placed unbaited in the burrow entrance (Fig. 24), where squirrels are trapped as they pass through. If you are using this type of trap within the range of the San Joaquin kit fox, you must place the trap in a covered box with an entrance no larger than 3 inches (7.5 cm) wide to exclude the fox, or you must spring the traps at dusk and reset them again in the morning.

Modified gopher box traps also are effective against ground squirrels (Fig. 25). The most attractive baits to use for ground squirrel traps are walnuts, almonds, oats, barley, and melon rinds. Grain baits are the most practical and least costly. Place the bait inside the trap well behind the trigger or tied to it. Bait the traps without setting them for several days until squirrels become used to taking the bait. Then put in fresh bait and set the traps. Box traps can be placed in pairs (Fig. 25), in groups, or inside larger boxes. Place the traps so that nontarget animals are not likely to be caught—for example, inside a larger box with openings no larger than 3 inches (7.5 cm) wide, which is just large enough for ground squirrels.

Fumigants. Fumigation can be very effective against ground squirrel populations. The best time to start fumigation is in the late winter or early spring, soon after squirrels have emerged from hibernation. At this time the squirrels are active and the soil is moist. Fumigation is not effective when squirrels are hibernating or aestivating, because at those times they seal off their burrows. Fumigation is much less effective when soil is dry, because more of the fumigant can escape from the burrows through cracks in the soil. When using a fumigant, be sure to treat all active burrow systems in and around the orchard. Recheck all areas a few days after fumigation and retreat any that have been reopened. For safety, do not use fumigants on burrow systems that are adjacent to or may open under buildings. FOLLOW LABEL DIRECTIONS CAREFULLY and understand the hazards when using fumigants.

Gas cartridges provide an easy and relatively safe way to fumigate ground squirrel burrows. They are available commercially and from some agricultural commissioners' offices. Use one or two cartridges for each burrow that shows signs of activity. More than two may be needed for a large burrow system. Quickly shove the ignited cartridges into the burrow, using a shovel handle or stick, and seal the burrow entrance with soil. Watch nearby burrow entrances; treat and seal any that begin to leak smoke. The larger and more complex the burrow system, the more smoke it takes to be effective.

Aluminum phosphide* is a very effective fumigant when used early in the spring when soil moisture is high. Application personnel should be trained in the material's proper use and on its potential hazards . When aluminum phosphide pellets come into contact with moist soil in the burrow they produce phosphine gas, which is highly toxic to any animal. When using aluminum phosphide, treat every active burrow, fill the entrance with a wad of newspaper, and cover with soil.

Baits. Poison bait is usually the most cost-effective method for controlling ground squirrels, especially for large populations. A bait consists of grain or pellets treated with a poison registered for ground squirrel control. To be effective, the bait must be used at a time of year when ground squirrels are feeding on seeds (see Fig. 23) and will readily accept baits. They are most effective in late spring or early summer. In fall months, squirrels cache a lot of seeds, so it may require more bait to control the population. Before you use baits, place small amounts of untreated grain near burrows to see if the squirrels will take it. If the grain is taken, proceed with baiting. If it is not taken, wait several days or a week and try again. As nuts reach maturity, the chance of bait acceptance decreases because squirrels strongly prefer nuts over grain baits. You may have to wait until the nut crop is harvested before the ground squirrels will take bait again, but by this time much of the crop may have been consumed. When using poison baits, be sure to follow label directions carefully to reduce hazards to nontarget species.

Multiple-dose anticoagulant baits can be applied in bait stations, as spot treatments near burrows, or broadcast over

Figure 26. Four-inch (10-cm) plastic pipe can be used to construct a bait station for ground squirrels. The 4-inch pipe must have its entrance restricted to 3 inches (7.5 cm) when used within the range of the San Joaquin kit fox. Place baffles inside the pipe to keep bait inside the station.

larger infested areas. For these baits to be effective, animals must feed on them over a period of several days. Anticoagulant baits are often the preferred method to use because they are very effective and do not produce bait shyness. Also, an antidote is available if accidental poisoning of domestic animals occurs.

Bait stations are most commonly used in orchards to provide bait for ground squirrels. Check the label to make sure that the bait you are using is registered for use in bait stations. Various kinds of bait stations can be used; all of them are designed to let squirrels in but to exclude larger animals. One design is made of PVC pipe (Fig. 26). Make the openings about 3 to 4 inches (7.5–10 cm) in diameter and incorporate baffles to keep the bait inside the station. Special types of stations must be used within the ranges of the San Joaquin kit fox or endangered kangaroo rats to ensure that these species are excluded (see Table 8 and Figs. 20 and 21). Place bait stations near runways or burrows and secure them so they cannot easily be tipped over. If squirrels are moving into the orchard from adjacent areas, place bait stations along the perimeter of the orchard where squirrels are invading, one station every 100 feet (30 m). Use more stations when the number of squirrels is high. Check bait stations daily at first, then as often as needed to keep the bait replenished. If bait feeding is interrupted, the bait's effectiveness is greatly decreased. Be sure to pick up any bait that is spilled and replace bait that is wet or moldy. Successful baiting usually requires 2 to 4 weeks. Continue to supply bait until feeding ceases and you observe no squirrels, and then properly dispose of unused bait.

*Restricted-use material. Permit required for purchase or use.

When specified on the label, anticoagulant baits can be applied as spot-treatments, which are the most economical and effective for small populations. Scatter bait (1 pound is about ten placements) evenly over 40 to 50 square feet (3.7–4.7 sq m) near active burrows. Reapply according to label directions to make sure there is no interruption in exposure to the bait. Scattering takes advantage of the ground squirrels' natural foraging behavior and minimizes risks to nontarget species. Never pile the bait on the ground; this increases the hazard to livestock and nontarget wildlife. After treatment, pick up and dispose of carcasses whenever possible to prevent poisoning of dogs or other scavengers. Burying is a good method as long as carcasses are buried deep enough to discourage scavengers.

Bait containing zinc phosphide*, an acute poison, can be applied as spot or broadcast treatments to control ground squirrels during the nonbearing season or outside of orchards. Spot-baiting involves scattering 1 tablespoon of bait (about 60 baits per pound) by hand over a 2- to 3-square-foot area around each burrow opening. Bait may also be broadcast over infested areas with a mechanical spreader at the rate of 6 pounds of bait per swath acre. Be sure to check for bait acceptance before you apply the bait. The primary hazard to nontarget animals is through eating the bait. Assess the potential hazard to humans, livestock, and wildlife before using this type of bait; if it is risky, use another method to control ground squirrels.

Consult the DPR World Wide Web site (http://www.cdpr.ca.gov/docs/es/index.htm) for the latest recommendations on using poison baits in areas that are within the range of endangered species.

Pocket gophers can damage almond trees by feeding on roots. Some of the roots of this tree have been eaten by gophers, greatly reducing the upper portion of the root system. Note the underground tunnel next to the trunk.

Pocket Gophers
Thomomys spp.

Pocket gophers are potentially serious pests, especially in young orchards. Herbaceous cover crops, especially legumes, are their preferred food, but they also feed on the bark of tree crowns and roots, girdling and killing young trees and reducing the vigor of older trees. Gophers sometimes damage drip irrigation lines. Burrow systems may divert irrigation water, causing water stress in younger trees and increasing soil erosion. The mounds of soil pushed out of burrows may interfere with harvesting, mowing, and other orchard activities.

Adult pocket gophers are 6 to 8 inches (15–20 cm) long with stout yellowish or grayish brown bodies and small ears and eyes. They rarely are seen above ground, spending most of their time in a system of tunnels they construct 6 to 18 inches (15–46 cm) beneath the surface. A single burrow system can cover several hundred square feet and consists of main tunnels with lateral branches used for feeding or for pushing excavated soil to the surface. Gophers are extreme-

Pocket gophers form conspicuous, fan-shaped mounds when excavating. Burrow openings are plugged. The presence of darker soil indicates fresh activity.

*Restricted-use material. Permit required for purchase or use.

ly territorial, so you rarely will find more than one gopher per burrow system. The conspicuous, fan-shaped mounds that are formed over the openings of lateral tunnels are the most obvious signs of gopher infestation. These tunnel openings are almost always closed with a soil plug. Gophers feed primarily on the roots of herbaceous plants, and they also clip small plants and pull them into their burrows to feed. Gophers may come above ground to feed on vegetation a few inches from the hole. When they have finished using these feeding holes, they plug them with soil.

Gophers breed throughout the year on irrigated land, with a peak in late winter or early spring. Females may bear as many as three litters each year. Once weaned, the young leave their mother's burrow and travel to find a favorable location for establishing their own burrow. Some may take over previously vacated burrow systems. Buildup of gopher populations in the orchard is encouraged by the presence of most cover crops, especially perennial clovers, which are a preferred food. When cover crops dry up, gophers may be forced to feed extensively on the bark of tree roots and crowns. Damage to orchard trees is always below ground and usually is not evident until trees show signs of stress.

Management Guidelines

The best times to check for gophers are in the fall and spring, when mound building activity is at a peak, and following irrigation. Monitor monthly in the spring and pay close attention to orchard perimeters to determine if gophers are invading the orchard. Monitor orchards with ground covers more closely, since they are more likely to support gophers and the presence of vegetation may make burrowing activity harder to see. Plan to monitor immediately after you mow or irrigate, when fresh mounds are easier to see. Look for darker-colored mounds that indicate newly removed soil.

Begin control as soon as you see any gopher activity in the orchard. For infestations that cover a limited area, use traps or hand-applied poison bait. For infestations that cover a large acreage, a mechanical burrow builder can be used to place bait. Trapping and hand-baiting can be used anytime during the year, but they are easier when the soil is moist and not dry and hard. Although mechanical burrow builders can be used anytime if soil conditions are right, the best time is after the first major rains in the fall, when the soil is moist enough to retain the shape of the mechanically constructed burrows. A second treatment can be made in the spring while there still is enough natural moisture in the soil to support an artificial burrow.

Clean cultivation of the orchard floor reduces the food supply and destroys some burrows, making the orchard less habitable for gophers. Clean cultivation also makes it easier to monitor gopher activity. Flood-irrigated orchards may have fewer gopher problems than orchards that are drip- or sprinkler-irrigated. In flood-irrigated orchards, watch for gopher burrowing activity on berms. Gopher control in adjacent areas reduces the potential for gopher problems in the orchard.

Traps. Traps are effective against small numbers of gophers, but are labor intensive and therefore relatively expensive. Either pincer-type or box-type kill traps can be used.

To place traps you need to probe near a fresh mound to find the main tunnel, which usually is on the lower side of the mound (Fig. 27a). The main tunnel usually is 8 to 12 inches (20–30 cm) deep, and the probe will drop quickly about 2 inches (5 cm) when you find it. Place two traps in the main tunnel, one facing each direction (Figs. 27b and 28). Be sure to anchor traps to a stake with wire. After placing the traps, cover the hole to keep light out of the tunnel. If a trap is not visited within 48 hours, move it to a new location.

Baits. Strychnine*, zinc phosphide*, and anticoagulant baits are registered for control of gophers. Bait must be applied below ground. Be sure to follow label directions carefully for application rates and safety considerations. For commercial

*Restricted-use material. Permit required for purchase or use.

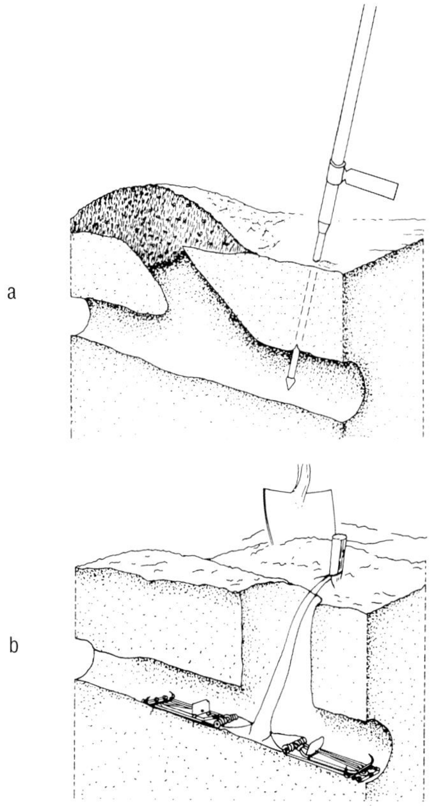

Figure 27. Use a probe to find gopher tunnels for placing traps (a). Begin probing 8 to 12 inches (20–30 cm) from the plug side of the mound. The probe will suddenly drop a few inches when you hit the main tunnel. Use a shovel to expose the main tunnel and place two traps in the tunnel, one in each direction (b). Tie the traps to a stake that is tall enough to be seen easily. Push each trap well back into the tunnel and cover the hole so that no light will enter.

Figure 28. When placing box traps for gopher control, be sure to cover or fill in the openings carefully so that no light will get in after the traps are pushed tightly against the tunnel openings. The traps in this drawing have not yet been set.

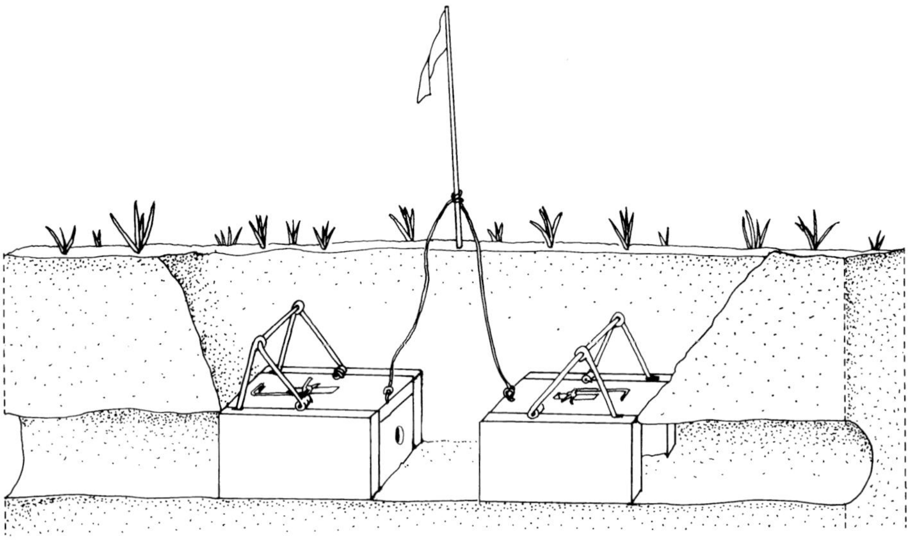

orchards, single-dose strychnine baits are by far the most cost-effective.

Apply baits by hand for small infestations or where the use of a mechanical burrow builder is not feasible. Use a probe to find the main tunnel next to a fresh mound or between two fresh mounds (see Fig. 27a). Once you find the main tunnel, enlarge the opening so you can drop bait into the burrow. After you place the bait, cover the hole to keep out light and prevent soil from falling onto the bait. Place bait in two or three places along the tunnel. This hand-application method can be used for single-dose or multiple-dose baits. Reservoir-type hand probes designed to deposit single-dose baits are available. Bait application is faster with these devices because they eliminate the need to place the bait by hand.

When large areas are infested, the most economical way to control gophers is to apply bait with a mechanical burrow builder. Burrow builders are tractor-drawn machines that construct artificial tunnels through infested areas, intersecting many gopher tunnels. Single-dose strychnine bait is dropped automatically into the artificial burrows, and gophers find the bait when exploring these new tunnels. Operate burrow builders down row middles within infested orchards and along orchard perimeters where gophers are moving in from adjacent infested areas. Adequate soil moisture is essential for successful burrow building. When operating the machine, periodically check to see that burrows are being formed properly. If possible, wait at least 10 days before running any other equipment over treated areas. Mechanical burrow builders may not work well on rocky soils; in some mature orchards that have not routinely been deep cultivated, surface roots may be damaged by the burrow builder. Do not use burrow builders in areas where gophers are not present: the constructed burrows may serve as travel ways for invading gophers.

After you use a mechanical burrow builder, follow up with a program of trapping or hand baiting to take care of any surviving or invading gophers. Begin about 10 days after mechanical baiting if there are signs of new gopher activity. If the orchard is free of gophers, concentrate your follow-up actions along orchard perimeters.

Fumigants. Most fumigant materials such as gas cartridges are not effective because gophers quickly seal off their tunnels when they detect the smoke or poison gases. However, aluminum phosphide* can be effective if applied in late winter or early spring when there is ample soil moisture to retain toxic gas, and before the gophers' major breeding period. Follow label instructions and follow all of the safety precautions given. To use aluminum phosphide, first probe to find the main burrow as with hand application of bait, then insert the number of tablets prescribed by the label into the burrow and seal the probe hole. As with other control methods, you need to keep monitoring for signs of renewed gopher activity. Re-treat the area if you find new mounds.

Eastern Fox Squirrel
Sciurus niger
Eastern Gray Squirrel
Sciurus carolinensis

In some almond-growing areas, two introduced species of tree squirrels, the eastern fox squirrel and the eastern gray squirrel, occasionally cause economic damage by feeding on nuts and collecting large quantities for storage. The eastern fox squirrel is much more common than the eastern gray squirrel, and is well established in city parks, residential areas, and adjacent agricultural lands, especially around Fresno, Sacramento, and coastal areas south of San Francisco. The eastern gray squirrel presently occurs only in a few locations in the San Francisco Bay Area, the San Joaquin Valley, and Calaveras County.

*Restricted-use material. Permit required for purchase or use.

Two tree squirrels are native to California: the western gray squirrel, *Sciurus griseus*, which in some almond-growing areas seems largely to have been displaced by the eastern fox squirrel and rarely is a problem to almond growers, and the Douglas tree squirrel, *Tamiasciurus douglasii*, which lives in forests and is not an agricultural pest.

In contrast to the ground squirrel, the tree squirrel has a bushier tail, and its fur is not flecked or mottled. Tree squirrels prefer to live in woody areas along creeks and river banks, from where they can invade adjacent orchards for food. Although they live mainly in trees, some tree squirrels, particularly the fox squirrel, spend considerable time foraging on the ground. The eastern gray squirrel feeds on a variety of green and ripe nuts, seeds, mushrooms, and fruits. The eastern fox squirrel has a similar diet but also feeds on bird eggs and insects. Both species usually feed only in localized areas of orchards adjacent to wooded areas. The eastern fox squirrel can be distinguished from the eastern gray squirrel by its reddish brown color; the eastern gray looks much like the native western gray squirrel.

Tree squirrels nest in tree holes or build nests of sticks and leaves in the branches of trees. Females of the eastern gray squirrel bear two litters per year of three to five young, one in late winter and one in late summer. Yearling females of the eastern fox squirrel bear one litter; older females have two litters per year, between January and April and between July and September.

Gray tree squirrels may only be controlled under a permit issued by the California Department of Fish and Game. Fox squirrels may be controlled without a permit; trapping and shooting are the only legal methods. Trapping is often sufficient to control tree squirrels that are damaging almond orchards. A ground squirrel box trap can be used. Fasten one or two box traps on a board and nail the board on top of a horizontal limb in a tree where you detect feeding damage. Place a handful of nutmeats well behind the trigger mechanism to attract the squirrels. You can remove a considerable number of tree squirrels with relatively few traps if you keep them in continuous operation while squirrels are feeding on almonds.

Tree squirrels may also be controlled by shooting, but it is illegal to poison them. Consult your local game official for regulations concerning the shooting or hunting of tree squirrels.

Voles (Meadow Mice)
Microtus spp.

Voles, also called meadow voles or meadow mice, damage trees by feeding on the bark around the base of the tree. Small trees are most susceptible to being completely girdled and killed by voles, but even large trees can be severely damaged or killed. Voles also can damage irrigation lines by chewing small holes in them. Voles are most likely to cause problems in orchards with year-round cover crops or where you allow dense vegetation to build up around the bases of trees. Vegetation management and, when necessary, the proper use of trunk guards on young trees usually keeps damage to a minimum. Baiting is used to control populations that reach harmful levels.

Adult voles are larger than house mice but smaller than rats. They are active both day and night and all year. Females bear several litters each year, with peaks of reproduction in spring and fall. Populations cycle, climaxing every 4 to 7 years and declining fairly rapidly. Grasses and other dense ground covers provide food and cover that favor the buildup of vole populations. You can recognize vole activity by the presence of narrow runways in grass or other ground cover connecting numerous shallow burrows with openings about 1½ inches (4 cm) in diameter. Voles seldom travel far from their burrows and runways.

Management Guidelines

In midwinter, begin to monitor monthly for active runways in cover crops or weedy areas. Look for fresh vole droppings and short pieces of clipped vegetation, especially grass stems,

The eastern fox squirrel, more common than the eastern gray, has a bushy tail and red-brown fur.

Voles (meadow mice) chew the bark off trees just above and below the soil line.

Voles are gray or brown, with a blunt nose, inconspicuous ears, and a short, slightly hairy tail.

in runways. Look for burrow openings around the bases of orchard trees. If you find any, remove the soil from around the base of the tree and look for bark damage. Voles usually start chewing on bark about 2 inches (5 cm) below the soil line and then move upward. If you do not check carefully, you may not notice damage until late spring or summer, when it may be too late to prevent significant injury to the trees. Be sure to monitor fence rows, ditchbanks, and other areas near the orchard where permanent vegetation is favorable for the buildup of vole populations.

Habitat Management. Ground cover provides voles with food and cover. Because voles travel only a few feet from their burrows, destruction of vegetation will cause them to abandon the site or die out. A vegetation-free zone 30 to 40 feet (9–12 m) wide between the orchard and adjacent areas helps reduce the potential for invasion by voles, but such a wide area is rarely practical. If you observe vole activity within the orchard, clean cultivation around the trees or good strip weed control may be sufficient to prevent extensive damage to trees.

Control the vegetation around tree trunks to reduce the likelihood of vole damage. Use hand-hoeing or herbicides to keep an area about 3 feet (90 cm) out from the tree trunk free of vegetation. If you maintain ground cover in the row middles, keep it mowed fairly short as it will be less attractive to voles.

Baiting. If you find damaging infestations within the orchard, poison baits can greatly reduce the vole populations. Baiting can also reduce populations in adjacent areas before they invade the orchard. Single- and multiple-dose baits are available. Restrictions related to endangered species are listed in Table 8. It is extremely important to understand and follow the label directions for use. For small infestations, scatter the bait in or near active vole runways and burrows according to the bait's label directions. For larger areas and where the bait label permits it, you can make broadcast applications. For noncrop land, apply bait in fall or spring before the reproduction peak. Note that label restrictions prohibit the application of bait during the nut-bearing season. Bait acceptance will depend on the amount and kind of other food available. One of the most effective baits for voles is crimped oat groats treated with the single-dose poison zinc phosphide.* Because this material has a high potential for creating bait shyness among survivors, do not apply it more than two times in one year, and space those applications several months apart.

Tree Guards. You can use wire or plastic trunk guards to protect young trees from voles and rabbits. An effective guard can be made of a cylinder of ¼-inch or ½-inch hardware cloth that is 24 inches (60 cm) wide and of sufficient diameter to allow several years' growth without crowding the tree. Bury the guards' bottom edge at least 6 inches (15 cm) below the soil surface to discourage voles from burrowing beneath them. Plastic, cardboard, or other fiber materials can be used to make trunk guards. These materials are less expensive, also provide sunburn protection, and are more convenient to use; however, they may not provide the same degree of protection against vole damage. If you use any of these other materials, check underneath them periodically for evidence of voles burrowing underneath them to gnaw on the tree trunk. Good basal or strip weed control improves the effectiveness of trunk guards.

Other Controls. Trapping is not practical for voles because so many individuals have to be controlled when they are causing problems in commercial orchards. Fumigation is not effective because of the shallow, open nature of vole burrow systems. Repellents are not considered effective in preventing damage.

Black-tailed Jackrabbit
Lepus californicus

Cottontail and Brush Rabbits
***Sylvilagus* spp.**

Rabbits may cause severe damage to young trees by chewing the bark off the trunk and clipping off branches within their reach to eat buds and young foliage. They may also gnaw on drip irrigation lines. Jackrabbits are the most common rabbit pests. Cottontail and brush rabbits may damage trees in orchards near the more wooded or brushy habitats favored by these species.

A jackrabbit is about the size of a large house cat. It has very long ears, short front legs, and long hind legs. Jackrabbits live in the more open areas of the Central Valley, coastal valleys, and foothills. They make depressions underneath bushes or other vegetation, where they remain secluded during

*Restricted-use material. Permit required for purchase or use.

the day. Jackrabbits are classified as hares, and as is characteristic with hares, the young are born fully haired, with open eyes, and become active within a few hours. Cottontail and brush rabbits are smaller than jackrabbits and have shorter ears. They build nesting areas where thick shrubs, woods, or rocks and debris provide dense cover. Their young are born naked and blind, and stay in the nest for several weeks.

Rabbits are active all year in almond-growing areas. They often live outside of orchards, moving in to feed from early evening to early morning. They damage trees primarily in winter and early spring, when other sources of food are limited. You can prevent damage with proper fencing or tree guards. You can also bait, trap, or shoot rabbits, depending on the species and the size of the population.

Management Guidelines

Periodically examine new plantings for rabbit damage. If you find damage, look for droppings and tracks that indicate rabbits as the cause. Voles also chew the bark from the trunk, but the bark damage caused by rabbits usually extends higher on the tree, and the tooth marks are distinctly larger. If you find damage, monitor the orchard perimeters in early morning or late evening to see where the rabbits are entering and to obtain an estimate of the number of rabbits involved. You can also estimate the number of rabbits at night by using a spotlight, which produces readily observed eyeshine. Once the orchard is 4 or 5 years old, rabbits usually do not present a serious problem.

Fencing. Rabbit-proof fencing is an effective tool for preventing damage to young orchards planted where rabbits are a major concern. Make the fence of woven wire or poultry netting at least 4 feet (1.2 m) wide and with a mesh diameter of 1 inch (2.5 cm) or less. Bend the bottom 6 inches (15 cm) of mesh at a 90-degree angle and bury it 6 inches deep, facing away from the orchard, to keep rabbits from digging under the fence. If you are building a fence to exclude deer, and rabbits are a potential problem, it is a good idea to add rabbit-proof fencing along the bottom. Unless you are already building a deer fence, the cost of a rabbit fence may be prohibitive for a large orchard when you only need it for a few years. Tree guards are an alternative.

Tree Guards. You can make tree guards from wire mesh (Fig. 29), hardware cloth, plastic, paper, or cardboard. Cylinders made from poultry netting, hardware cloth, or some

Rabbits can kill young trees quickly by chewing the bark off their trunks. This damage is similar to that caused by meadow mice, but it appears higher on the tree trunk.

Figure 29. Wire mesh cylinders secured with stakes prevent rabbit damage to young trees. If mesh is small enough and buried several inches, the cylinders will protect trees from voles as well; if tall enough, they will prevent deer damage.

plastics, and secured with stakes or wooden spreaders, offer good protection against rabbits. Make the cylinders at least 2½ feet (75 cm) high to keep jackrabbits from reaching foliage and limbs by standing on their back legs.

If you need protection against voles as well, use smaller-mesh wire and bury the bottom of the cylinder (see the section above on managing voles). The tree guard offers a practical way to prevent damage when you are replanting a few trees in an established orchard.

Baiting. Poison baits may be a practical means of control for large numbers of jackrabbits or for jackrabbits that are damaging trees over a large area. Baits cannot be used for cottontail or brush rabbits. Multiple-dose baits for jackrabbit control are available from most county agricultural commissioners' offices. Follow label directions carefully. These baits are placed in bait stations specifically designed for rabbits (Fig. 30). Place bait stations containing 1 to 5 pounds (0.5–2 kg) of bait near trails and secure them so they cannot easily be tipped over. Use as many stations as is necessary to ensure that all jackrabbits have easy access to bait, spacing them 50 to 200 feet (15–60 m) apart along the perimeter where they are entering the orchard. Inspect bait stations every morning for the first several days to keep bait supplies replenished; it may take this long before jackrabbits become accustomed to feeding at the stations. Increase either the amount of bait in the stations or the number of stations if all the bait is consumed in a single night. Replace bait that becomes wet or moldy. It usually takes 2 to 4 weeks or longer before results are seen with multiple-dose bait. Continue baiting until feeding ceases and you no longer observe jackrabbits. Because of the open nature of this bait station, be sure to take precautions so domestic animals or wildlife do not have access to the bait. Be sure to dispose of unused bait properly at the end of the baiting program, and bury or burn the rabbit carcasses on a regular basis. See Table 8 for restrictions related to endangered species.

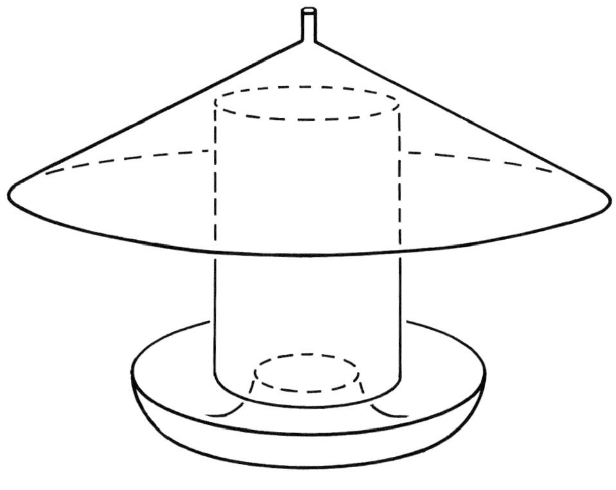

Figure 30. Self-feeding bait stations designed for jackrabbits consist of a covered container, which holds and dispenses anticoagulant bait.

Other Methods. Shooting, applying repellents, and trapping may provide effective control for small populations of rabbits, or it may be used to temporarily reduce damage until other measures such as fencing or tree guards can be put in place.

You can shoot all three types of rabbits if they are causing damage. If small numbers of rabbits are involved, this may be all that is necessary to prevent significant damage while trees are young. For best results, patrol the orchards early in the morning and late in the evening.

Repellents may provide temporary control for rabbit damage. They are sprayed or painted on tree trunks in the dormant season or on foliage and trunks during the growing season. Labels specify the proper application timing. Repeat applications as needed to protect new growth and to replenish repellent that is washed off by rain or irrigation.

Box-type or similar traps can provide effective control for small populations of cottontail or brush rabbits. Trapping generally is not effective against jackrabbits, because they do not readily enter traps.

Mule Deer
Odocoileus hemionus

Deer can be serious pests of newly planted trees in some foothill and coastal orchards and in Central Valley orchards near riparian habitats. Young trees can be severely stunted, deformed, or killed when deer browse on new shoots. Deer also feed on new growth on older trees, but this seldom causes significant damage. Bucks occasionally break limbs or injure bark when they use trees to rub the velvet off their antlers.

Management Guidelines

Deer feed mostly at night. To confirm their presence, look for tracks and fecal pellets in the vicinity of damaged trees. You may also use spotlights to check for deer at night. If deer are causing significant damage, deer-proof fencing provides the most effective and lasting control. It is costly, but if you are planting orchards where deer will present continuing problems, it will likely pay for itself in the long run.

Fencing. Fencing is most effective at excluding deer when it is put in place before you plant the orchard. Fencing must be at least 7 feet (2.1 m) high to exclude deer. On sloping terrain, an 8-foot (2.4-m) or taller fence may be necessary. Woven wire fences are used most often in California; however, electric fences and mesh fences made of polypropylene have gained some popularity in the past decade. Your choice of fence will be influenced by factors such as the potential severity and cost of deer damage, the duration over which you require protection, and the topography of the area.

Woven Wire Fences. Fences made of woven wire can effectively exclude deer if they are tall enough (Fig. 31). You can use a 6-foot (1.8-m) fence of woven wire with several strands of smooth or barbed wire along the top to extend the height to 7 or 8 feet (2.1–2.4 m). Be sure the fence is tight to the ground, or deer will crawl under. Check the fence periodically to make sure it is in good repair and that no areas have washed out, allowing deer to crawl under the fence. A smaller-mesh fencing installed and properly buried along the bottom of the fence will exclude rabbits as well as deer (see the section above on rabbits).

Wire mesh cylinders around individual trees may be effective where a few new trees are being planted in a location subject to deer damage. Make the cylinders at least 4 feet (1.2 m) tall and of large enough diameter to keep deer from reaching over them to the foliage. Secure the cylinders with wooden stakes so they cannot be tipped over.

Electric Fences. Electric fencing is less expensive to install than woven mesh fencing, but it costs more to maintain. High-tensile wire is the best choice because it is more resilient than other types; it can absorb the impact of deer, falling limbs, and farm equipment without stretching or breaking. Use a high-voltage, low-impedance power source that provides sufficient voltage to repel deer while being less likely to short out when vegetation touches the wires. Control vegetation around the base of the fence; in wet weather, contact with wet foliage can drain enough voltage from the fence to render it ineffective.

Other Controls. Habitat management usually is not an option for deer control because the deer travel long distances to reach food sources. Repellents may offer some protection to tree foliage, at least for a short time; they must be reapplied after rains and as new foliage emerges. Noisemaking devices may be effective for a few days, but deer quickly grow accustomed to them. If only a few deer are involved, having someone patrol newly planted orchards at night with a spotlight to frighten deer away may prove effective. The California Department of Fish and Game can issue depredation permits to allow you to shoot deer when they are causing damage. This may be necessary if a deer gets inside a fenced orchard and is not able to escape. Shooting will not solve a serious deer problem, but may prevent damage long enough to allow you to construct a fence.

Coyotes
Canis latrans

Coyotes occasionally cause damage in almond orchards by chewing on irrigation lines. Coyotes are classified as predators and can be taken whenever they are causing a problem. Contact your agricultural commissioner for details about regulations and control options.

Figure 31. A wire mesh fence is an effective means of excluding deer from an orchard if the fence is tall enough and kept tight to the ground.

Birds

Birds that cause significant damage to almond crops are crows (*Corvus brachyrhynchos*), scrub jays (*Aphelocoma coerulescens*), and in some areas, the yellow-billed magpie (*Pica nuttalli*). Occasionally during the dormant season, house finches (*Carpodacus mexicanus*) and to a lesser extent crowned sparrows (*Zonotrichia* spp.) may cause some disbudding in trees; this damage generally occurs in isolated areas and is a minor problem statewide. Table 9 briefly describes the most common species that cause damage in almonds, explains their legal status, and lists available management options.

Crows, scrub jays, and magpies can subsist on many kinds of food, including carrion, but their natural foods are insects, acorns, wild berries, and fruits. Cultivated fruits and nuts are often easy to obtain, and the birds take full advantage of these food sources. Scrub jays instinctively store food in the fall and carry away more fruit and nuts than they can eat. Thus, a single bird can account for considerable crop loss. Jays usually are solitary birds and occur singly or in pairs; rarely do they flock. In some orchard situations, with the ready availability of food, they have been known to occur in groups. Crows, on the other hand, join into flocks for most of the year and establish a central roost, which can consist of thousands of birds. Each day crows fly from the roost to feeding areas. Magpies generally occur in smaller groups of a few to several dozen birds. Crows, scrub jays, and magpies cause the greatest amount of damage during the period from hull split to harvest. After harvest, these birds can remove mummy nuts left in the orchard; however, allowing undisturbed access after harvest may contribute to increased bird damage the following year by improving survival and encouraging nearby nesting in the spring.

Table 9. Common Bird Pests of Almonds, Legal Status, and Management Options.

Bird name and silhouette	Description	Legal status and restrictions	Control options
Crowned sparrow (*Zonotrichia* spp.) 	Two species, 6 to 7 in (15–17 cm). Typical sparrow coloring of brownish on black with grayish breast. Adult white-crowned sparrow has three white and four black stripes on head. Golden-crowned sparrow has dull gold crown with black border. Small or large flocks feed on dormant flower buds.	Classified as migratory nongame bird. Can be controlled under the supervision of the county agricultural commissioner.	frightening, trapping
Crow (*Corvus brachyrhynchos*) 	Large, chunky, black bird, 17 to 21 in (43–53 cm). Heavy black bill and feet. Groups of a few birds or large flocks feed on mature nuts.	Classified as migratory nongame bird. Can be killed by landowners, tenants, or persons authorized by landowners or tenants when damaging crops.	frightening, shooting
European starling (*Sturnus vulgaris*) 	7½ to 8½ in (19–22 cm). with short tail. Bill is yellow in spring and summer, dark in winter. Plumage is iridescent black or purplish, heavily speckled with white. Large flocks feed on mature nuts.	Classified as nongame bird that may be killed at any time. No federal restrictions.	frightening, trapping, shooting
House finch (*Carpodacus mexicanus*) 	5 to 6 in (12–15 cm). Male has rosy red or orangish head, rump, and breast, with brownish back and wings, and brown streaks on sides. Female lacks the red or orange coloration. Small to large flocks feed on dormant flower buds.	Classified as migratory nongame bird. Can be controlled under the supervision of the county agricultural commissioner.	frightening,[1] trapping
Scrub jay (*Aphelocoma coerulescens*) 	Aggressive bird 10 to 12 in (25–30 cm). Head, wings, and tail are blue, underparts and back are gray, throat white. No crest. Usually solitary birds, sometimes groups, feed on mature nuts.	Classified as migratory nongame bird. May be killed with deperedation permit from U.S. Fish and Wildlife Service.	frightening, shooting
Yellow-billed magpie (*Pica nuttalli*) 	Large, noisy bird, 16 to 20 in (40–50 cm). Distinct black and white markings on body with very long tail. Small groups feed on mature nuts.	Classified as migratory nongame bird. No permit required to control birds damaging crops.	frightening, shooting

1. Most methods are not very effective for this species.

The head, wings, and tail of the scrub jay are blue; the breast and back are gray. A single scrub jay can account for considerable crop loss.

Crows are the most serious bird pest of almonds, feeding on nuts from hull split to harvest. During winter they may remove large numbers of mummy nuts that remain on the tree after harvest.

House finches and crowned sparrows move into almond orchards from nearby cover or resting areas, such as undeveloped brushy areas and power lines, to feed on flower buds before they open. Unless birds are observed feeding on buds or trees are inspected closely, damage may not be noticed until bloom, when damaged trees will have fewer blossoms in the upper canopy. Damage tends to be localized in orchard margins near bird habitat; orchards surrounded by other orchards usually sustain little or no damage.

European starlings (*Sturnus vulgaris*) and blackbirds (*Agelaius* spp., *Euphagus* spp.) sometimes invade orchards to feed on nuts. Rarely, flocks of band-tailed pigeons (*Columba fasciata*) will consume large numbers of immature almonds. Acorn woodpeckers (*Melanerpes formicivorus*), Lewis woodpeckers (*M. lewisi*), and northern flickers (*Colaptes auratus*) occasionally damage almond trees.

Monitoring

Regular monitoring will help you determine when damage actually starts so you can start control actions early. Birds are much more difficult to control once they have become used to feeding in a particular orchard. Become familiar with damaging species so you can distinguish them from the many nonpest birds that may frequent orchards. The illustrations in Table 9 will help you identify the most common pest species, and several bird identification guides are listed under "Vertebrates" in the suggested reading.

To monitor the occurrence of bird damage, it is easier to watch for the movement of birds into or within the orchard than to see the damage itself. This is particularly true of bud damage in winter, which is hard to see and may go undetected until bloom. Bud damage usually occurs in the upper parts of trees on the margins of an orchard, next to brushy or wooded habitat from which flocks of house finches or crowned sparrows move into the orchard. Plan to watch for bird activity once a week in locations where you anticipate this type of injury.

Bird counts can help you decide when to take control actions and whether the controls you have used are having an effect. Once nuts begin to develop, plan to drive through the orchards twice a week during the time of day you expect most bird activity and count the birds you see. Where birds are moving into the orchard from adjacent habitat, they can be counted by a stationary observer. Hand-held counters are useful for making these counts. Bird counts will tell you how severe the problem is and if it is increasing. Keep records of bird species, counts, and locations for each season; they will help you plan control actions in advance. This way you can have control devices on hand and in working order when they are needed.

Management Guidelines

Remove brush piles in or near orchards; they offer refuge and loafing sites for birds and may exacerbate problems. Removal of brushy vegetation adjacent to orchards may also reduce the potential for bird problems. Frightening devices—noisemakers and visual repellents—are the primary means of controlling bird damage in almond orchards. To be effective, several different methods should be used in rotation, and you should start using them as soon as birds appear. Once birds have become used to feeding in an orchard, they are much harder to frighten away. Trapping can be effective against house finches and starlings.

Frightening. The most effective way to frighten birds from the orchard is to use a combination of noisemakers and visual repellents. If your orchard is near an urban area, check to see if there are any local ordinances that may restrict the use of noise-making devices. For maximum effectiveness, rotate from one type of frightening device to another; otherwise, birds will become used to it. Monitor bird activity while using frightening devices and switch to a different type if birds appear to be getting used to the technique you are using.

Roving patrols that fire shell crackers, bird bombs, or whistler bombs are among the most effective ways to frighten birds from orchards. Stationary noisemakers such as gas cannons (propane exploders) and electronic noisemakers are most effective if you use at least one device for every 5 acres (2 ha) and elevate them above the level of the tree canopy. Move the devices to new locations every 3 to 5 days so the birds will take longer to get used to them.

The most commonly used visual repellents are mylar streamers and large "scare-eye" balloons. Attach balloons to poles so they are above the tree canopy and hang mylar streamers in trees. Use visual repellents in combination with noisemakers to increase their effectiveness. Use one type of visual repellent with a each type of noisemaker and switch to a different visual repellent when you switch noisemakers.

The most effective bird frightening program makes use of as many different kinds of noisemakers and visual repellents as practical. Shooting is usually used to reinforce the various frightening methods:

First week:	Patrol orchard firing cracker shells.
Second week:	Use stationary propane exploders.
Third week:	Use stationary propane exploders and also patrol orchard firing cracker shells.
Fourth week:	Install electronic noisemakers with "scare-eye" balloons or mylar streamers near trees where damage is most severe.

Shooting. Birds such as scrub jays and magpies that usually invade orchards in small numbers can often be controlled by shooting. Check with local Fish and Game officials before shooting any birds. A depredation permit is required if you want to shoot scrub jays. Permits are not presently required for shooting crows, magpies, or starlings that are causing damage, but it is a good idea to check with authorities because regulations may change. Where permissible, shooting occasionally at a few birds will increase the effectiveness of your other noisemaking techniques, because the birds will begin associating loud noises with the real hazards of firearms.

Trapping. Trapping can be an effective way to control house finches and starlings, especially if it is conducted over a relatively large area; for instance, if several adjacent growers conduct a trapping program. The most effective trap for these species is the modified Australian crow trap. Diagrams and details of its use are given in *Vertebrate Pest Control Handbook,* listed under "Vertebrates" in the suggested reading. Successful trapping must take into account the behavior patterns of the birds being controlled. Place traps in suitable locations with adequate food and water to keep the trapped birds alive. Traps need to be checked regularly (at least once a day), and trapped birds should be removed and euthanized humanely. Trapping is best carried out by someone experienced with the technique. For house finches and crowned sparrows, trapping must be conducted under the supervision of the county agricultural commissioner.

Insects and Mites

Numerous insects and mites inhabit almond orchards. Most cause little or no damage to the tree or crop. Many play a beneficial role in the orchard system by performing activities such as pollination, feeding on organic debris, or feeding on pests. Only a small fraction cause economic damage to the crop. Of these, navel orangeworm, peach twig borer, San Jose scale, webspinning mites (twospotted and Pacific spider mites), and ants have the greatest potential for economic impact on the tree or crop throughout the almond-growing regions in California.

Navel orangeworm larvae overwinter in mature nuts, or mummies, that remain in the tree after harvest. Adult moths emerge in spring and lay first-generation eggs on mummy nuts in the orchard. The second generation may also develop on mummies or damaged new crop nuts; but at hull split, egg laying switches to the new crop, and the population begins a period of rapid growth. Larvae feed on all parts of the new nuts—the husk, shell, and nutmeat. Two cultural methods, effective removal of mummy nuts and a rapid, early harvest, provide the most effective control of navel orangeworm. In orchards isolated at least ¼ mile (0.4 km) from external sources of infestation, or in localities where area-wide orchard sanitation is practiced, these two procedures will reduce navel orangeworm damage below economic levels and eliminate the need for in-season chemical treatments for this pest. Some damage will occur in orchards where a substantial number of mummies remain from the previous season, regardless of the chemicals used or the area of the state in which the crop is grown. Hence, orchard sanitation, whenever possible, is essential to navel orangeworm management.

Peach twig borer, also a moth, is another major pest of almonds. Immature larvae overwinter in small chambers, or hibernacula, primarily excavated in crotches of 1- to 4-year-old limbs. In spring, they leave their hibernacula and feed on leaf and blossom buds, and inside the tips of branches. Unlike navel orangeworm, twig borer larvae can enter green nuts before hull split, although this occurs very rarely. The major period of crop infestation is during and after hull split, when twig borer larvae may enter the kernel and feed on nutmeats. Properly timed applications during bloom of the bacterial insecticide *Bacillus thuringiensis* (Bt) or the selective material spinosad can control peach twig borer.

San Jose scale does not directly damage nutmeats but can damage the tree, eventually causing yield reductions. It feeds on plant juices and contributes to an overall decline in vigor, growth, and productivity. Severe infestations can kill young trees and twigs and branches of older trees, reducing vigor and productivity. In some areas, San Jose scale populations are becoming resistant to a number of pesticides traditionally used to control this pest.

Application of superior-type oil during the dormant season, at proper rates with good spray coverage, controls immature San Jose scales and other scale pests, as well as the eggs of European red mite and brown mite. In many cases, dormant oil treatments and resident natural enemy populations will keep San Jose scale from reaching damaging levels provided that broad-spectrum insecticides are not used during the growing season. Addition of insecticide to the dormant oil may be needed to control heavy infestations because mature scales are not killed by oil alone. Applying insecticide during the dormant season is less harmful to natural enemy populations, although there is the threat that runoff will contaminate rivers or creeks. Dormant oil plus insecticide will also control peach twig borer, although bloomtime application of Bt or spinosad is the preferred control method if scales are not a major problem.

When adequate cultural controls against navel orangeworm are practiced, dormant oil sprays are used for scales, and bloomtime treatments are made for peach twig borer, in-season sprays for insect control may not be necessary. Some insecticides kill the natural enemies of spider mites and scales and cause flare-ups of these pests. If you don't spray during the growing season for navel orangeworm, peach twig borer, or San Jose scale, natural enemies will have a better chance of maintaining pest populations below damaging levels.

Several species of mites can cause economic damage in almond orchards. The occurrence of these mites on almonds and their relative importance varies regionally. Two of these species, twospotted and Pacific spider mites, are important throughout the Central Valley. These two species, along with the strawberry spider mite, are referred to as webspinning mites. They cause a similar type of damage, have similar life histories, are difficult to distinguish in the field, and are controlled in the same manner. Mite outbreaks are often induced by foliar insecticides; use of pyrethroids either during the season or in dormant sprays is especially likely to cause outbreaks of mites. Certain orchard conditions such as dust from roadsides, water stress, and frequent cultivation or flailing also tend to promote mites. If not carefully monitored and treated when necessary, webspinning mites can defoliate trees and reduce the following season's growth and yield. Trees that are stressed, for example, those receiving insufficient irrigation, will suffer greater defoliation from the same level of mite feeding than trees that are not stressed.

In addition to webspinning mites, European red mite and brown mite (also called brown almond mite) are occasionally present in almond orchards in large numbers. Both species are controlled adequately by good coverage with a high label rate of dormant oil spray and usually do not require in-season treatments. Low to moderate numbers of these mites in spring are beneficial because they serve as a food source for mite predators. If mite predators have an adequate food supply early in the season, their populations can build up sufficiently to control webspinning mites later in the summer.

Ants are mostly a problem in the San Joaquin Valley, although they can cause damage in all almond-growing regions. The pavement ant and southern fire ant are the most important economically. They feed on nutmeats at harvest when the nuts are drying on the ground and may feed on mature nuts still on the tree. Some ants tend scales and aphids, protecting these pests from natural enemies and allowing them to increase. Ants can sometimes be suppressed by hull split sprays, but severe infestations need to be controlled with a soil-applied insecticide. More effective and less disruptive baits are available; they must be applied well in advance of harvest to be effective.

Other insect pests of almonds include oriental fruit moth, several borers, olive scale, European fruit lecanium, boxelder bug, leaffooted plant bug, stink bugs, fruittree leafroller, tent caterpillars, tenlined June beetle, and leafhoppers. Many of these insects occur only in localized areas, or they occur infrequently and infestations are not predictable. Of these pests oriental fruit moth, the plant bugs, and some of the borers have the greatest potential for causing economic damage.

Oriental fruit moth sometimes injures terminal shoots of trees in the same way as peach twig borer, and shoot strikes caused by these two pests are easily confused. Young trees can be damaged severely by shoot dieback, especially during training when they have a limited number of terminal shoots. Oriental fruit moth can also be found mining in almond hulls. Although most of their feeding occurs between the hull and shell, larvae occasionally feed on the nutmeats as well.

Peachtree borer, American plum borer, prune limb borer, and Pacific flatheaded borer can severely damage or kill a young tree and greatly reduce the vigor of an older one. Peachtree borer has long been a pest in the Santa Clara Valley and Contra Costa County and occasionally causes problems in almonds in the San Joaquin Valley. Larvae enter trees at or near ground level and tunnel in the wood of the trunk. American plum borer and prune limb borer occur in the Central Valley. They invade young trees and mine the layer of wood just beneath the bark of the scaffold crotch. As a result, scaffolds frequently break off during windy periods.

Shothole borer is a beetle that infests trees already weakened by some other stress. If the infestation is not severe and the primary source of beetles is treated or removed, trees can usually outgrow a shothole borer invasion when they regain vigor.

Two species of scale, olive scale (an "armored" scale) and European fruit lecanium (a "soft" scale), occur sporadically throughout the almond-growing regions. Neither insect damages trees as seriously as the San Jose scale. Both are controlled by natural enemies, if not disrupted by insecticides applied for other pests, or by a dormant oil spray and rarely cause any significant damage.

Occasionally an orchard may be invaded in the spring by true bugs that migrate into almonds and feed on the crop until a more suitable host becomes available. Boxelder bugs generally occur in the Sacramento Valley, stink bugs are more of a problem in the San Joaquin Valley, and leaffooted plant bugs are more common in the Fresno area although they do cause problems in other areas. Feeding by these bugs causes small nuts to abort and creates small black depressions on the kernels of older nuts. Although many nuts can be affected, these bugs are usually not treated; their occurrence is unpredictable, and by the time damage is noticed treatments will not prevent much additional damage.

Tenlined June beetle occurs on almonds in a few locations of the San Joaquin Valley. Larvae of this beetle feed on tree roots and slowly kill infested trees. The only available control is to remove affected trees and replant after fumigating the soil.

Fruittree leafrollers and tent caterpillars feed on the buds and foliage of almond trees. The leafrollers occasionally get into the green nuts but do not normally cause serious economic damage.

Monitoring Insect and Mites

Monitoring for insects and mites is essential to a good pest management program. Use monitoring results, together with field history and weather data, to identify potential problem areas or "hot spots," to determine whether or not a spray is necessary, to time sprays, to determine the effectiveness of a previous treatment, and to forecast the need for additional control measures. Monitoring programs are available for navel orangeworm, peach twig borer, San Jose scale, mites, ants, oriental fruit moth, and some of the tree borers. You must correctly identify pests, their associated damage, and their natural enemies; even closely related species may require different control measures. Photos and descriptions in this manual will assist you in the identifications. For additional help, consult your UCCE farm advisor or pest control adviser (PCA) and the UC IPM World Wide Web site (www.ipm.ucdavis.edu).

Monitor for insect and mite pests at the times indicated in Figure 32. Begin monitoring before damage occurs and continue through the pest's damaging stages. Some pests are best monitored during the dormant period, others during the

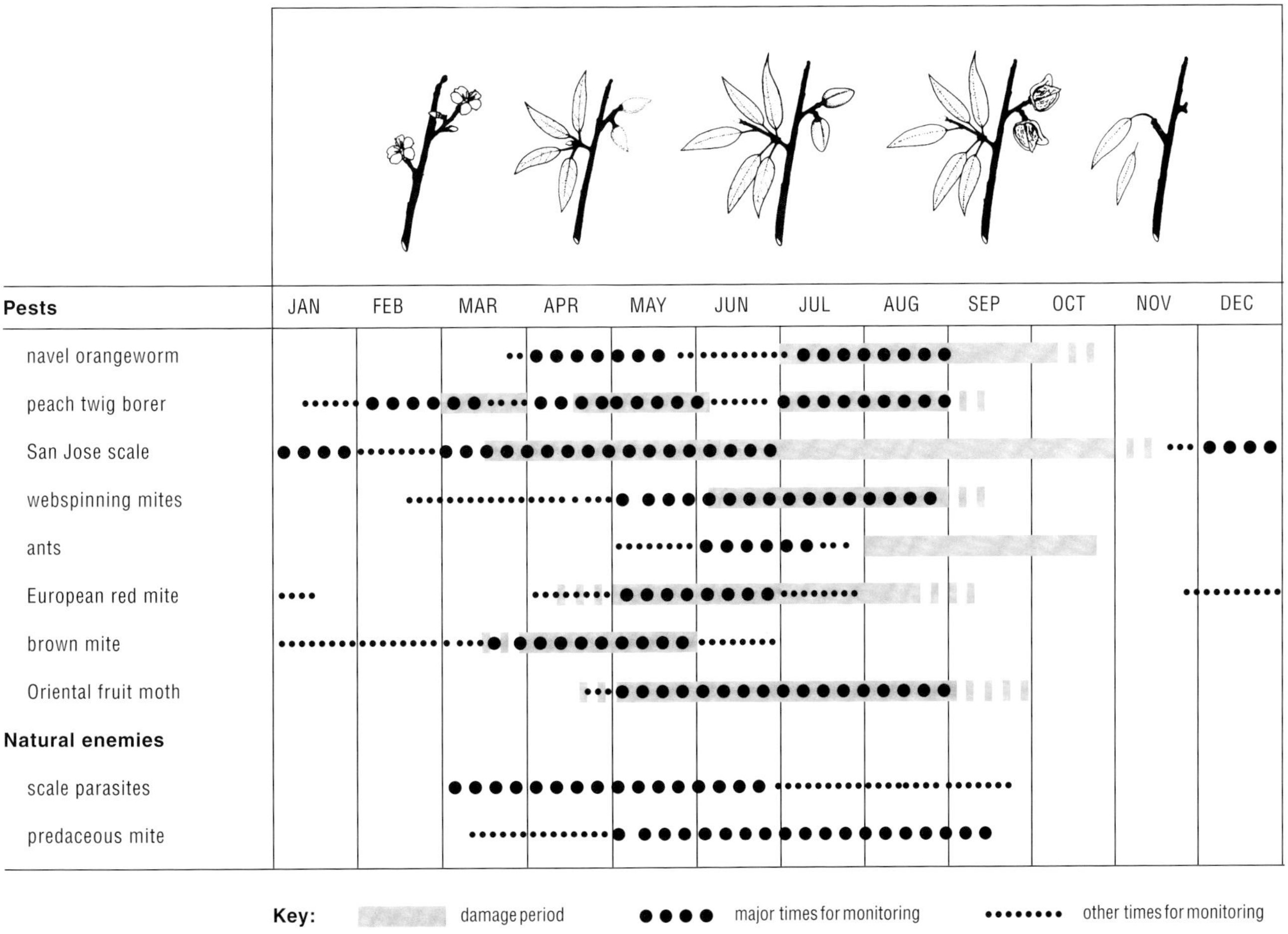

Figure 32. Monitor insect pests, mite pests, and natural enemies when they are likely to cause damage or at other times indicated. Developmental times for the almond branch across the top of the figure are for an early cultivar. Events may occur up to 1 month later in late cultivars.

growing season. Check every orchard at least once a week during the growing season and more often during critical periods.

When you monitor insects and mites, keep in mind that different species or different life stages prefer certain parts of the tree or the orchard. For example, brown mites feed on leaves only during the coolest part of the day and are found on woody areas of the tree at other times. Infestations of webspinning mites are most likely to begin in dusty locations, usually roadside trees, and toward the center of individual trees early in the season. European red mite infestations frequently develop in shaded areas with high humidity. Pest development may vary among neighboring orchards due to different cultural practices or the age and cultivar of the trees.

Keep a record of all your monitoring results. Make separate records for each orchard, noting special conditions and pest problems. Record your results on a sampling form, such as the one in Figure 33 (for insects monitored with traps) or Figure 47 (for mites monitored by visual counts).

In addition to counting the actual number of insects trapped, keep records of temperatures and use this information to help you follow insect development using degree-days. You may use data from your own weather station or use weather data from a local station available from the UC IPM Web site (www.ipm.ucdavis.edu). Programs are available at this Web site to calculate degree-days for specific pests using data from local weather stations.

Monitoring Methods

The sampling area, or block, will depend upon the site of the orchard, the insect monitored, and the ability to make individual treatment decisions on portions of the orchard. Conditions within each area should be uniform, and all sections of the orchard, except hot spots, should be included in the sampling scheme. Hot spots, such as the edges of an orchard where dust or insecticide drift may lead to a buildup of mites, areas with shallow soils, or stressed areas, should be monitored and treated separately. At each sampling area, look for natural enemies of mites as well as for pest species. Sufficiently high populations of natural enemies can rapidly bring a potentially damaging mite population under control. As a general guideline, divide large orchards into 80-acre (32-ha) sampling blocks or blocks that you plan to spray as a unit. Orchards that exceed 1,000 acres (404 ha) can be divided into larger sampling blocks if conditions within each block are uniform. Within each block, choose four sampling areas and place one peach twig borer pheromone trap and one San Jose scale pheromone trap in each area. If you anticipate the need to schedule a treatment for navel orangeworm, place two egg traps in each area. If you are monitoring for oriental fruit moth as well, place one oriental fruit moth trap in each area. For convenience, hang the peach twig borer, San Jose scale, oriental fruit moth, and navel orangeworm traps on two or more adjacent trees. Place the second navel orangeworm trap about 10 trees away from the first. In orchards that are 80 acres (32 ha) or less, select four sampling areas and place one navel orangeworm, peach twig borer, and San Jose scale trap in each area. If you are also monitoring for oriental fruit moth, use one trap at each location for this moth as well. You can use trees with traps and adjacent trees to monitor other pests, such as mites and wood borers.

The above guidelines represent the ideal. The actual number of traps used in a particular orchard will also be influenced by other criteria, such as whether or not the sampler is monitoring several orchards in the same vicinity, thus reducing the number of traps needed in each orchard. Always use at least two traps in smaller orchards. More detailed information on sampling is given in the sections on specific pests in this chapter.

Visual Sampling. Some pests can be monitored by thoroughly examining leaves, trunks, twigs, or nuts for eggs, larvae, adults, or damage. Check for scale insects in the tops of trees by examining prunings for the presence of live scales during the dormant season. To detect tree borer infestations, examine the tree trunk and scaffold crotch carefully for entry holes or frass. The type of borer being monitored will determine the location on the trunk to look for evidence of an infestation. Examine the orchard floor for ants.

Assess mite populations by sampling leaves to determine the presence or absence of mites and mite predators or by brushing and counting the number of mites per leaf (see the monitoring section for mites in this chapter). Mites may first appear in hot spots, such as dusty border areas of the orchard or areas with coarser or shallower soils that are more likely to become stressed. Do not base a treatment decision for the entire orchard on samples taken from these areas. Spot-treating localized infestations preserves natural enemies in other areas of the orchard. Mass releases of predatory mites may be used for localized infestations of pest mites, but conservation of natural predator populations is always a good idea.

Although egg traps are used to follow seasonal development of navel orangeworm (see below), the severity of navel orangeworm infestation in soft-shelled almond cultivars is correlated with the presence of mummy nuts. These are easily seen in winter.

Traps. Monitoring programs for navel orangeworm, peach twig borer, San Jose scale, and oriental fruit moth use traps that attract insects by scent. These traps can help you detect the presence and seasonal development of these pests in the orchard. You can use trapping information to time pest control actions more effectively. The number of traps needed depends on the size of the orchard and the insect being monitored. For convenience, place the traps for the four species in adjacent trees at a sampling site.

INSECT MONITORED ______________________________

ORCHARD LOCATION ______________________ TRAPPING PERIOD ____________ TO ____________

TRAP BOTTOMS CHANGED ______________________ PHEROME CAPS OR BAIT CHANGED ____________

COMMENTS ______________________________

____

____

		INDIVIDUAL TRAP COUNTS					AVERAGE TRAP COUNTS
Date	Accumulated degree-days	Trap 1	Trap 2	Trap 3	Trap 4	Trap 5	Pests/trap/day

Figure 33. Record data from traps on a record sheet, such as this one, or on a computer spreadsheet. Use a separate form for each insect—that is, navel orangeworm, peach twig borer, San Jose scale, and oriental fruit moth—or you can adapt a form for the different species you may wish to monitor.

Egg-laying activities of navel orangeworm can easily be monitored with egg traps, small plastic containers with two large windows that are covered with nylon organdy screening. The bait is a mixture of almond press cake and almond oil that attracts female moths, which lay eggs on the outside of the trap. Traps painted with flat black spray paint are more effective than white traps.

Egg traps can help you determine when navel orangeworm flights are beginning so that degree-day calculations can be used to time a hull split spray properly, if needed. Traps should not be used to determine if a spray is necessary: the number of eggs found on traps has not been correlated to the subsequent magnitude of navel orangeworm damage. However, experience with traps may allow you to conclude that navel orangeworm pressure is particularly high or low in a given year.

Pheromone traps are used to monitor peach twig borer, San Jose scale, and oriental fruit moth activity. Pheromones are attractants produced by animals to communicate with individuals of their own species. Pheromones currently being used in almond IPM programs are sex pheromones released by females to attract the males for mating.

Most commercially available pheromone traps are made of cardboard with one or more surfaces coated with a sticky substance. Attached to the trap is a replaceable plastic or rubber lure (cap) impregnated with a pheromone. The scent is slowly released from the lure, and attracted insects get stuck on the sticky surface of the cardboard. Sex lures are species-specific, although other closely related species are sometimes caught. If you are catching other nonrelated species, contact the manufacturer. Pheromone traps require regular changing of the lures and sticky liners, but they are easier to service than most other trap types, such as blacklight traps or liquid bait traps, and are more economical.

Traps can be used to keep track of pest development. They can also be used to schedule monitoring activities and control actions when used together with degree-day calculations. However, they should not be used to assess damage potential and the need for control action. Flight activity varies with environmental conditions, and trap catches may vary with weather. For example, cool temperatures or rainy weather may result in low trap catches. Be sure to correlate trap catches with weather information. Trap catch also varies with the number of traps used for a given acreage.

Prevention and Management

Cultural Practices. Orchard sanitation is an important management tool for reducing the buildup and spread of insects. Removing mummy nuts from the trees and destroying them deprives the navel orangeworm of overwintering sites and egg-laying sites in the spring. Also, removing fruit during the winter from nearby alternate hosts such as walnut, prune, fig, and peach can reduce navel orangeworm populations by eliminating these potential overwintering sites. Harvesting early reduces navel orangeworm infestations because larvae invade only nuts with split hulls; early harvest reduces the time susceptible nuts are on the tree, and navel orangeworm moths will not lay eggs on nuts that are on the ground. Adequate irrigation enables mite-infested trees to withstand feeding by greater numbers of mites before economic loss occurs than mite-infested trees that are not receiving sufficient irrigation. Finally, pruning trees to remove damaged or dead limbs will reduce invasions by wood-boring insects.

Table 10. Natural Enemies of Insect and Mite Pests of Almonds.

	NATURAL ENEMIES	
PEST	Parasites	Predators
navel orangeworm	*Goniozus legneri*, *Copidosoma plethorica*	lacewings, minute pirate bugs, assassin bugs, damsel bugs, *Phytocoris* spp., spiders
peach twig borer	*Bracon gelechiae*, *Euderus cushmani*, *Macrocentrus ancylivorus*, *Copidosoma varicornis*, *Spilochalcis*, *Toxophoroides*, tachinid fly (*Erynnia* sp.)	gray ant, grain itch mite, lacewings, minute pirate bugs, assassin bugs, damsel bugs, *Phytocoris* spp., spiders
webspinning mites, European red mite	—	western predatory mite and other predatory mites, spider mite destroyer, sixspotted thrips, minute pirate bugs, lacewings, western flower thrips, *Phytocoris* spp.
San Jose scale	*Aphytis* spp., *Encarsia (Prospaltella) perniciosi*	lady beetle (*Chilocorus orbus*), lacewings, *Cybocephalus californicus*, *Phytocoris* spp.
Parthenolecanium scales	*Aphytis* spp., *Coccophagus* spp., *Encarsia* spp., *Metaphycus* spp.	lady beetles (*Chilocorus orbus*, *Hyperaspis* spp., *Rhyzobius lophanthae*), lacewings, *Cybocephalus californicus*, *Phytocoris* spp.
oriental fruit moth	*Ascogaster quadridentata*, *Apanteles* sp., *Bracon* sp., *Macrocentrus ancylivorus*	lacewings, minute pirate bugs, assassin bugs, *Phytocoris* spp., spiders

— = no data

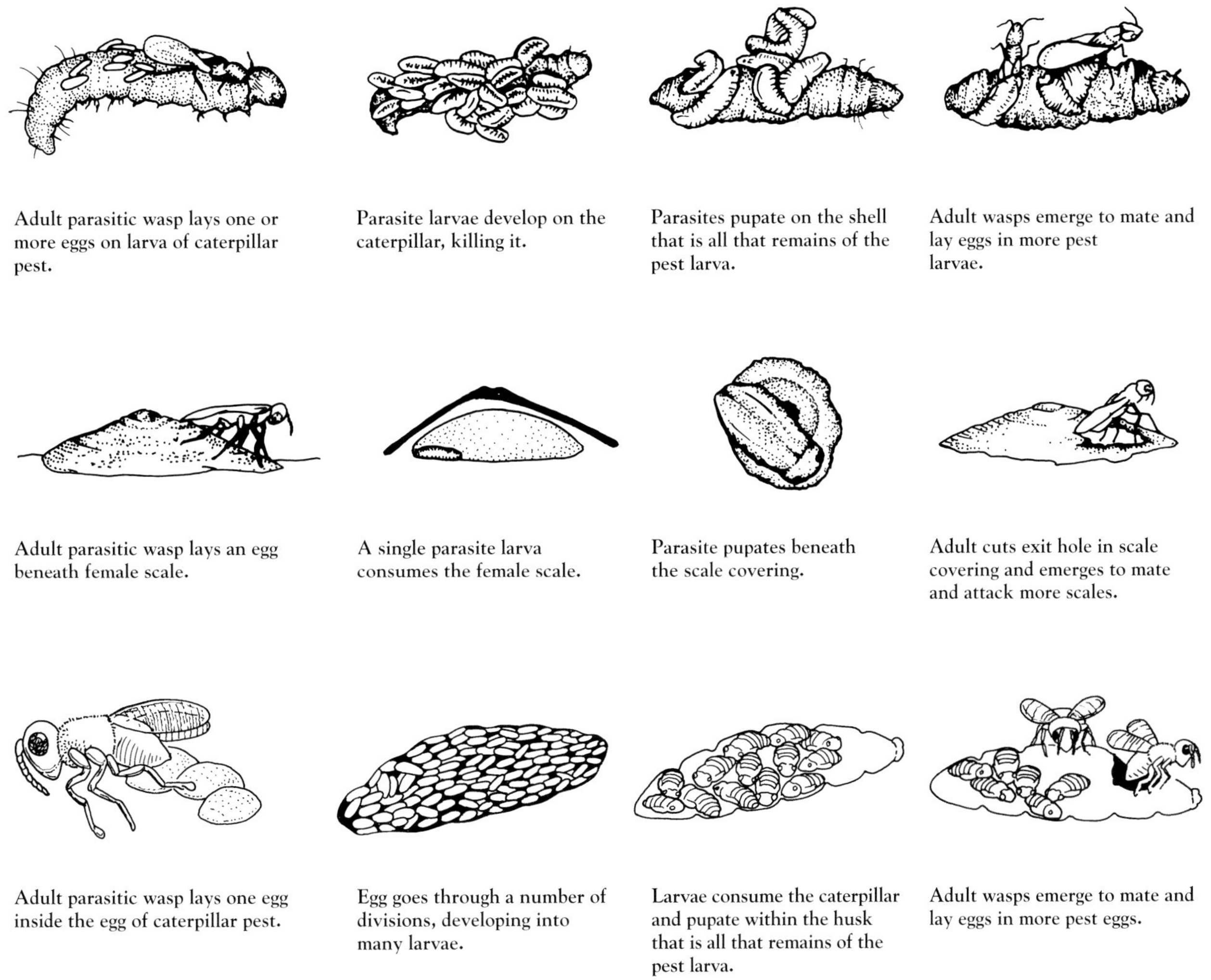
Adult parasitic wasp lays one or more eggs on larva of caterpillar pest.
Parasite larvae develop on the caterpillar, killing it.
Parasites pupate on the shell that is all that remains of the pest larva.
Adult wasps emerge to mate and lay eggs in more pest larvae.
Adult parasitic wasp lays an egg beneath female scale.
A single parasite larva consumes the female scale.
Parasite pupates beneath the scale covering.
Adult cuts exit hole in scale covering and emerges to mate and attack more scales.
Adult parasitic wasp lays one egg inside the egg of caterpillar pest.
Egg goes through a number of divisions, developing into many larvae.
Larvae consume the caterpillar and pupate within the husk that is all that remains of the pest larva.
Adult wasps emerge to mate and lay eggs in more pest eggs.

Figure 34. Three examples of parasite life cycles. Top: *Goniozus legneri* on navel orangeworm. Middle: *Aphytis* sp. on San Jose scale. Bottom: *Copidosoma (Pentalitomastix) plethorica* on navel orangeworm.

Biological Control. Potential pest populations in almond orchards may be controlled by other organisms inhabiting the orchard. These biological control agents include predators, parasites, and pathogenic microorganisms. Numerous native predators and parasites of almond pests occur throughout the Central Valley. In addition, exotic natural enemies of the navel orangeworm (those found in its natural home range) have been collected and introduced into the Central Valley by University of California scientists.

Natural enemies are very important in the control of mites and scales but do not always provide economic control of other major almond pests. Natural control of scales and mites is easily disrupted by in-season chemical treatments, resulting in pest outbreaks. You can reduce chemically induced mite outbreaks by carefully selecting the types and rates of insecticides used and timing applications so that they are least disruptive to natural enemies. Sprays that reach a cover crop may destroy natural enemies residing there. Use directed sprays and spot sprays as much as possible to minimize harm to nontarget species. (Table 10 lists natural enemies of almond pests, and Figure 34 illustrates the life cycles of three different parasites.)

Pheromone Mating Disruption. Growers can use mating disruption with pheromones, also called pheromone confusion, to control populations of peach twig borer. This technique also works well for oriental fruit moth; however, because this insect is not predictably a pest of almonds, its use would rarely prove economical. Mating disruption may be available in the future for controlling navel orangeworm. The success of mating disruption requires correct timing and proper placement of dispensers. The technique is most effective in orchards with lower pest populations and orchards that are isolated from untreated hosts of the pests being controlled. Area-wide programs may be necessary for mating disruption to be successful in almond orchards. Mating disruption does not work well in windy locations, and control is poor in the upper parts of orchards situated on sloping terrain. In orchards with high pest populations, it is necessary to use higher rates of pheromone and more point sources, and to place dispensers high in trees. New methods for dispensing pheromones are continually being developed to improve the technique. More specific information on materials and rates can be found in *UC IPM Pest Management Guidelines: Almond*, listed under "Pesticide Application and Safety" in the suggested reading and available at the UC IPM web site (www.ipm.ucdavis.edu). Be sure to follow manufacturers' instructions closely with regard to the placement and number of dispensers.

Oviposition Disruption. Navel orangeworm moths are attracted to nuts for egg laying (oviposition). Volatile compounds produced by the nuts are responsible for this attraction. If these compounds are applied to leaves and branches of almond trees, females will lay eggs there, and the larvae that hatch from these eggs will not feed and develop successfully. Commercial products that stimulate this egg-laying behavior are being produced, and they may prove useful for selectively managing navel orangeworm.

Pesticide Sprays. Correct application of a needed pesticide is essential to a successful integrated pest management program. Choose pesticides that have the least impact on natural enemies to improve your overall IPM program. To achieve effective pest control, you must choose the correct rate, coverage, ground or air speed, gallonage, and type of equipment (see the section on pesticides in the chapter "Managing Pests in Almonds.") For information on the type and rate of pesticide to use, see *UC IPM Pest Management Guidelines: Almond*, listed under "Pesticide Application and Safety" in the suggested reading and available at the UC IPM Web site (www.ipm.ucdavis.edu).

Dormant Sprays. Applying an oil spray during the dormant season controls immature stages of San Jose and other scales, and also controls eggs of mite species that overwinter in the orchard as eggs. The oil can be applied any time after the first heavy rain and after the leaves have fallen in the winter; however, application just before or at bud swell usually is most effective. A heavy rain before oil application ensures that the trees are not moisture stressed, thus avoiding possible phytotoxic effects of the oil on the trees. Absence of leaves allows best coverage with the spray.

Adding an organophosphate insecticide to the oil spray has been used for many years to control mature stages of scale and the overwintering stage of peach twig borer. The oil helps the insecticide penetrate to target areas and is responsible for killing mite eggs. There is serious concern about the runoff of organophosphate insecticides into surface water during the dormant season. Best management practices (BMP's), such as maintaining a cover of vegetation on the orchard floor, using filter strips, and proper disposal of waste water, are recommended to help reduce runoff. Special care in using these materials must be exercised in sensitive areas particularly prone to runoff. Pyrethroids or spinosad, a more selective material less harmful to natural enemies, can be used in place of organophosphates, but they do not control scales. In addition, pyrethroids can be more harmful to predatory mite populations. Avoid using conventional insecticides with dormant oil applications, or use them sparingly based on results of monitoring scale and peach twig borer populations. They are generally unnecessary if scale populations are low, and bloom sprays or pheromone disruptants can be used to manage peach twig borer.

Bloom Sprays. Carefully timed sprays of *Bacillus thuringiensis* (Bt) or spinosad during bloom are the preferred method for control of peach twig borer in an IPM program. These materials are much less harmful to natural enemies and honey bees than conventional insecticides, although care must be taken not to apply spinosad when bees are present in the orchard. Guidelines for using bloom sprays are given in the section below on peach twig borer. Bt and spinosad do not control scales, so other control actions for scales may be needed where bloom sprays are used.

In-Season Sprays. Postbloom in-season sprays are not required in many orchards in many years if sanitation practices and dormant and/or bloom sprays are used. However, some orchards in some years will have high pest populations in spring and may require sprays in May or June. Sprays later than June are rarely recommended.

Use the guidelines in this chapter to determine which in-season sprays, if any, you will need to apply. The optimal spray timing for a particular pest will vary with temperature conditions from year to year; it can also vary in regard to the optimal spray timing for other pests. Figure 35 illustrates examples of optimal early season spray timings for major almond pests, based on degree-day accumulation techniques outlined in this chapter, for three different years. To effectively time pesticide applications, always monitor the pests you need to control.

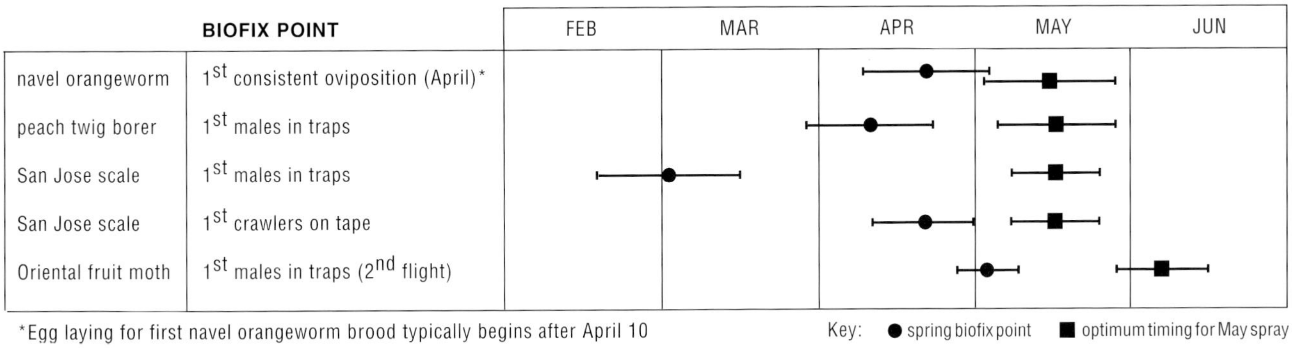

Figure 35. Spring biofix points and optimal May spray timings for navel orangeworm (NOW), peach twig borer (PTB), San Jose scale (SJS) and oriental fruit moth (OFM). The bars around these points indicate the variation in timing between warmer years, when they occur earlier, and cooler years, when they occur later. Note the synchrony of the four species to each other. San Jose scale has two biofix points, depending on the trapping method used. May sprays for these insects are recommended only when high populations warrant their use. These sprays are generally more disruptive than dormant or bloom sprays or sprays applied for later generations.

Navel Orangeworm

Amyelois transitella (Pyralidae)

Navel orangeworm is the primary insect pest of almonds in California. A native of the southwestern United States and Mexico, it was first discovered feeding on damaged navel oranges in Arizona in 1920. Although this was the origin of its name, navel orangeworm is not an economic pest of oranges. It feeds on the dried, decaying, or damaged fruits of many trees, including citrus, stone fruits, figs, and pomegranates, as well as a number of noncrop plants such as acacia and locust. In almonds, walnuts, and pistachios, however, navel orangeworm attacks healthy nuts after hull split and is a major pest.

Description

Adult moths have irregular silvery-gray and black patterns on the forewings and legs. The navel orangeworm belongs to the pyralid, or snout moth, family and has a small, snoutlike projection formed by a pair of palps in front of the head. In cool, wet spring weather adults may live 2 to 3 weeks; as the season progresses and days become hot and dry, they survive less than a week. Most female moths mate within 2 days of emergence and lay an average of 85 eggs per female over a period of 2 to 8 nights, depending upon the temperature. No eggs are laid when night temperatures drop below 55°F (13°C).

First-generation and some second- and third-generation eggs are laid singly on the surface of mummy nuts; occasionally they are also found close to the nut on twigs or nut stems. Most of the second-, third-, and fourth-generation eggs are laid on the new crop nuts. When first laid, the eggs are creamy white, flattened, and oval with irregular reticulations on the surface. The color changes from white to pink and then to reddish orange just before hatching. At this time, the black head capsule can also be seen inside the egg. Eggs hatch in 4 to 23 days, depending on temperature, with eggs laid in summer hatching in 4 to 8 days. Development of all life stages is temperature-dependent and occurs more rapidly during warmer weather.

Newly hatched larvae are the same reddish orange color as the mature egg. After feeding begins, larval color is influenced by the nature of the food and can range from milky white to pink. The head capsule is dark reddish brown in all instars. A pair of crescent-shaped marks on the second segment behind the head distinguishes navel orangeworm from most other larvae except carob moth (covered next in this chapter). Upon reaching maturity, larvae spin a closely woven cocoon and pupate.

Navel orangeworm pupae are light to dark brown and are found inside nuts or between hulls and shells, where the larvae have been feeding. The pupal stage lasts from 6 to 10 days in hot weather to several weeks in late winter or early spring.

Seasonal Development

Moths of a given generation may be active and lay eggs over a period of about 2 months, depending on the temperature. Those activity periods are called flights. The navel orangeworm has three and sometimes four flights each season. The fourth flight occurs in years when temperatures are warmer than usual and degree-day units are accumulated more rapidly than in cooler years. Figure 36 shows the typical seasonal development of the navel orangeworm.

Navel orangeworms overwinter as larvae in almond mummy nuts that remain in the orchard after harvest. Larvae do not usually enter diapause, so feeding and development may occur during the warmer periods of winter. Larvae pupate within the mummies in early spring, and moths of the overwintering brood begin emerging in late March or April, marking the beginning of the first flight. Emergence and egg laying by the overwintering generation usually peak in May and are completed by early to mid June. First-generation eggs are laid on mummy nuts remaining in the trees, and larvae develop exclusively in these nuts. Egg traps are used to monitor for the laying of first-generation eggs, which establishes the biofix used for degree-day calculations.

First-generation female moths that emerge in mid to late June for the second flight lay their eggs on mummy nuts or occasionally on green nuts damaged by peach twig borer or oriental fruit moth. Navel orangeworm larvae can then enter these damaged nuts before hull split and go directly into the kernel to feed.

At the onset of hull split in the Nonpareil cultivar, egg laying begins on the new crop nuts. As the infestation develops, moths prefer to oviposit on nuts previously infested by navel orangeworm or peach twig borer, although they readily attack uninfested nuts. In soft-shell cultivars larvae penetrate the shell and feed on the nutmeats, but they are unable to penetrate the shell of hard-shell cultivars. Larvae can also feed and complete their development in the hull. Although a single larva in an almond of the new crop is most common, there may be two or three, and as many as 17 have been found in a single nut.

Second-generation larvae that have developed on the Nonpareil crop during July and August may emerge as moths (third flight) in time to lay eggs on and reinfest this crop—hence the importance of a timely harvest. They can also infest later-maturing crops of pollenizer cultivars, such as Monterey or Fritz, still to be harvested. Navel orangeworm populations can grow rapidly through the remainder of the season because the new crop nuts are a much better and more plentiful food supply than mummies from the previous year, and because temperatures are warmer during July and

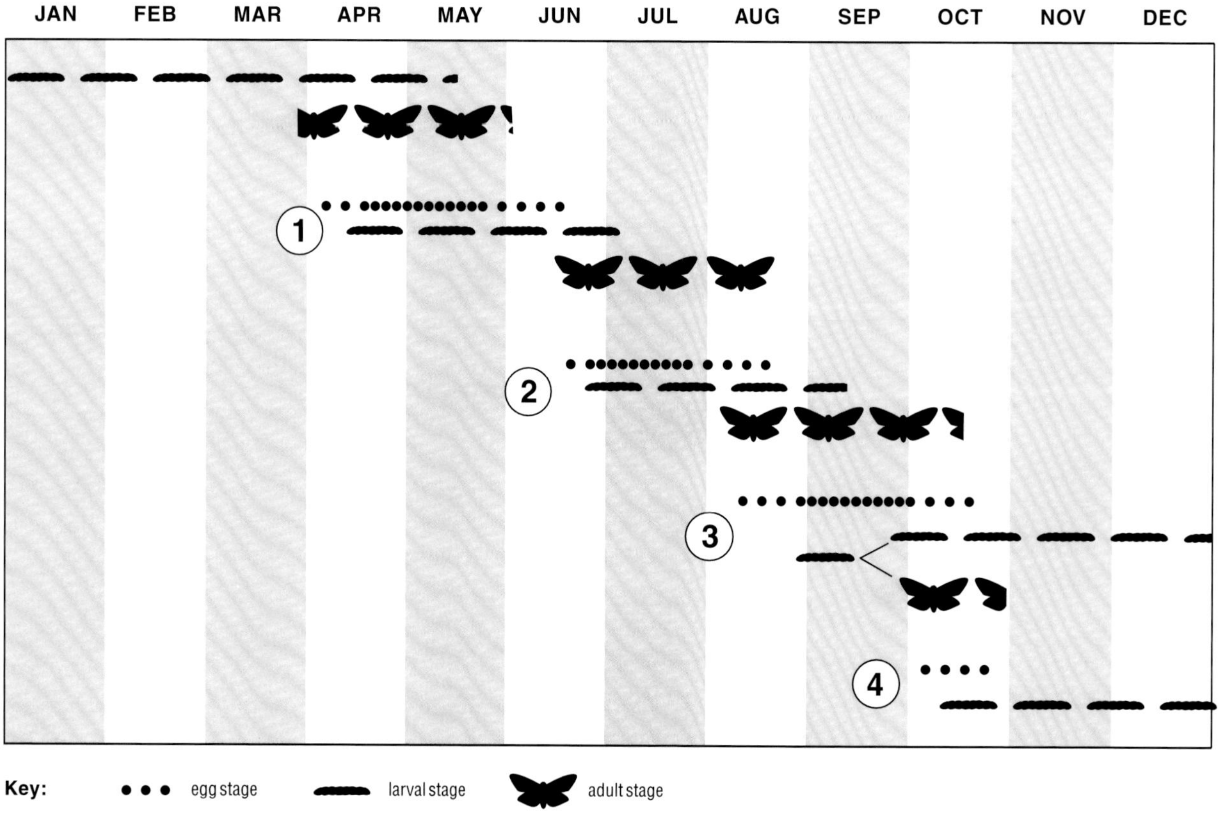

Figure 36. Seasonal development of the navel orangeworm. Egg laying by navel orangeworm varies from year to year and from district to district. For the egg stage, the denser dots indicate the period during which egg laying typically peaks. The date hull split begins can vary from one year to the next, depending on temperature. The fourth flight and the partial fourth generation are more common in southern portions of the Central Valley, especially in years with above average temperatures.

Adult navel orangeworms are dull, grayish brown moths with irregular silver-gray and black patterns on the forewings and legs. A pair of palps in front of the head form a snoutlike projection.

Navel orangeworm eggs are creamy white when first laid. The color changes from white to pink and then turns reddish orange just before hatching.

NAVEL ORANGEWORM

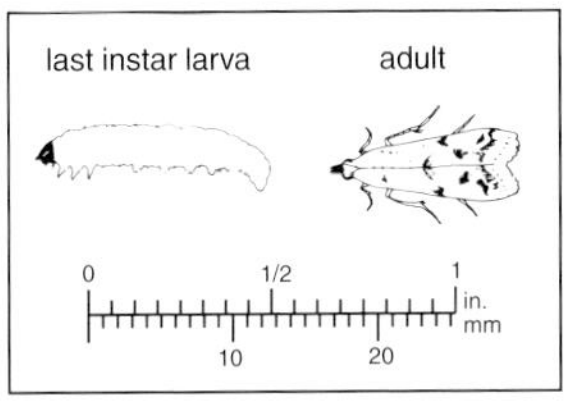

Older navel orangeworm larvae have a dark red-brown head capsule and range in color from milky white to pink.

Newly hatched navel orangeworm larvae are the same red-orange color as the mature egg.

August than in the late spring and early summer. As a result, the second generation of moths (third flight) is significantly larger than the overwintering or first generation, and because they lay eggs for the third generation, they have the potential to cause the most nut damage.

In unusually warm years, some of the early third-generation eggs may be able to complete development and give rise to a small moth flight. This mostly occurs in the central and southern portions of the San Joaquin Valley in years of above-average temperatures. These moths lay eggs for a fourth generation. Eggs hatch and larvae overwinter along with some of the older third-generation larvae.

Damage

Navel orangeworm larvae damage almonds by feeding directly on the nutmeats. First-instar larvae that bore directly into the nutmeat leave a minute quantity of frass at the point of entry, or pinhole. Larvae that feed first on the hull or outer parts of the shell and bore into the nutmeat in the second or later instar do not produce a pinhole, but

The navel orangeworm larva has a pair of crescent-shaped marks on the second segment behind the head. The presence of these marks distinguishes it from other larvae such as the oriental fruit moth, but not from the closely related carob moth.

rather a larger entry hole. As the larvae feed and get larger, they consume most of the nut and produce large amounts of frass and webbing.

Nuts infested in July are often completely consumed by harvest. Most of the completely consumed nuts are blown out by the pick-up machines and hullers and are not recorded as worm-damaged in harvest samples. Eggs laid on nuts just before harvest will hatch, and the larvae will develop in harvested nuts that are not fumigated immediately. Besides damaging nutmeats and leaving frass and webbing, navel orangeworms cause nuts to be more susceptible to fungal infections that can lead to aflatoxin contamination, a serious threat to the marketability of any crop. The almond industry has established strict grading standards to control the risk of aflatoxin for consumers.

Some cultivars of almonds are more susceptible to navel orangeworm damage than others. Shell hardness and seal and the time and duration of hull split all influence susceptibility. Soft-shell cultivars with a poor seal are more susceptible to navel orangeworm infestations than hard-shell cultivars with tight seals. Later-maturing soft-shell cultivars with a lengthy hull split period are also very susceptible to attack because the nuts are still on the trees when third-generation moths emerge in August and September to lay eggs. Therefore, late-maturing cultivars with a lengthy hull split period, soft shells, and poor seals show the highest levels of navel orangeworm damage. In recent years industry-wide nut rejection levels, which are primarily due to navel orangeworm damage, have been highest in Jordanolo, Merced, Thompson, and Carrion cultivars, followed by Nonpareil, Ne Plus Ultra, and Fritz, in that order. Ruby, Peerless, Butte, and Price are relatively resistant, while Mission, Padre, and Carmel are the most resistant to navel orangeworm.

Shell hardness, seal, and length of hull split period (and therefore, the susceptibility of cultivars) may vary somewhat from year to year, depending on growing conditions. Also, navel orangeworm infestations may vary according to crop set. A light crop may have a heavier infestation than a heavy crop because navel orangeworm oviposition is concentrated on fewer nuts. The result may be a higher percentage of damaged nuts in a given year, even though the size of the navel orangeworm population is the same as in previous years.

Navel orangeworm larvae consume the almond and produce large amounts of frass and webbing. A dark brown pupa can be seen encased in the webbing and frass in the center of the photo.

Mummies, or nuts that remain on the tree after harvest, provide overwintering sites for navel orangeworm larvae.

Management Guidelines

Two cultural methods—effective removal of mummy nuts in fall or winter and rapid, early harvest—provide the most effective control of navel orangeworm. Insecticide treatments are needed only when these practices are not carried out properly or where infested trees remain nearby. If sprays are needed, they are timed to coincide with egg laying in spring or at hull split as indicated by egg traps hung in the orchard. Insecticides provide, at best, only 50% control. Two introduced wasps can be found in almond orchards, but they cannot be relied on to provide effective control alone.

Orchard Sanitation. Navel orangeworm overwinters and spends the first generation in spring in mummy nuts. The most effective way to prevent economically destructive populations of navel orangeworm is to remove mummy nuts from the trees by February and destroy them. Best control is achieved when an average of less than two mummies per tree remains after February 1. Mummies on the ground should be destroyed by March 15. When a good orchard sanitation program is carried out in an orchard at least ¼ mile (0.4 km) from infested trees, together with an early harvest, sprays are not usually needed for navel orangeworm. Area-wide sanitation programs are even more effective at keeping navel orangeworm below damaging levels. Such programs should be instituted wherever possible. Even when chemical control programs are planned, partial removal of mummy nuts will reduce navel orangeworm damage.

Remove mummy nuts from the tree either as part of normal harvest operations or during the winter (December and January). If mummies are removed as part of harvest, returns from recovered nuts may help offset the cost of cleaning.

Mummy nuts can be removed from the tree by mechanically shaking the tree or by hand-poling. If an average of two

Shakers can be used to remove mummy nuts during winter.

Mummy nuts also can be removed by knocking them from trees with poles.

or more mummies per tree remain in January, pole or shake again. Walk through the orchard, selecting at least 2 trees per acre at random, and counting the mummies to get an average. Generally, trunk shaking is cheaper than hand-poling and is better for trees over 12 feet (3.7 m) tall, if the orchard soil is not too wet for the shakers. Loss of a few fruit buds from mechanical shaking does not significantly affect the following season's yield. Hand-poling may be efficient for trees less than 20 feet (6.1 m) tall and with fewer than 50 mummies per tree, but it should be done before bud swell or it may remove a significant number of buds. Bud removal may slightly reduce crop potential—especially on Merced and Thompson cultivars, where many of the mummy fruits and buds are located along main limbs. After removing mummies, be sure to spot-check your work by counting as described at the beginning of this paragraph. Perform further removal if the average for the orchard is above two per tree.

Remove mummies in the winter under the wettest possible conditions, preferably after 1 or 2 days of wet fog or drizzle. Moisture softens the fibers holding the mummies on the tree and also soaks into the mummies, making them heavier and easier to remove. Fog and moisture during the winter months may be limited, making cleanup hard to complete. When optimal conditions occur, first shake or pole the cultivars that are most susceptible to navel orangeworm infestation and most difficult to clean, such as Merced and Thompson. These cultivars are prime overwintering sites for navel orangeworm. Less-susceptible cultivars can be knocked later. Some growers conducting winter cleanup take maximum advantage of the number of hours of fog or heavy dew by shaking trees at night or in the early morning.

Other cultivars such as Nonpareil, Ne Plus Ultra, and Price are easier to clean. Mummies on Mission trees should also be removed because navel orangeworm larvae can overwinter between the hull and the shell of these nuts. Mummies on Peerless trees generally fall off the trees; cleaning this cultivar is not essential, but it will help to concentrate bird activity on other mummies in the orchard that might harbor navel orangeworm.

Birds feed on mummy nuts and often remove substantial numbers during the winter. If you clean the orchard in early December, the birds will concentrate their activities on any remaining mummy nuts over a longer period of time. Birds may be significant in the final stages of orchard cleanup in orchards near river bottoms, in flyways, or adjacent to other crops where they often rest or feed. In orchards that are in areas of large, solid blocks of almond plantings, however, bird activity may be inconsequential.

Mummy nuts remaining on the ground after harvest or winter nut removal should be destroyed by the time moths start emerging in mid-March. You can still benefit from mummy destruction as late as May 1, but for best results, mummies should be destroyed by March 15. Discing or flail mowing between rows, a normal operation in many orchards, along with wet conditions that are conducive to rotting, will kill most navel orangeworms present. On bare ground where herbicides are applied over the entire orchard floor or on herbicide-treated strips, berms, or permanent irrigation checks, blow or sweep nuts to the middle of the row before discing or flailing. Although mummy nuts will decompose adequately without flailing in orchards with dense ground cover, bare areas in the ground cover will allow navel orangeworms to survive.

Unkempt orchards and backyard fruit and nut trees are a source of reinfestation when they are near a commercial orchard and should be cleaned up if possible. Uncleaned commercial orchards are a source of infestation for neighboring orchards. Sanitation is most effective when it is carried out on an area-wide basis. For maximum effectiveness from a sanitation program, an orchard should be at least ¼ mile (0.4 m) from external sources of infestation. Even in nonisolated orchards, however, sanitation is beneficial and economical.

Rapid, Early Harvest. Getting Nonpareil and other soft-shelled almonds off trees, picked up, hulled, and to the handler as early and quickly as possible is essential for effective navel orangeworm control. Even in orchards that are cleaned and that have received navel orangeworm sprays, damage can be high if early-maturing soft-shelled almonds are not harvested as early as possible. This is because egg laying by the third navel orangeworm flight coincides with the optimal time of harvest.

For an early harvest, you must take into account not only the degree of hull split and the percent of nut removal but also the start of the third navel orangeworm flight as well. To detect the beginning of this flight, use egg traps and inspect green nuts for the first new eggs, which will indicate that the third flight has begun. You can use a degree-day model to predict the start of egg laying by the third flight in the same way as described below for the second flight under "Hull Split Sprays." If possible, do not delay Nonpareil harvest beyond the beginning of the third flight; a second shake can be performed at less cost than the additional insecticide treatment that is needed if harvest is delayed, and damage will be lower on the nuts harvested at this time.

The earlier you harvest your soft-shelled crop, the less time it will be exposed to the damaging third-generation larvae, and the lower your infestation will be. However, if harvest is too early, problems with embedded shells can be increased, reducing nutmeat quality. When 95 to 100% of the fruit at eye level are visibly split (Stage 3 or more, see Fig. 6), shake a few trees to determine if nut removal is satisfactory. If too many nuts remain in the tree, try again in a few days. When at least 95% of the nuts fall when trees are shaken, begin harvest. Hulls will still be green but they will dry rapidly on the ground. Navel orangeworm moths do not lay eggs on nuts on the ground.

If early-maturing, soft-shelled almonds are removed from the orchard in adequate time, much of the navel orangeworm population will be removed with them, reducing the number of moths in the third flight. Early harvest of later-maturing soft-shelled cultivars also prevents overlap of generations on the same crop.

If you can't harvest an entire orchard as early as desired, first harvest those blocks that had a higher mummy load at the beginning of summer or are near external sources of navel orangeworm. Harvest susceptible soft-shelled cultivars, such as Nonpareil, Thompson, Merced, and Ne Plus Ultra, as early as possible.

Almonds can be picked up after harvest as soon as the hulls are dry enough for hulling. The length of time for almonds to dry on the orchard floor depends on a variety of factors such as temperature, the degree of shading, the cultivar of the tree, how soon before harvest the trees were irrigated, and how soon after hull split the nuts were harvested.

A good rule of thumb to determine hullability and field storage is to hold the hull where it is split from the shell between the thumb and index finger and bend it back on both sides. If it snaps, it will hull properly. If the hull just bends back, the nuts are not ready for the huller. Check 10 randomly selected nuts from several locations in the orchard. If at least 80% of them snap, the nuts can be delivered to the huller.

If you are unable to harvest before the beginning of the third flight, pick up the almonds as soon as possible after they are dry and immediately hull or fumigate them. Fumigation at this time destroys larvae while they are small and before they damage the kernel. Otherwise, navel orangeworms already present in those almonds will continue to develop and cause damage. More importantly, adults can emerge from these nuts to infest later-maturing pollenizer crops still in the trees. Rapid removal of nuts at harvest will also help reduce ant and squirrel damage. If possible, fumigate or hull worm-susceptible cultivars first. Until the hulls and nuts are separated, eggs and small larvae on the hulls remain a threat to the nutmeats.

On-Farm Fumigation. Consider on-farm fumigation immediately after harvest if navel orangeworm infestations were a problem during the season. Although almonds are fumigated as a standard practice upon receipt by handlers, the interval between harvest and handler fumigation can be several weeks. During this time, navel orangeworm and peach twig borer larvae already present in the nuts can continue to develop and cause more damage. Even if the crop is sent directly to the huller, enough time may elapse that on-farm fumigation will provide significantly less insect damage. On-farm fumigation is simple, inexpensive, and will halt activity of navel orangeworm and other insects. Follow the guidelines for fumigation in *Fumigation of Inhull Almonds on the Farm*, listed under "Pesticide Application and Safety" in the suggested reading.

Biological Control. Several parasitic wasps have been introduced into California and are commonly found parasitizing navel orangeworm larvae in commercial almond orchards.

The encyrtid wasp, *Copidosoma* (*Pentalitomastix*) *plethorica*, was introduced into California from Mexico in the 1960s and is now widespread in California. This parasite lays its egg inside the navel orangeworm larva, and each egg develops into a large number of larvae that consume the host and pupate inside the exoskeleton of the dead orangeworm. The bethylid wasp, *Goniozus legneri*, was imported from Uruguay and Argentina and released in California orchards in 1979. This wasp lays one or more eggs on the surface of navel orangeworm larvae, and each egg develops into a single parasite larva that consumes the host from the outside. Both wasps coexist in almond orchards, but *Goniozus legneri* is found primarily from Stanislaus County south. Together these two parasites appear to cause a greater mortality of navel orangeworm than either one alone. Although both often reach high densities, natural populations do not reliably provide effective control of navel orangeworm in all orchards. *Goniozus legneri* is very common in orchards with high numbers of mummy nuts, and therefore navel orangeworms, but it does not reach high densities in orchards with low densities of navel orangeworm. One reason is that it overwinters poorly compared to navel orangeworm. Augmentation of natural *G. legneri* populations with releases of insectary-reared wasps can increase parasite numbers. However, such releases should not be relied upon as the sole control technique. They should be used in conjunction with good sanitation, early harvest, and monitoring of navel orangeworm activity.

Although birds are generally considered orchard pests in the summer, they do remove mummy nuts from the trees following harvest and are considered beneficial during the winter months. Their effectiveness is determined by the orchard's proximity to areas where birds normally congregate in the winter months.

The parasitic wasp *Copidosoma (Pentalitomastix) plethorica* lays its egg inside the navel orangeworm larva. Each egg develops into a large number of larvae that eventually pupate inside the exoskeleton that is all that remains of the dead larva, as shown in this photo.

The parasitic wasp *Goniozus legneri* lays one or more eggs on the surface of navel orangeworm larvae. Parasite larvae consume the orangeworm from the outside.

Monitoring and Insecticide Sprays. Manage your orchards to avoid spraying for navel orangeworm; sprays can cause serious outbreaks of mites and destroy natural enemies of the navel orangeworm, San Jose scale, and other insect pests. In years with a good crop set, an orchard that meets all the criteria listed below will probably not need an in-season spray to deliver a crop with a tolerable level of orangeworm damage (up to 4%) on Nonpareils. (In years when the crop set is light, however, damage may be higher because progeny of overwintering navel orangeworms will be concentrated on fewer new nuts, and the larger nuts of a small crop will have sutures that are open wider, making them more susceptible to worm damage.) Sprays may not be needed in an orchard:

- That has an average of less than two mummies per tree on February 1. Count the number of mummies on 2 randomly selected trees per acre (10 per ha) to determine the average number of mummies per tree. Check a minimum of 15 trees per block.

- In which mummies on the ground are destroyed by March 15.
- That is isolated by at least ¼ mile (0.4 km) from external sources of infestation or is in an area-wide sanitation program.
- That will be harvested promptly, that is, shaken just as the crop reaches 100% hull split or at least before the third flight of moths, which developed from the second egg hatch, emerges.

In almond orchards where an adequate sanitation program has not been carried out, or where there is a nearby external source of navel orangeworm, an insecticide application may be necessary at hull split and/or during the spring period of egg laying in May. Usually only one treatment is necessary, and it is best applied at hull split. The hull split spray is easier to time than the spring spray, and it does not disrupt natural control of mite pests in the early season. Also, Bt may be applied for navel orangeworm at hull split, but more than one application is necessary to cover the period of egg hatch. If you also need to control peach twig borer, however, apply your spray in spring and be sure to monitor mite populations afterwards.

Soft-shelled cultivars that are not harvested early may need an additional spray before harvest. This additional spray, however, is not as cost-effective as early harvest and orchard sanitation. Also, harvest interval requirements will restrict the use of certain materials.

Using Egg Traps. If in-season sprays are needed, they should be timed relative to egg laying, which is monitored with navel orangeworm egg traps. Black egg traps work best, but their caps do not need to be black. The bait usually used to attract female navel orangeworm to lay eggs on the traps is a mixture of almond press cake and 10% (by weight) crude almond oil. If kept moist, the bait can be used up to 8 weeks without change; replace it when it gets wet from rain or becomes lumpy and moldy. The bait will last indefinitely when kept in cool storage.

Use a black egg trap baited with a mixture of almond press cake and almond oil to monitor navel orangeworm egg laying. The trap's cap does not need to be black.

Place egg traps in orchards by the first week of April. Use at least four traps per orchard, and in large orchards use an average of one trap per 10 acres (4 ha). Divide large orchards into 80-acre (32-ha) sampling blocks or blocks that will be sprayed as a unit. Each block should be subdivided into four sampling areas, and two traps should be hung in different trees in each area. Orchards that exceed 1,000 acres (404 ha) can be divided into larger sampling blocks if conditions within each block are uniform.

Hang egg traps at head height on the north side of Nonpareil trees, 1 to 3 feet (30–90 cm) inside the drip line of the tree. Do not place traps in border trees. In orchards with sprinklers, keep the bait dry by installing low-angle sprinkler heads in trap areas (which will also help control some diseases by reducing the wetting of branches and foliage) or hanging the traps over the sprinkler heads.

Preceding the major spring flight, a low level of egg laying activity may occur sporadically between late March and mid-April. To monitor orangeworm activity, check the traps twice weekly and graph the number of eggs per trap per night, as shown in Figure 37. The actual start of major spring activity is characterized as a consistent, increasing trend in egg laying (75% of traps with continuous egg laying), which typically begins after April 10 and coincides with warming temperatures. Watch the graph and remove any eggs laid on the traps before a definite increase in egg laying. (A toothbrush works well for this purpose.) When egg laying increases, circle freshly laid white eggs on two or three of the traps with a white indelible marker, remove the bait from these traps, hang them in varied locations in the tree canopy, and watch the eggs to see when they hatch. With typical spring weather, eggs will hatch about 10 days after they were laid. If the weather turns cold, hatching will take longer. If a spring spray is needed for navel orangeworm, the correct time to apply it is when marked eggs begin to hatch. Watch eggs closely to make sure they are actually hatched; they may be attacked and consumed by predators such as *Phytocoris* sp., which suck the contents of eggs without leaving a visible hole.

Meanwhile, to confirm that a definite spring brood has begun, continue to record egg laying from the rest of the traps, and graph egg trap data until the peak period of egg laying is clearly under way.

Egg traps may not work as well during the second, third, and fourth moth flights because splitting nuts of the new crop are more attractive to females. Egg laying on traps may not show a clear increase even though moth populations are increasing.

Hull Split Sprays. If needed, a summer application for navel orangeworm is timed relative to the beginning of hull split and the occurrence of egg laying. The best time to spray for

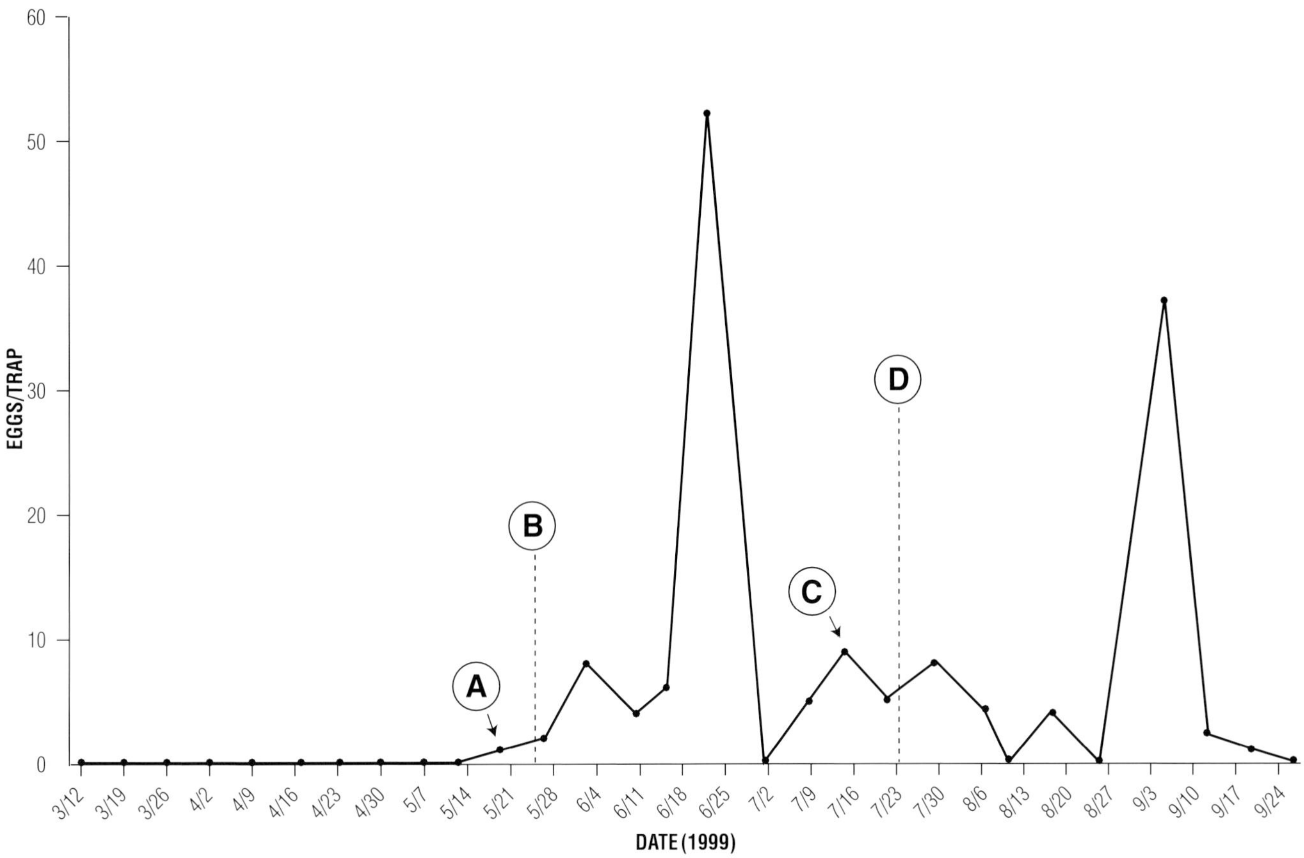

Figure 37. Graph of navel orangeworm egg trap catches. Egg laying for navel orangeworm typically begins after April 10 and is characterized by a definite increasing trend in egg laying. Using the beginning of egg laying as a biofix (point A on the graph—the first of two consecutive monitoring dates with an increase in the number of eggs on traps), the degree-day model for navel orangeworm predicts egg laying by the second flight to begin after 1,056 degree-days (D). Eggs laid by the first flight are predicted to begin hatching about 100 degree-days after the biofix (B). In this orchard, the beginning of hull split occurred on July 14 (C). If a hull split spray were applied, it should be applied at the beginning of egg laying by the second flight, which occurred after the beginning of hull split. The degree-day accumulation for this orchard is shown in Figure 38.

navel orangeworm is when hull split begins on sound fruit in the tops of the trees and eggs are being laid on traps. At this time, the nuts at eye level will be less mature than those at the top and have only a deep ridge in the hulls at this time. Nuts in the top southwest quadrant of the tree split first. Use long-extension pruning shears to cut small branches from this top portion of five or six trees in the orchard to check for initiation of hull split. Check for eggs on egg traps. If eggs are not being laid, wait until egg laying starts after hull split begins.

Timing of egg laying can be predicted using the navel orangeworm degree-day model accessible from the UC IPM World Wide Web site (www.ipm.ucdavis.edu). This model uses a lower developmental threshold of 55°F (12.8°C) and an upper threshold of 94°F (34.4°C) with a horizontal cutoff. The biofix for the start of a degree-day accumulation using this model is the beginning of a consistent increase in egg laying on egg traps. When at least 75% of the egg traps in a given location show increases in the number of eggs on two consecutive monitoring dates, the biofix point is the first of those two dates—for example, point A in Figure 37. A degree-day accumulation, using weather data from a station near the orchard illustrated in Figure 37, is shown in Figure 38. Egg laying by the second flight of moths is predicted to begin 1,056 DD after the first biofix—point C in Figure 37. If hull split begins before egg laying is predicted, apply the hull split spray at the beginning of egg laying. If hull split begins after egg laying is predicted, apply the spray at the beginning of hull split. If protection of late-maturing cultivars may be needed, egg-laying by the third moth flight is predicted 723 DD after a biofix for the beginning of egg laying by the second flight, established as described above for the first biofix.

A phenology model using a vertical cutoff and the same lower and upper developmental thresholds also can be used to predict egg laying by the second and third moth flights. Using this model, egg laying by the second flight is predicted 1,153 DD after the beginning of egg laying by the first flight, and egg laying by the third flight is predicted 797 DD after the beginning of egg laying by the second flight.

Although a phenology model will provide a good reference, it is best to confirm that egg laying has started by checking egg traps in your orchards. Remember that the number of eggs laid on each trap during hull split will be very reduced compared to earlier in the season because the female moths are more attracted to the splitting crop of new nuts.

DATE	AIR TEMPS (F) MIN *	MAX *	DEGREE-DAYS DAILY		DEGREE-DAYS ACCUMULATED
May 19 1999	47	78	9.27	1	9.27
May 20 1999	48	72	6.66		15.93
May 21 1999	46	82	10.96		26.89
May 22 1999	55	91	18.00		44.89
May 23 1999	55	84	14.50		59.39
May 24 1999	55	84	14.50		73.89
May 25 1999	55	88	16.50		90.39
May 26 1999	59	91	20.00	2	110.39
May 27 1999	56	86	16.00		126.39
May 28 1999	53	86	14.71		141.10
May 29 1999	49	79	10.16		151.27
May 30 1999	50	82	11.85		163.12
May 31 1999	53	74	8.76		171.88
Jun 01 1999	51	74	8.22		180.10
Jun 02 1999	53	65	4.35		184.46
Jun 03 1999	46	67	4.13		188.58
Jun 04 1999	51	70	6.30		194.88
Jun 05 1999	51	82	12.12		207.00
Jun 06 1999	52	82	12.41		219.40
Jun 07 1999	49	77	9.21		228.61
Jun 08 1999	44	79	9.21		237.82
Jun 09 1999	45	81	10.31		248.12
Jun 10 1999	47	84	12.12		260.24
Jun 11 1999	48 1	82 1	11.38		271.62
Jun 12 1999	47 1	84 1	12.12		283.73
Jun 13 1999	53	84	13.72		297.45
Jun 14 1999	58	88	18.00		315.45
Jun 15 1999	56	84	15.00		330.45
Jun 16 1999	53	87	15.21		345.66
Jun 17 1999	55	92	18.50		364.16
Jun 18 1999	57	91	19.00		383.16
Jun 19 1999	55	88	16.50		399.66
Jun 20 1999	52	92	17.35		417.01
Jun 21 1999	63	89	21.00		438.01
Jun 22 1999	62	96	23.79		461.80
Jun 23 1999	65	96	25.28		487.08
Jun 24 1999	60	92	21.00		508.08
Jun 25 1999	57	86	16.50		524.58
Jun 26 1999	50	83	12.34		536.92
Jun 27 1999	54	87	15.57		552.50
Jun 28 1999	59	94	21.50		574.00
Jun 29 1999	60	98	23.44		597.44
Jun 30 1999	64	100	25.94		623.38
Jul 01 1999	65	98	25.90		649.29
Jul 02 1999	59	90	19.50		668.79
Jul 03 1999	54	84	14.08		682.86
Jul 04 1999	53	84	13.72		696.58
Jul 05 1999	54	93	18.57		715.15
Jul 06 1999	54	91	17.57		732.72
Jul 07 1999	55	90	17.50		750.22
Jul 08 1999	56	96	20.81		771.03
Jul 09 1999	58	93	20.50		791.53
Jul 10 1999	60	97	23.13		814.66
Jul 11 1999	64	89	21.50		836.16
Jul 12 1999	73	92	27.50		863.66
Jul 13 1999	72	97	29.05		892.72
Jul 14 1999	66	93	24.50		917.22
Jul 15 1999	59	88	18.50		935.72
Jul 16 1999	56	85	15.50		951.22
Jul 17 1999	54	86	15.08		966.29
Jul 18 1999	54	89	16.57		982.86
Jul 19 1999	56	84	15.00		997.86
Jul 20 1999	52	81	11.91		1009.78
Jul 21 1999	54	84	14.08		1023.85
Jul 22 1999	56	89	17.50		1041.35
Jul 23 1999	58	83	15.50	3	1056.85
Jul 24 1999	53	81	12.23		1069.08
Jul 25 1999	54	90	17.07		1086.15
Jul 26 1999	57	91	19.00		1105.15
Jul 27 1999	58	86	17.00		1122.15
Jul 28 1999	55	84	14.50		1136.65
Jul 29 1999	54	86	15.08		1151.73
Jul 30 1999	56	89	17.50		1169.23

1. biofix date
2. egg hatch predicted
3. second flight egg laying predicted

Figure 38. Example of a degree-day accumulation for navel orangeworm using the phenology model available at the UC IPM web site (www.ipm.ucdavis.edu), weather data from a station near the orchard illustrated in Figure 37, and egg trap data from that orchard.

Fruit on border trees will split first; check trees in the main part of the orchard to determine if hull split has begun. Check for hull split by examining fruit from the tops of trees. Sample for initiation of hull split twice a week. The hull does not need to be completely open to be considered split, just enough so that a visible opening is present without squeezing the fruit (Stage 2, or disappearance of Stage 1; see Fig. 6).

Spraying at hull split provides a partially protective residue on the nuts. A large portion of the second-generation eggs is laid within the split hull, protected from the hull split spray. However, a treatment applied at hull split provides about 40 to 60% control overall. This spray reduces the number of early-maturing, second-generation moths and thus reduces the amount of egg laying that occurs in the early part of the third generation. Suppressing the early egg laying period of the third generation reduces the amount of navel orangeworm damage inflicted before nuts can be harvested. Because hull split sprays are only partially effective, the best control approach remains winter sanitation combined with an early harvest.

A properly timed hull split spray provides control of navel orangeworm and some control of peach twig borer. Bt can be used for the hull split spray. It does not harm mite predators or other natural enemies, but two applications are necessary to protect susceptible nuts because the residual activity lasts only a few days. Good coverage is essential because Bt is a stomach poison and the larvae must ingest the material for it to be effective. If you have a history of mite problems and significant numbers of mites are present in the orchard, use a miticide with the hull split spray if you are using a broad-spectrum insecticide instead of Bt. When application is timed to Nonpareil hull split, insecticides do not provide protection for the unsplit soft-shelled cultivars in the same orchards. Leaving later-maturing cultivars unsprayed at Nonpareil hull split reduces insecticide costs and provides reservoirs for predators of pest mites, but it should be followed by good sanitation to remove mummy nuts in winter.

Commercial products that mimic the volatile compounds in nuts that attract female navel orangeworm moths are being produced for application to almond foliage and branches to disrupt normal egg laying. If these products prove useful, they could be applied as an alternative to insecticide hull split sprays.

Monitor later-maturing soft-shelled cultivars closely for third-generation navel orangeworm activity when the hulls on these cultivars begin to split. Crack the nuts and check for larvae. Consider a preharvest spray to protect these cultivars if damage seems likely. Remove all harvested nuts from the ground, however, before spraying later-maturing cultivars to avoid violating preharvest intervals.

Spring Sprays. Spring sprays for navel orangeworm are suggested primarily for situations in which the dormant or bloom spray was not applied for peach twig borer and in which mummy counts are high. A single May spray can be applied to target both pests effectively in some, but not all, situations, saving considerable expense. Peach twig borer populations often peak a few days after the navel orangeworm egg hatch begins, making the single spray feasible. If two sprays are required, it is best to time the spring spray for peach twig borer and treat navel orangeworm at hull split. It is important to monitor for both insects to tell if a single spray will be effective.

The best time to apply a spring spray for navel orangeworm is just after the first eggs of the spring brood hatch. This time will vary according to year and location. You can use the same degree-day accumulation described above to estimate when egg hatch will begin. Egg hatch is expected when 100 DD have accumulated (for example, point B in Fig. 37). You can also observe eggs deposited on traps to determine when egg hatch begins. However, egg trap counts will not tell you if a spray is needed. Evaluate your orchard using the criteria listed above to determine if you need to spray.

Evaluating the Effectiveness of a Control Program. Table 11 lists estimates of the effects of different control practices on the reduction of navel orangeworm damage. You can use these estimates along with estimates of the cost of different control practices and crop value to estimate the cost-effectiveness of various control programs.

Carob Moth

Ectomyelois ceratoniae (Pyralidae)

The carob moth is closely related to the navel orangeworm. This species was discovered in southern California in the early 1980s, and is now found as far north as the southern San Joaquin Valley. It occurs in the Mediterranean region, where it is a pest of almonds and other crops such as dates, citrus, carob, fig, quince, and pomegranates. The carob moth not only looks like the navel orangeworm, it has a similar life cycle and host range, and it causes the same type of damage to the almond crop. It lays eggs directly on mummies or hull split fruit, where both larvae and pupae can be found.

It is easiest to distinguish the carob moth from the navel orangeworm when it is in the pupal stage. The tan colored carob moth pupa has a raised dark ridge in the middle of the thorax that extends from the head to abdomen. In addition, each abdominal segment has two short, dark spines, and two small anal hooks protrude from the posterior end.

To distinguish carob moth larvae from navel orangeworm larvae, look for the characteristics illustrated in Figure 39. Because the carob moth is not currently found in most almond-growing areas of California, notify your agricultural commissioner if you think it is present in your orchard.

Table 11. Effects of Different Insect Management Practices on Reduction of Crop Rejects Caused by Navel Orangeworm.

Insect management practice[1]	Estimated reduction in navel orangeworm rejects (%)[2]
1. winter cleanup of mummies (to an average of 2 per tree or less)	
a. orchards near external source of infestation	50–70
b. isolated orchards	70–80
2. early harvest (100% hull split)	70
3. 1a and 2 combined	75
4. 1b and 2 combined	80–90
5. properly timed May spray	40–60
6. properly timed hull split spray	40–60

1. It was assumed that a dormant spray of an insecticide with a narrow range oil or bloomtime spray was applied for peach twig borer.

2. Use these percentages as guidelines only. Results obtained by growers may differ slightly. Adjust the control percentages to reflect the results typically obtained in your own orchard with each of the insect management practices listed.

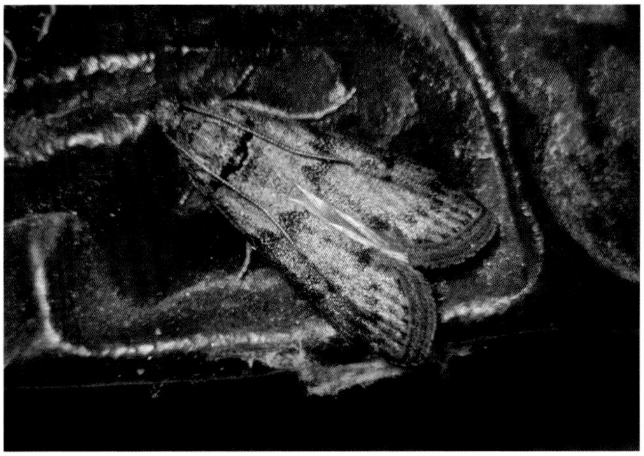

THOMAS D. EICHLIN

The carob moth is silver with black wavy patterns on its forewings.

MAX E. BADGLEY

The carob moth pupa (top) has a raised dark ridge in the middle of the thorax and two short, dark spines on each abdominal segment. These features are not present on navel orangeworm pupae (bottom).

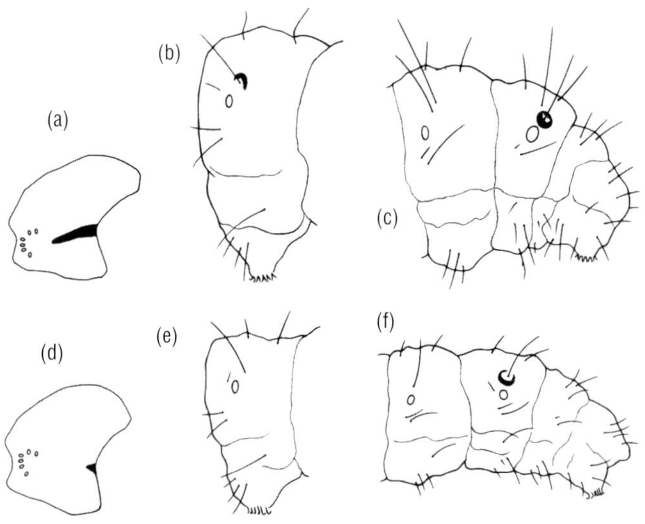

Figure 39. Because distinguishing characteristics can be variable, look at more than one feature to distinguish carob moth from navel orangeworm larvae. Carob moth larvae have a darkened band on the side of the head capsule (a); navel orangeworm larvae do not have this band (d). The hair just above the spiracle on abdominal segments one through seven has a pigmented area above its base in carob moth larvae (b), but not in navel orangeworm larvae (e). On the eighth abdominal segment, the pigmented area at the base of the hair usually forms a full circle in carob moth larvae (c), but only a horseshoe-shaped crescent in navel orangeworm larvae (f). (Adapted from a drawing by Thomas Eichlin, California Department of Food and Agriculture.)

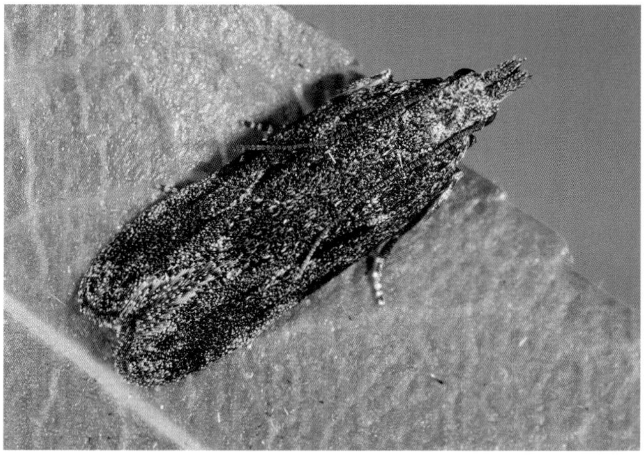

The peach twig borer adult is a small, mottled gray moth with a snout-like projection from the head.

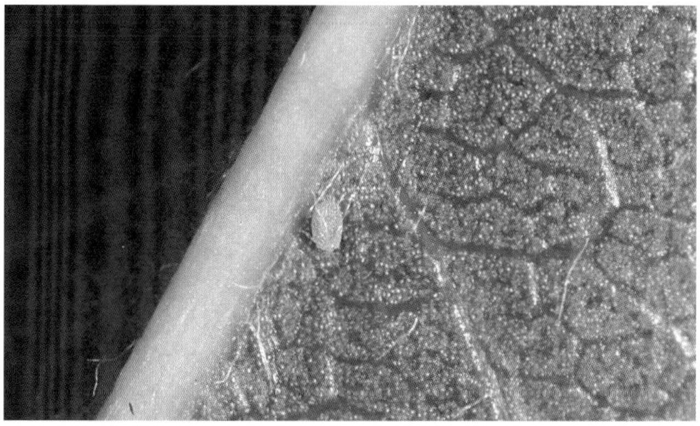

Peach twig borer eggs are white when first laid. The color changes to orange as the embryo develops inside.

Peach Twig Borer
Anarsia lineatella (Gelechiidae)

Peach twig borer is a major pest of almonds and stone fruits. Native to Europe, the insect was first reported in California in the 1880s and is now found throughout the state.

Until arrival of the navel orangeworm in the 1940s, peach twig borer was the most serious insect pest of almonds. Today there is a relationship between the two insects: fruit attacked by peach twig borer are preferred over sound fruit by navel orangeworm for egg laying.

PEACH TWIG BORER

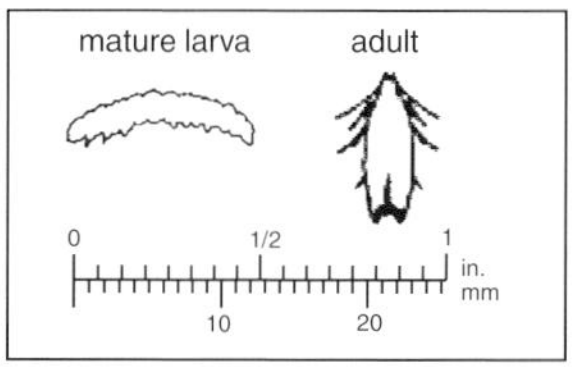

Description

Moths of the peach twig borer are nocturnal and rest on the undersides of large limbs and in bark crevices during the day. The moths have steel-gray, mottled forewings.

Each female lays about 90 to 100 eggs on shoot terminals and fruit in spring and summer and on young branches in fall. Eggs are also laid on the undersides of leaves next to veins or the midrib. When first laid, the oval eggs are white with a hexagonal pattern on the surface. As the embryo develops inside, the color changes to orange. Depending upon temperature, eggs hatch in 4 to 18 days.

When they first hatch, peach twig borer larvae are light brown with a black head and prothorax. As they grow, the head and prothorax remain black, but the body turns chocolate brown, and white portions between each body segment give the appearance of bands. The larvae have four to five instars (overwintering larvae have the additional instar) and take 10 days to 4 weeks to mature, depending on temperature.

Larvae primarily bore into twigs, but they also feed on nuts. When they are ready to pupate, the larvae leave the shoot or nut they have been inhabiting, making an exit hole as they leave. The pupal stage lasts 4 to 11 days and usually occurs in crevices, curled leaves on the tree, or debris on the ground, or for later generations, between the hull and shell. The pupae are dark brown and are not found in a cocoon.

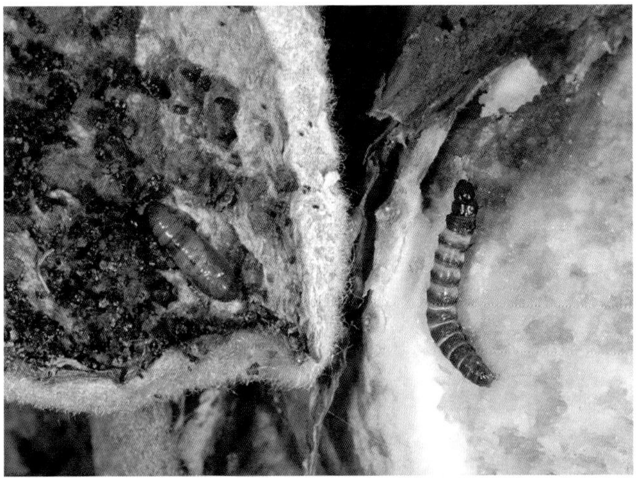

The peach twig borer larva is a small, distinctive caterpillar with a black head and thorax and a dark brown body with white portions between each body segment that give it a ringed appearance.

The overwintering site for peach twig borer larvae can be located by looking in the limb crotches of first- or second-year wood.

The peach twig borer larva forms a chimneylike pile of frass when boring a hibernaculum into the bark.

This shoot terminal has been mined by a peach twig borer larva, and the affected portion of the twig has wilted, causing a symptom known as "flagging" or a "shoot strike."

Seasonal Development

Peach twig borer overwinters as first- or second-instar larvae in cells (hibernacula) bored under the thin bark of limb crotches or bark cracks in 1- to 4-year-old wood. Larvae are inactive when it is cold but they do not diapause. They may feed within hibernacula exposed to the sun on warm winter days. To locate a hibernaculum, look for minute, chimneylike piles of frass and sawdust that larvae cast out when they are constructing the cells. It is easier to see hibernacula in fall before the rains wash away the frass chimneys, on warm winter days when feeding resumes, or in early spring when activity picks up and new particles of wood and frass are evident. When the trees bloom in spring, larvae leave their hibernacula and feed on flower buds or young nutlets and foliage. They can damage several clusters of leaves or buds before settling down and mining into the interior of a twig. The mine extends 1 to 2 inches (2.5–5 cm) from the shoot tip, and the affected portion of the twig wilts, resulting in the characteristic damage symptom called "flagging" or a "shoot strike."

There are usually four generations of peach twig borer each year in almonds. Overwintering larvae develop mostly in green shoots during spring, and most of the moths emerge from mid-April through May and lay first-generation eggs. Larvae of the first generation develop in green shoots and occasionally in immature fruit. First-generation moths emerge from mid-June through July and lay second generation eggs

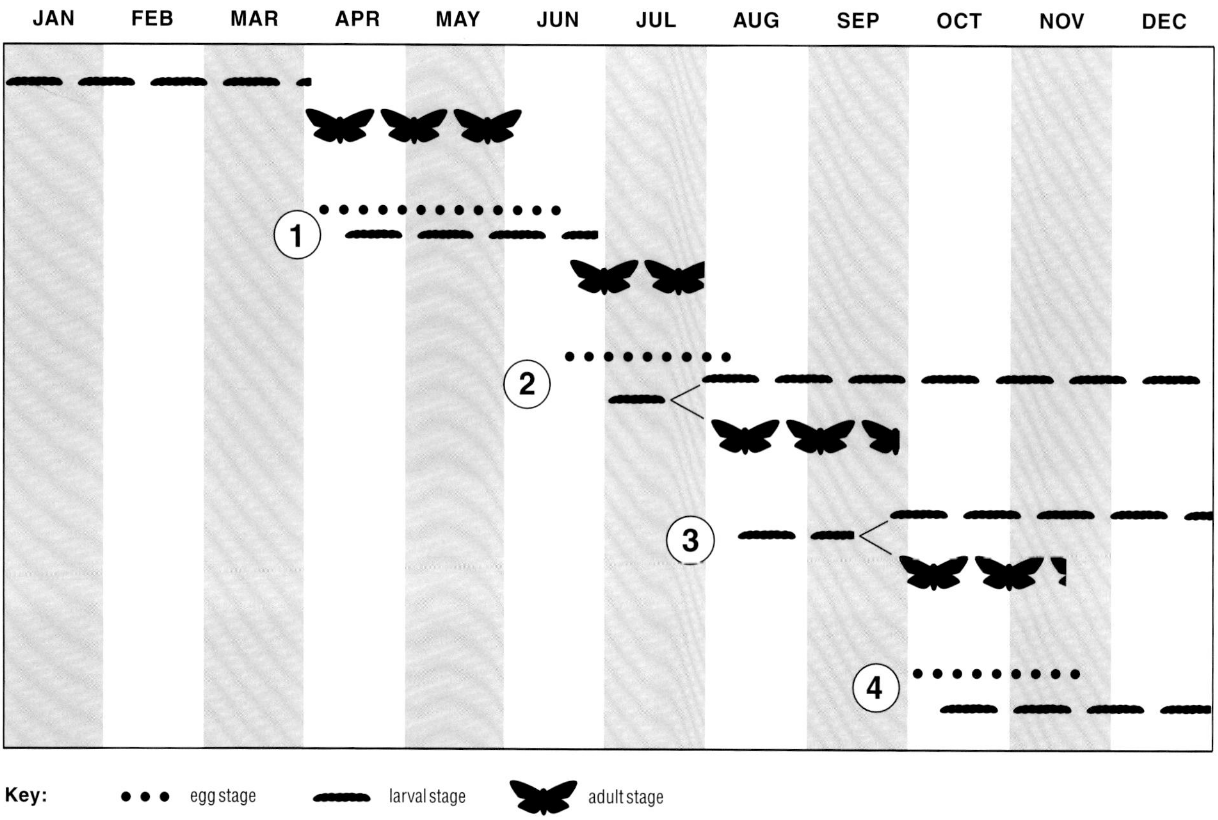

Figure 40. Seasonal development of the peach twig borer. The timing of moth flights varies from one season to the next and from one growing area to another. Symbols represent the time during which a moth flight usually will occur. In any given location, the actual timing and length of a moth flight varies.

Feeding by peach twig borer larvae on almond nutmeats causes shallow channels and surface grooves on the kernels.

on both green fruit and leaf shoots. Figure 40 outlines the general seasonal development of peach twig borer. When second-generation moths begin to emerge in August, they lay eggs on maturing fruit. Some of the larvae of this generation and the previous one can go into hibernacula for overwintering; others continue to develop on nuts or in shoots, and third-generation moths generally emerge in October. These moths lay fourth-generation eggs, and the larvae build hibernacula soon after hatching.

Damage

Peach twig borer larvae can damage both the growing shoots and the nut crop. They mine the tender shoots of the almond tree, killing the tips and causing lateral twig growth. This can be serious in young trees because shoot damage can make it difficult to train the trees during pruning. On older trees this type of damage is unimportant.

Although Oriental fruit moth larvae also mine the growing tips of shoots, they have white or pink bodies with brown head capsules, in contrast to the distinctly segmented, dark brown bodies and black head capsules of peach twig borers. They usually kill more of the shoot tip than peach twig borer.

Like the navel orangeworm, twig borer larvae can feed on the hulls, between the hull and shell, or directly on the nutmeats; but twig borer larvae differ greatly in appearance from navel orangeworm larvae and do not produce webbing. Also, navel orangeworm bores into the nutmeat, and several larvae may feed in a single nut, whereas peach twig borer causes only shallow channels and surface grooves on the kernels and normally feeds one larva to a nut. Peach twig borer and navel orangeworm often infest the same nut. Because peach twig borer feeding is often masked by navel orangeworm feeding, peach twig borer infestations are usually greater than that detected in the harvest sample.

Nuts of certain cultivars seem to be more severely damaged by peach twig borer than others. Susceptibility to peach twig borer damage is generally highest in the Merced and Thompson cultivars, followed by Nonpareil, Fritz, Price, and Ne Plus Ultra.

Management Guidelines

Peach twig borer has about 30 species of natural enemies. Among those most commonly found in California are two chalcid wasps (*Copidosoma* [*Paralitomastix*] *varicornis* and *Euderus cushmani*), the grain or straw itch mite (*Pyemotes ventricosus*), green lacewings (*Chrysoperla* and *Chrysopa* spp.), and the California gray ant (*Formica aerata*). In some years these natural enemies destroy a significant portion of larvae, but by themselves they do not reliably maintain peach twig borer populations below economically damaging levels. No effective cultural controls for peach twig borer are known.

It is not necessary to spray for peach twig borer where this insect has been managed effectively and in orchards where populations are always low, but adequate monitoring is necessary. Be sure to monitor for peach twig borer hibernacula during the winter to determine if moderate to high densities are present; also monitor for larvae during bloom and in shoot strikes during April. Look for shoot strikes, cut them down, and examine them to see if they were caused by peach twig borer or oriental fruit moth. If it is not difficult to find four or more shoot strikes per tree in a mature orchard, control action may be needed during the season or in the following dormant season or bloom.

Sprays for peach twig borer can be applied during dormancy, during bloom, or during the growing season in spring or at hull split. Normally, applications should be required at only one of these times. The preferred time to spray is during bloom with Bt, which is not harmful to natural enemy populations, or with spinosad. Bt also can be used during the season, but it is not as effective and multiple applications are needed. Spinosad can be used effectively as a dormant spray and is much less harmful to natural enemy populations than organophosphates or pyrethroids. Alternatively, peach twig borer can be managed using pheromones applied as mating disruptants during the season. Different monitoring procedures must be followed to properly time different types of treatment. However, even if you treat during dormancy or at bloom and are not planning on a spring spray, it is a good idea to continue to track the peach twig borer life cycle with pheromone traps and monitor for shoot strikes.

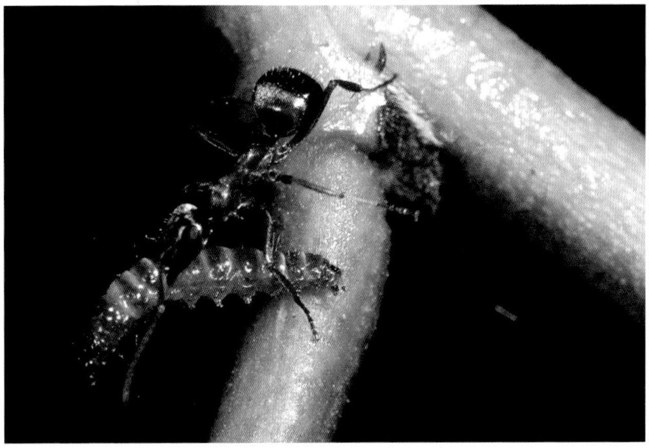

The California gray ant, *Formica aerata*, is an important predator of peach twig borer larvae in Central Valley orchards.

To monitor for peach twig borer emergence, cut out small wedges of bark around the small, reddish piles of frass that mark the entrances of hibernacula (arrow).

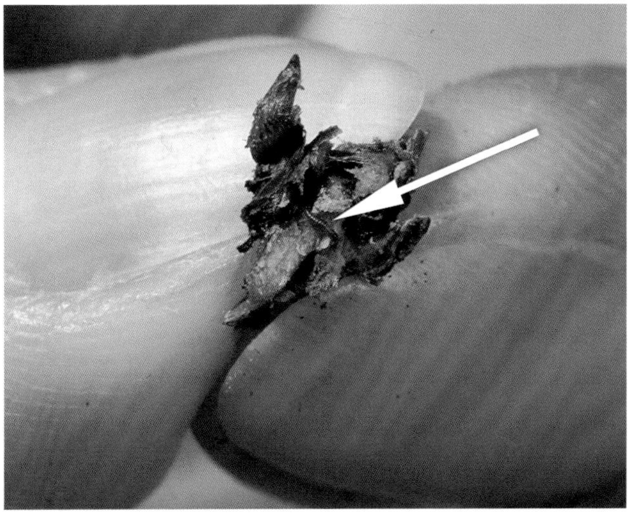

Pinch each wedge of bark to open the hibernaculum and use a hand lens to look for a peach twig borer larva (arrow).

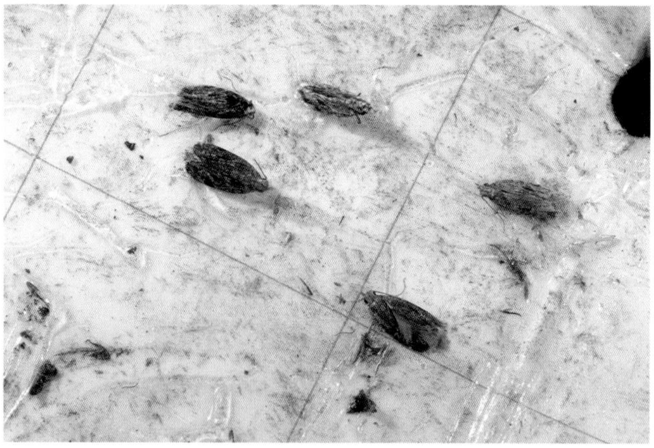

Counts of male peach twig borer moths caught in pheromone traps like the one shown here are used to monitor the development of successive peach twig borer generations in an orchard.

Bloom Sprays. Peach twig borer can be controlled during bloom with well-timed applications of Bt or spinosad. This treatment is a preferred alternative to traditional dormant applications of conventional insecticides, which may have negative environmental impacts. The bloomtime treatments target the overwintering larvae, which feed on exposed shoots and buds before boring into the shoots. The larvae must consume the Bt; once larvae begin feeding inside shoots, Bt treatments are ineffective. Treatments applied before new shoot tissue is present (e.g., during the dormant season) are also ineffective.

For best control, make two applications timed to larval emergence. Monitoring the percentage of larval emergence from hibernacula is time-consuming, but it allows the most accurate timing of Bt applications and determines the length of the emergence period, which can be affected by temperature. Starting at the popcorn stage(see the Appendix), examine at least 10 hibernacula each week in an orchard that has not been treated with a dormant insecticide to determine the percentage of larvae that have emerged. Make the first application when 20 to 40% of the larvae have emerged, and make the second application 7 to 10 days later or when emergence has reached 80 to 100%. If emergence is spread out, a third spray may be needed when emergence finally reaches 80 to 100%.

Alternatively, Bt applications can be timed according to bloom, but this approach is less precise than monitoring emergence. Make the first treatment at early bloom, just beyond the popcorn stage. Make the second treatment 7 to 10 days later. A third application may be needed at petal fall if bloom is extended due to cool weather.

Ground applications give better coverage than aerial applications. When aerial sprays are not properly applied over orchards, as much as 90% of the material applied can fall to the ground instead of being deposited on tree surfaces. If aerial sprays are necessary because ground applications cannot be made, make sure the applicator uses proper nozzles and flies sufficiently high above the canopy—approximately 20 feet (6.1 m)—to allow maximum deposition on the upper tree branches.

Dormant Sprays. A traditional time to treat for peach twig borer is during the dormant season with an oil spray that contains insecticide. Spraying during tree dormancy has several advantages: coverage of the bark is more complete; beneficial insects and mites are less affected (except when using pyrethroids); other pests such as San Jose scale, brown mite, and European red mite are simultaneously controlled; and there are fewer conflicts with other cultural operations in the orchard. The dormant spray kills overwintering twig borer larvae in their hibernacula. If you apply the spray properly, twig borer populations can be reduced over 95%. However, organophosphates used in the dormant season have been detected in some waterways following rainfall, and there is concern about their use in locations or under conditions that may result in their entry into waterways. Other insecticides applied as dormant sprays also can result in toxicity to aquatic life if they enter waterways, so best management practices that prevent their movement out of orchards should always be practiced. Spinosad, a biologically derived insecticide, also can be used as a dormant spray with oil to control peach twig borer, and preliminary research indicates that it is relatively nontoxic to aquatic invertebrates.

A dormant spray of oil without insecticide will control immature scales and mite eggs, but it will not control peach twig borer. However, dormant oil sprays appear to enhance the effectiveness of Bt applied during bloom.

Spring Sprays. If you did not spray during the dormant season or at bloom and have a history of peach twig borer damage, you may need to apply a spring spray. In orchards that are between 1 and 3 years old, spring sprays for peach twig borer are recommended because young trees are attractive to this pest and a few shoot strikes can damage developing scaffolds. When trees are older than 3 years, monitor shoot strikes in mid to late April to determine the need to spray. Always be sure to confirm that the strikes were caused by peach twig borer and not oriental fruit moth by looking for larvae in the strikes. The spring spray is directed against the hatching larvae of the first generation. In most years this should effectively control peach twig borer until the crop is harvested. If your control emphasis is peach twig borer, time the spray using peach twig borer pheromone traps and degree-day calculations. If your control emphasis is navel orangeworm or both navel orangeworm and peach twig borer, time the spray based on data collected from monitoring navel orangeworm egg traps.

Place pheromone traps in the orchard by March 20 in the southern areas of the Central Valley and by April 1 in the northern areas to detect moth emergence. Use an average of one trap per 20 acres (8 ha) to monitor the orchard; place the traps at least 1 acre (209 ft, or 64 m) apart (see Monitoring Methods at the beginning of this chapter for additional information). Hang the traps 6 to 7 feet (2 m) high in the northeast quadrant of the tree, 1 to 3 feet (30–90 cm) from the outside of the canopy. Count and remove the moths at least twice weekly during major flight activity. Record and graph the average number of moths per trap per night (Fig. 41). Replace pheromone dispensers according to the manufacturer's recommendations, and replace trap bottoms when they become obviously dirty or after 100 to 150 trapped moths have been removed.

Flight activity of the peach twig borer is greatest between the hours of midnight and 2 A.M. and at temperatures above 60°F (15.5°C). When temperatures fall below this threshold, or if it's windy, very little flight activity takes place. Pay close attention to nightly temperatures if your trap data indicates that a major flight is occurring. If the temperature drops below 60°F at midnight, your trap catches will decline, but they will increase when the weather turns warm (Fig. 41). This increase does not necessarily signify the start of the next peach twig borer generation: it could signify the resumption of the interrupted one. Rely on degree-day calculations to determine when the next generation has started.

Begin calculating degree-days in the spring when male moths are detected in the traps on at least two consecutive sampling dates. A degree-day model for peach twig borer can be found at the UC IPM World Wide Web site (www.ipm.ucdavis.edu). This program will calculate degree-days using temperature data you provide or data from a weather station in your area. An example of such a calculation is shown in Figure 42. About 220 DD accumulate from first male moth catch to first egg hatch in the next generation. The optimal timing for first-generation or spring larval treatment, called the "May spray," is between 400 and 500 DD after the start of the spring flight in April.

It is best to control peach twig borer in the dormant season or during bloom, because sprays of broad-spectrum insecticides in spring disrupt natural controls of other pests, especially spider mites. Pyrethroid insecticides in particular should not be used as spring sprays.

Hull Split Sprays. Control of peach twig borer at hull split is difficult and not generally recommended because the timing is so critical. The hull split spray is directed at the hatching larvae of the second generation. To be effective, it must be

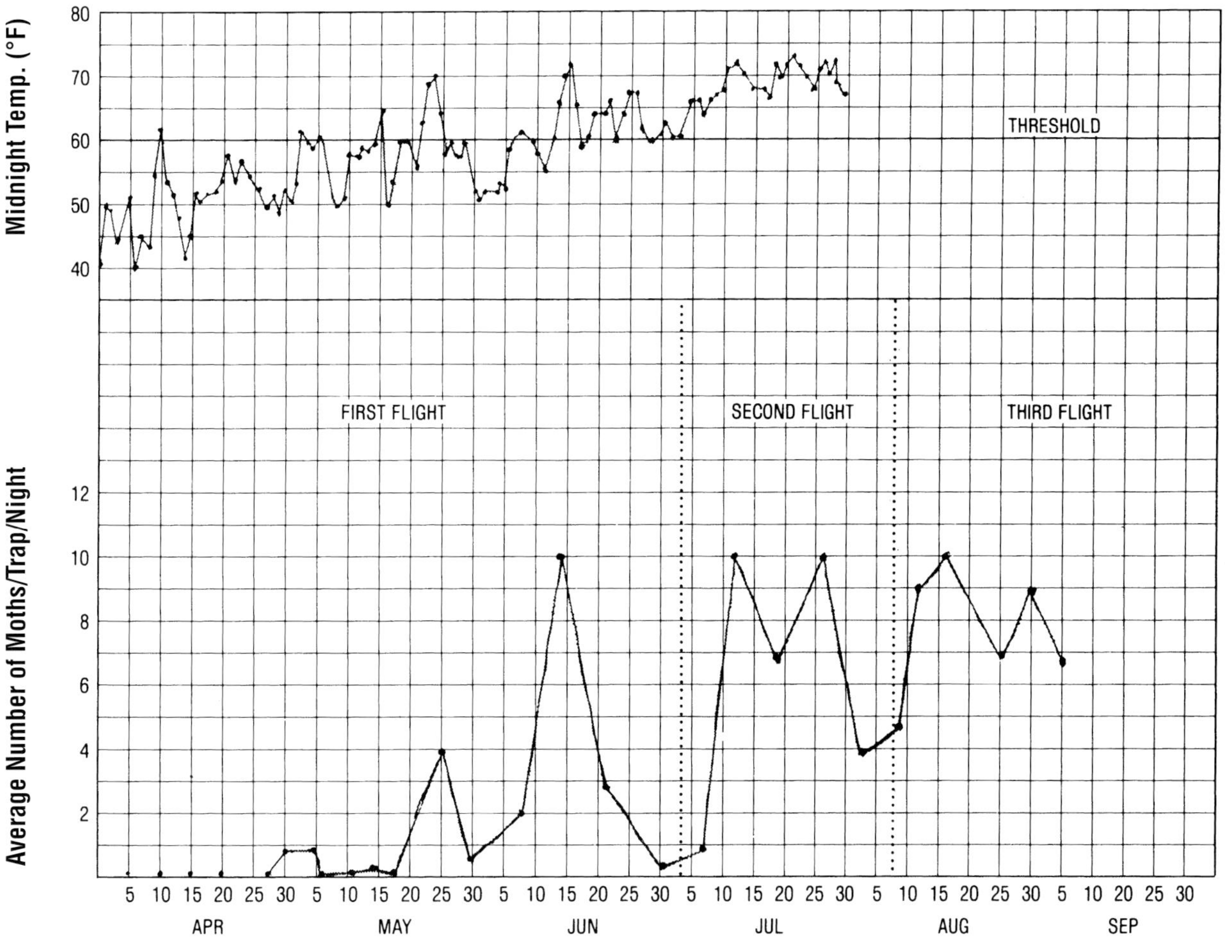

Figure 41. Graph of peach twig borer trap catches and daily midnight temperatures. Note that trap catches decline when temperatures fall below the threshold of 60°F (15.5°C).

applied at very early hull split, when about 1% of the hulls of sound fruit in the treetops have split.

Pheromone Mating Disruption. Mating disruption with sex pheromones has been used successfully to control peach twig borer in almond orchards, and research is under way to improve the reliability of the technique and methods of application. Pheromone mating disruption is most effective in orchards with low moth populations that are not close to other untreated peach twig borer hosts, and when it is used on an areawide basis. Areawide application is more important with peach twig borer because this species is more migratory than other species such as oriental fruit moth. The effectiveness of mating disruption is reduced by small orchard size, uneven terrain, reduced pheromone application rates, and improper timing. With the continued use of pheromone mating disruption there will be a continual reduction of the peach twig borer population as long as moths don't migrate in from outside the treated area.

Place pheromone dispensers in orchards in mid-March or when you begin to catch the first moths in pheromone traps. Follow the manufacturer's recommendations for placement, number of dispensers, and replacement intervals. Use higher load rates and more dispensers for orchards with heavy infestations. Reapply the pheromones at the recommended timing for the second flight. Pheromone traps placed within treated orchards will help monitor the effectiveness of the mating disruption. If you are catching more than 5 moths per pheromone trap per week within one generation of harvest, treat with an insecticide rather than replace dispensers.

When using pheromone mating disruption, monitor the orchard regularly for shoot strikes to verify that control is being achieved. Also monitor nuts from the tops of trees regularly for signs of larvae or damage; monitor more frequently during the final 8 weeks before harvest. Treat with an insecticide if there are more than two shoot strikes per tree after the first moth flight or if larvae are found in green fruit.

DATE	AIR TEMPS (F) MIN ±	MAX ±	DEGREE-DAYS DAILY	DEGREE-DAYS ACCUMULATED
Apr 15 1998	39	67	6.06	6.06
Apr 16 1998	41	74	9.55	15.61
Apr 17 1998	42	79	12.12	27.73
Apr 18 1998	45	79	12.83	40.56
Apr 19 1998	48	88	18.19	58.75
Apr 20 1998	50	93	20.77	79.51
Apr 21 1998	52	95	22.28	101.79
Apr 22 1998	55	86	20.50	122.29
Apr 23 1998	53	62	7.50	129.79
Apr 24 1998	50	73	11.50	141.29
Apr 25 1998	48	74	11.24	152.53
Apr 26 1998	51	84	17.50	170.03
Apr 27 1998	51	90	20.31	190.34
Apr 28 1998	49	93	20.34	210.68
Apr 29 1998	54	91	22.13	232.81
Apr 30 1998	57	83	20.00	252.81
May 01 1998	58	74	16.00	268.81
May 02 1998	54	69	11.50	280.31
May 03 1998	57	74	15.50	295.81
May 04 1998	54	73	13.50	309.31
May 05 1998	50	74	12.00	321.31
May 06 1998	57	69	13.00	334.31
May 07 1998	57	77	17.00	351.31
May 08 1998	53	69	11.00	362.31
May 09 1998	44	66	6.37	368.68
May 10 1998	47	72	9.95	378.63
May 11 1998	50	62	6.00	384.63
May 12 1998	49	60	4.63	389.26
May 13 1998	49	64	6.61	395.87
May 14 1998	50	69	9.50	405.37
May 15 1998	47	73	10.44	415.81
May 16 1998	46	64	5.82	421.63
May 17 1998	41	69	7.24	428.87
May 18 1998	51	82	16.50	445.37
May 19 1998	46 1	75 1	11.14	456.51
May 20 1998	42 1	72 1	8.80	465.31
May 21 1998	48 1	73 1	10.74	476.06
May 22 1998	47	79	13.39	489.45
May 23 1998	55	78	16.50	505.95
May 24 1998	55	72	13.50	519.45

Figure 42. Example of a degree-day calculation for peach twig borer to determine the optimal time for a May spray. If treatment is needed, it should be made when 400 to 500 DD have accumulated since pheromone traps first begin to consistently trap peach twig borer moths. This example was obtained using the peach twig borer model at the UC IPM World Wide Web site (www.ipm.ucdavis.edu).

San Jose Scale

Quadraspidiotus perniciosus (Diaspididae)

San Jose scale was accidentally introduced into California from China during the 1870s. It now is widespread and known to infest over 700 species of plants. The most susceptible of these hosts are deciduous fruit and nut trees such as pears, plums, apples, peaches, nectarines, and almonds. Severe infestations kill fruit wood, reduce yields, and eventually kill the tree.

Description and Seasonal Development

San Jose scale emerges as nymphs; hence there is no visible egg stage. Three stages occur during the first instar: the "crawler," the "white cap," and the "black cap." The crawler stage is the only one in which a scale infestation can spread. All other developmental stages, except adult males, are immobile and attached to a single feeding site under a protective scale covering. The bright yellow crawler is about the size of the sharp end of a pin (0.2 mm) and resembles a mite. After leaving the protection of the female's cover, it relocates by crawling; by being carried on equipment or field workers, or on birds or insects that land on branches where the crawlers are active; or by wind. Within 8 to 24 hours of birth it settles down on the bark, inserts its mouthparts into the host, and begins to feed on the tree's sap. As it feeds, a white waxy covering (the white cap) is secreted. After about a week, a band of dark wax appears around the periphery of the white cap, marking the beginning of the black cap stage. Eighty percent of San Jose scale overwinter in the black cap stage; during mild winters, the remaining 20% are mostly mated adult females. These females will not survive a severe winter, however. The crawler, white cap, and black cap stages are susceptible to dormant oil sprays.

When the weather warms in February and the tree's sap flows, black cap nymphs resume their growth and undergo the first molt. The male's scale covering elongates in the latter part of the second instar, and they become distinguishable from females, whose covering remains circular. Females molt twice before reaching maturity, while males molt a total of four times.

Following the second molt, males pass through two nonfeeding instars in which the scale cover remains elongated, is gray, and has a small, indistinct knob near the larger end. After the fourth molt, the male emerges from under the scale cover as a delicate, minute, two-winged, yellow-brown insect. Adult males can mate immediately after emerging and most only live 6 to 8 hours in warm weather.

Females live under the scale covering for the remainder of their lives. At maturity the cover is gray-brown and has a tiny white knob in the center. The insect body under the covering is bright yellow and has no appendages.

The life cycle of armored scales such as San Jose scale is illustrated in Figure 43.

When females are mature, they produce a sex pheromone to attract males. The males' spring flight period coincides with the receptivity of the females in March and April. Within 5 to 6 weeks, females begin producing first-generation crawlers. Depending upon temperature, a female can produce crawlers for 6 to 8 weeks at a rate of about 10 per day.

Crawlers that emerge in May give rise to the first-generation male flight in June. Two more generations follow in August-September and October-November. Crawlers produced in October and November overwinter as black caps and produce the overwintering flight the following spring. Under optimal conditions of hot, dry weather, a generation can be completed in 7 to 8 weeks.

Damage

San Jose scale sucks plant juices from the trees and injects a toxin, contributing to death of twigs and limbs and to overall decline in tree vigor, growth, and productivity. The scale feeds on any thin-barked wood. On 1-year-old wood,

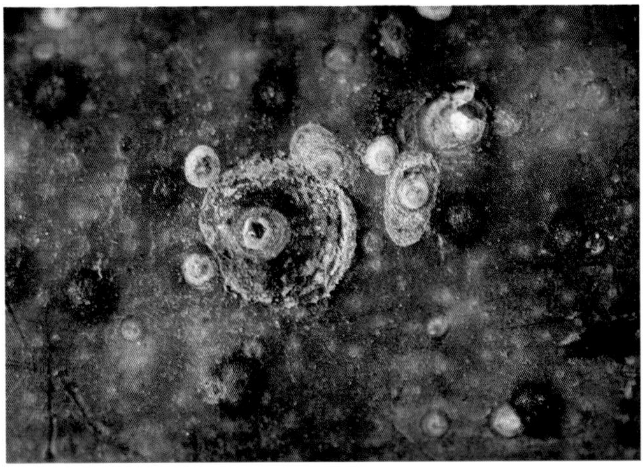

Following the first molt, the male San Jose scale's covering is elongated, while the female's remains circular. White cap and black cap stages can also be seen in this photo.

If you remove the covers from San Jose scales, you can see the bright yellow bodies of the male and female underneath.

Tiny, bright yellow crawlers of the San Jose scale migrate to new feeding locations. After crawlers settle down to feed, they secrete a white, waxy covering. This stage is known as the white cap stage.

SAN JOSE SCALE

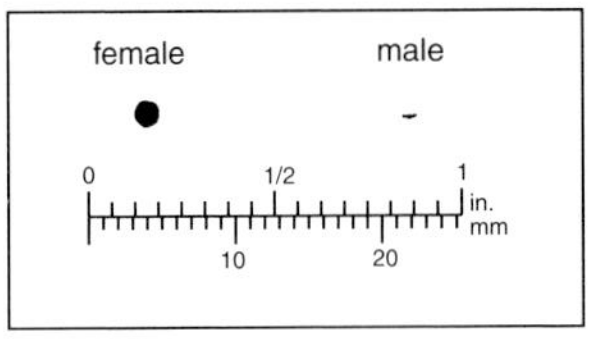

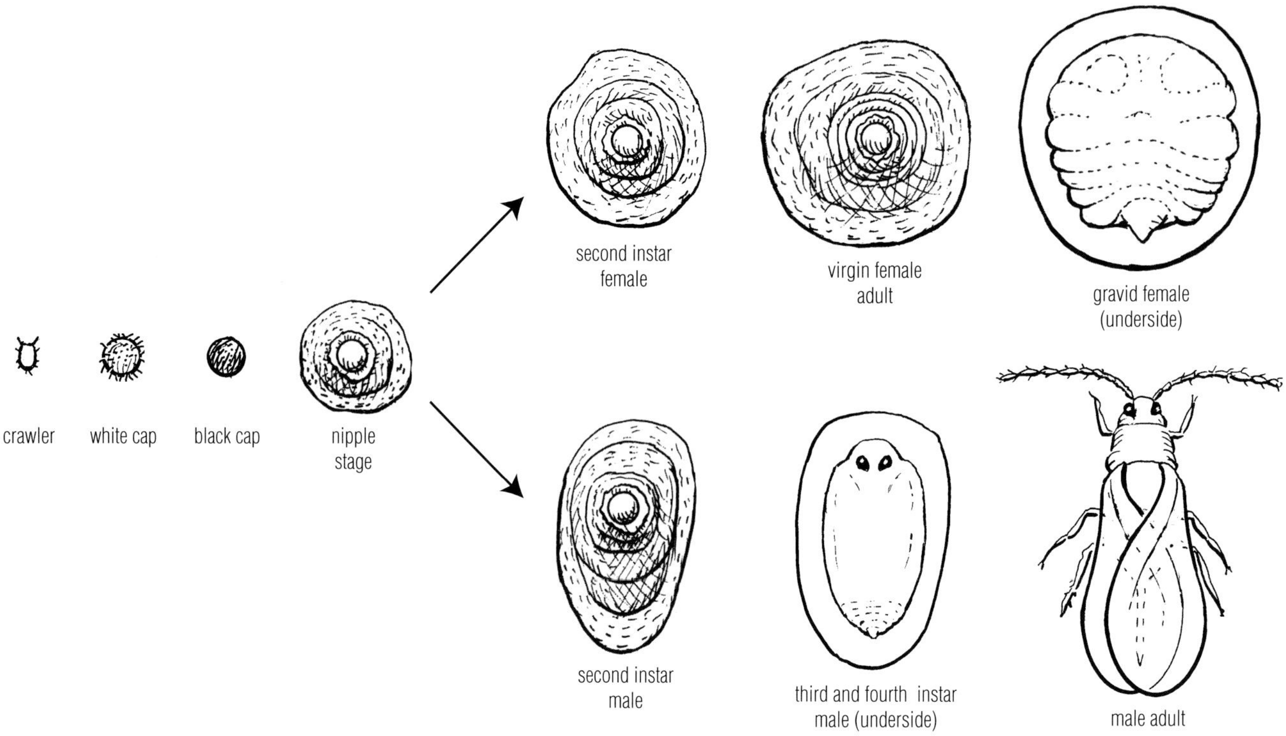

Figure 43. Life cycle of the San Jose scale, a typical armored scale. Eggs hatch into tiny crawlers that soon settle and secrete a cottony covering (white cap), then a dark waxy covering (black cap). Later, a more solid covering is formed (nipple stage). After the first molt, males begin to develop an elongated covering, while the female cover remains round. Females molt three times; the final stage, the mated female, remains legless, wingless, and immobile beneath the rounded cover. Males molt four times, developing eyespots in the third and fourth instar, which can be seen by turning over their elongated covers. Adult males have legs and two wings.

it produces a red halo around the feeding site. San Jose scale often infests the fruit of apples, pears, and stone fruit, but it usually is not found on almond hulls. Infested trees appear water stressed, and extremely heavy populations can cause the bark to crack and exude gum. Fruit spurs and branches may be killed, permanently affecting structure, vigor, and productivity of mature orchards.

Management Guidelines

San Jose scale has many natural enemies that can frequently keep the pest under control if not disrupted by application of broad-spectrum insecticides. Many orchards that have not used broad-spectrum sprays for 2 to 3 years do not have San Jose scale problems. Low to moderate levels of scales can be managed with sprays of oil during the dormant season. High populations require the addition of more broad-spectrum insecticides to the dormant spray.

Biological Control. Natural enemies that feed on San Jose scale include two predaceous beetles: the twicestabbed lady beetle *Chilocorus orbus* (Coccinellidae) and another small beetle, *Cybocephalus californicus* (Nitidulidae). Several small wasps parasitize this scale. The most important are the Encyrtidae species *Encarsia (Prospaltella) perniciosi* and *Aphytis* spp. Populations of these parasites may quickly build to high levels when insecticides are not used during the dormant season and when less-harmful materials such as Bt or spinosad are used at bloom or during the season. In most orchards where broad-spectrum insecticides have been eliminated, parasites have reached densities that keep San Jose scale at low levels within 2 to 3 years.

Monitoring. Examine the orchard during dormancy to spot San Jose scale infestations. Twigs or branches that have been killed by the scale retain their leaves during the winter. Scale populations are often irregularly distributed within orchards and may be hidden beneath loose bark in older trees. Be aware that scale populations may be higher after mild winters, because mated adult females, normally killed by cold weather, are likely to survive. Frequently, infestations first appear in the tops of trees where it is hardest to get good spray coverage. The best way to detect infestations is to examine prunings during the dormant season, especially those from the treetops. If scales are found on the prunings of the previous season's growth in several orchard locations, a dormant spray should be considered. Use pheromone traps or sticky tapes (see the section on spring sprays, below) to detect San Jose scale infestation during the growing season and to time a spring spray, if needed. Pheromone traps can also be used to monitor parasitic wasp populations because these wasps are attracted to the pheromone produced by female scale. Learn to distinguish parasites from male scales in the traps. Also look for emergence holes in scale covers that indicate parasitization.

CYBOCEPHALUS

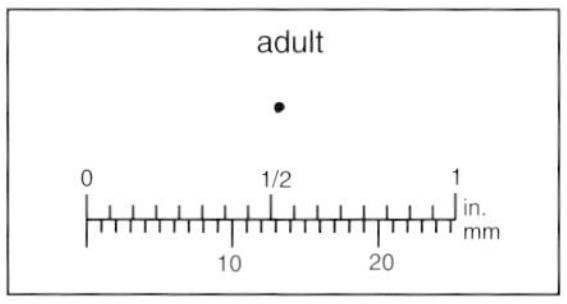

TWICESTABBED LADY BEETLE

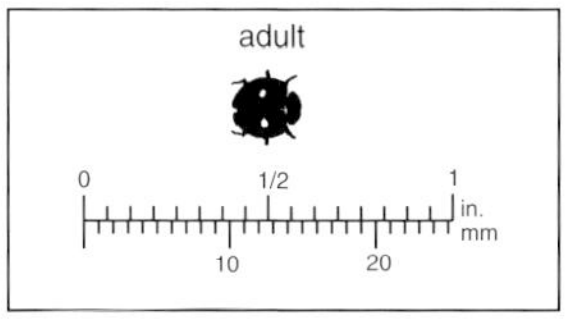

The adult male San Jose scale has a dark band across its back (left). The tiny yellow wasp *Aphytis melinus* parasitizes San Jose scale. It resembles the adult male scale but lacks the dark band across the back (right).

Red halo around the feeding site is evidence of San Jose scale on young wood.

The twicestabbed lady beetle adult is black with two prominent red spots on the back. Larvae and adults of this species feed on mealybugs and scale.

The tiny nitidulid beetle, *Cybocephalus californicus,* is an important predator of armored scale insects such as San Jose scale.

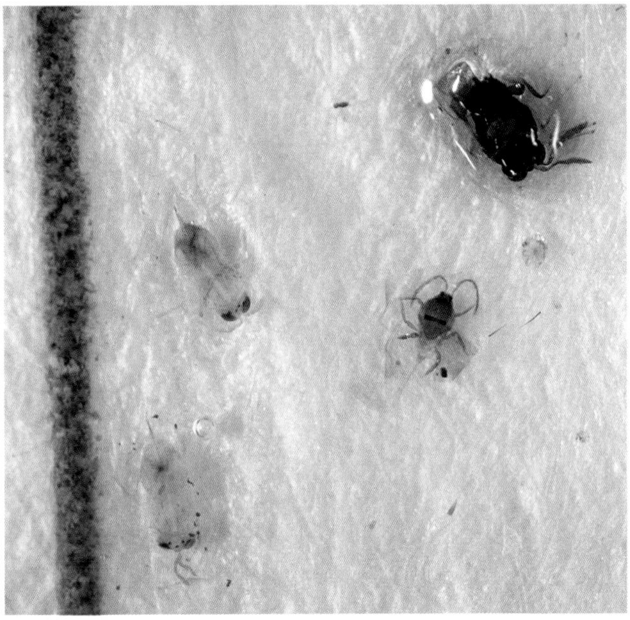

San Jose scale pheromone traps will trap parasites as well as male scales, all of which are attracted to the pheromone. Shown here are two *Aphytis* sp. (left), a male scale (center), and *Encarsia perniciosi* (upper right).

Adult parasites leave exit holes in the coverings of female San Jose scales. Checking for scales with these holes can give an estimate of the activity of scale parasites.

Dormant Sprays. A properly applied dormant spray of oil without insecticide controls immature stages of San Jose scale. This is the preferred management strategy for this pest. Use high label rates of oil when scale is present to ensure adequate coverage and best control. Oil can damage trees when applied to trees that are stressed for water. Apply high label rates of oil only when soil moisture is ample following winter rains or irrigation. Avoid using organophosphate more often than absolutely necessary, and take care to avoid runoff into water. If monitoring of prunings indicates high density of scale, inclusion of an organophosphate in the dormant spray may be warranted. In orchards where broad-spectrum pesticides are not used, oil sprays and natural enemies may be sufficient to keep San Jose scale below damaging levels.

Spring Sprays. If a dormant spray is not applied, the best alternate time to spray is during crawler emergence, which is in May in the southern portion of the state and in late May or early June in northern areas. Do not try to control San Jose scale and navel orangeworm with the same spray, however. Not only are they best controlled by different materials, but the spray timing may differ.

Seasonal flights of male scales and populations of parasites can be monitored with pheromone traps. Place the tent-shaped traps 6 to 7 feet (2 m) high in the north or east side of the trees by February 25 in southern areas and by March 15 in northern areas. Securely fasten them to tree limbs with a wire hanger. This prevents traps from being blown out of the trees by wind. Use at least three to four traps, regardless of orchard size. In orchards larger than 60 acres (24 ha), use one trap per 20 acres (8 ha) (see the section on monitoring methods at the beginning of this chapter for additional information).

Inspect the traps twice weekly for male scales. Because male scales are so small, use a hand lens to inspect the traps. Look for the dark band on the back of the male scale to distinguish it from *Aphytis* spp., Encyrtidae scale parasites that are similar in appearance to male scales. Record your data on a form similar to the one in Figure 33, and clean the trap. Replace traps whenever they become too dirty for accurate identification and counting of male scales; replace pheromone caps according to the manufacturer's recommendations.

After the first male scales have been trapped, begin calculating degree-days to predict crawler emergence. A model that calculates degree-days for San Jose scale is available at the UC IPM World Wide Web site (www.ipm.ucdavis.edu), or you can keep track of daily temperatures and use a degree-day table (also available at the IPM web site under "San Jose scale" in the almond *Pest Management Guidelines*). Crawler emergence begins about 405 DD after the first male flight. The optimal timing for a May crawler spray is 600 to 700 DD after the first male scale is collected in March or April. Crawlers can be controlled with properly applied oil sprays.

Because male scales are weak flyers, erratic spring weather can greatly influence their flight activity and, thus, the effectiveness of pheromone traps. Sticky tape traps can be used to monitor for scale crawlers as an alternative or in addition to using pheromone traps for male scales. Sticky tape traps consist of clear plastic tape that is wrapped around scaffold limbs (Fig. 44). Emerging males and crawlers that are migrating to a new location stick to the surface of the tape. The optimal spray timing using sticky tape traps is 200

Figure 44. When applying double-sided sticky tape to monitor for San Jose scale crawlers, choose scaffold limbs that are 1 to 2 inches (2.5–5 cm) in diameter and be sure the tape is wrapped tightly around the wood so the crawlers cannot get underneath.

The olive scale is purple underneath its covering. The color helps distinguish it from San Jose scale, which is yellow.

to 300 DD after the first crawlers are trapped. Consult your UCCE farm advisor for additional information.

If a late-season scale population has flared up in your orchard because the previous dormant spray was missed or improperly applied, do not try to treat the infestation during the growing season, except during the May crawler period. If this period has passed, wait until the following dormant season to apply a spray. Late-fall or postharvest treatments are not effective against San Jose scale because of overlapping generations. Only crawlers, white caps, black caps, and emerging males are susceptible to sprays.

Olive Scale
Parlatoria oleae (Diaspididae)

The olive scale is another armored scale (family Diaspididae) that occurs in almond orchards. Its covering is similar in shape, size, and color to San Jose scale, but the insect's body underneath is purple. Olive scale is not widely distributed in almonds but occurs in localized areas in some counties. Like San Jose scale, it secretes a tough, waxy covering that protects and camouflages the insect. In spring, olive scale lays eggs beneath its shell. Eggs begin to hatch in late April and crawlers scatter to new sites. If undisturbed by insecticides applied for major almond pests, natural enemies provide good control of olive scale. This scale is also adequately controlled by oil sprays applied during the dormant season.

European Fruit Lecanium
Parthenolecanium corni (Coccidae)

European fruit lecanium, formerly called brown apricot scale, occurs throughout the Central Valley but is rarely a problem on almond unless disrupted by insecticides. It is a soft scale (family Coccidae), larger and more bulbous in shape than armored scales. Soft scales excrete sticky honeydew upon which sooty mold can grow; armored scales do not.

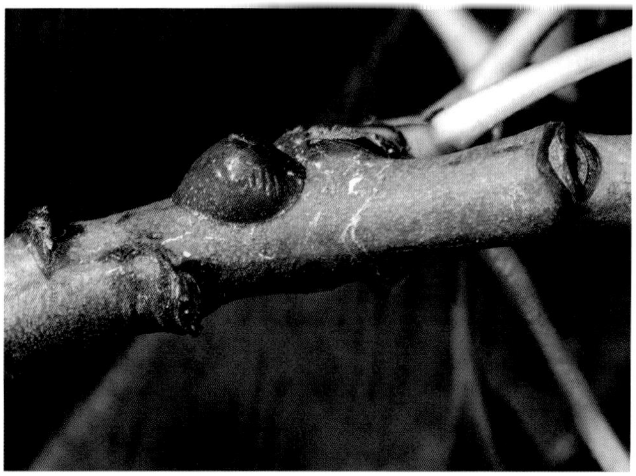

The domed, shiny brown shell of the European fruit lecanium has several ridges along the back.

European fruit lecanium passes the winter as a nymph. Control with oil is best at this stage (before February 1) because in late winter females harden, develop a rounded shell, and are no longer effectively controlled with oil sprays. The domed shell is shiny brown, about ⅜ inch (10 mm) in diameter, and has several ridges along the back. Eggs are laid in the spring, and the adult female's vital organs waste away as eggs accumulate under the arched canopy of the shell. Hatching occurs from May to July; there is only one generation per year. The young develop through the remainder of the season during which they produce copious amounts of honeydew.

The nymphs can be controlled in spring by oil once they have emerged from under the female's domed cover, but timing this spray is extremely difficult. Wait until winter when the leaves are off the trees for best coverage and control with an oil spray. Lecanium has many natural enemies, including parasitic wasps and lady beetles. Look for emergence holes in scales for evidence of parasitization.

MITES

Four species of mites can cause economic damage in Central Valley almond orchards. Their pest status depends upon various cultural practices, the weather, and the level of predation against them. The use of pyrethroids or carbamates for insect pests often causes mite problems because these insecticides are particularly harmful to mite predators.

Two of these mite species, twospotted mite and Pacific spider mite, occur throughout the Central Valley. In the early season, these spider mites, referred to in this manual as webspinning mites, are found in the lower central portion of the tree; once the weather gets hot, their numbers grow rapidly and they disperse throughout the tree. European red mites and brown mites (brown almond mites) are generally found in the northern San Joaquin and Sacramento valleys early in the season; however, European red mites can also occur in high numbers in the central and southern San Joaquin Valley. Both mites are found on leaves throughout the tree canopy and are potentially damaging early in the growing season. Populations of European red mites usually decrease with the onset of hot weather but may remain high throughout the growing season.

Most of these mites overwinter in almonds. European red mites and brown mites overwinter as red eggs that are laid on spurs and in limb crotches. Twospotted and Pacific mites overwinter as red or orange diapausing adult females on ground litter, orchard floor vegetation, and under bark scales of the tree. Figure 45 illustrates the periods of activity for these pest mites.

Webspinning Spider Mites

Pacific Spider Mite, *Tetranychus pacificus* (Tetranychidae)

Twospotted Spider Mite, *Tetranychus urticae* (Tetranychidae)

Pacific and twospotted mites are called webspinning spider mites because dense webbing occurs on leaves and branch terminals when populations of these mites reach high numbers. Adults of these species are difficult to distinguish from each other in the field. Both may occur in the same orchard, but it is not known if they differ in the effect their feeding has on almond trees or if different almond cultivars vary in their susceptibility to damage by these mites. Because they have similar life histories and are controlled in the same manner, however, there is no practical reason at this time to distinguish the two species when monitoring orchards.

Description and Seasonal Development

Webspinning mites overwinter as red or orange adult females on ground litter, on orchard floor vegetation, and under rough bark scales of almond trees. During the growing season the color of the oval-shaped females ranges from light yellow-green to almost black, depending upon age and host food. All have dark spots on both sides of the body. Adult male mites are smaller and less common than female mites. Male mites do not overwinter.

When the weather warms in the spring, females become active, moving up the tree to begin feeding on young almond leaves. They lay their eggs on the undersides of leaves. The spherical eggs are translucent when first laid but turn opaque before hatching. Immature mites molt three times. First-stage mites have only six legs, but later stages and adults have eight. Early in the growing season, webspinning mites are typically found on foliage in central or lower portions of trees as they radiate out from overwintering sites. Spring populations are often localized in hot spots either along roadsides or on the orchard's periphery. They

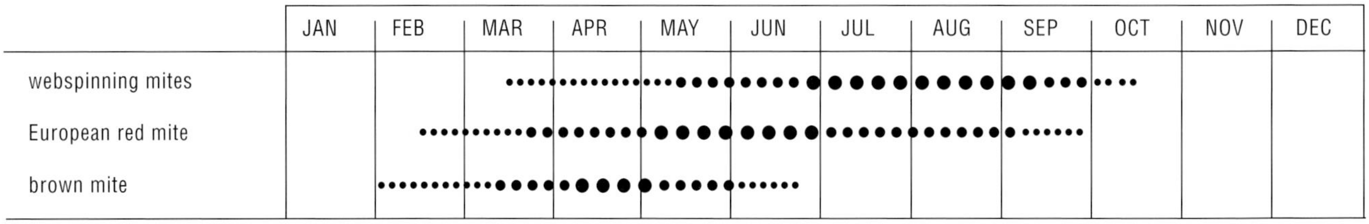

Figure 45. Typical periods of activity for pest mites in almond orchards. The larger the circle, the greater the potential for mite activity.

Overwintering female webspinning mites have red or orange bodies with dark spots on both sides of the body.

As the season progresses, webspinning mite adults generally are light colored with dark spots on both sides of the body, but their color depends on the quality of the foliage. Eggs are spherical and translucent. The smaller mite in the center right is a male.

Feeding by webspinning mites first appears as a light stippling on the leaves, as seen in the leaf on the left side of this photo. The leaf in the center has moderate stippling.

SPIDER MITE

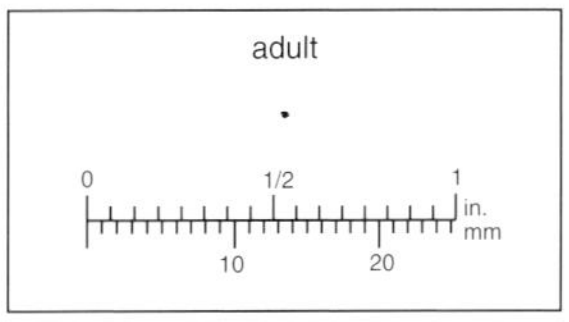

may also occur in high numbers in early spring in orchards that received pyrethroid applications the previous season. By May they are more generally distributed throughout tree canopies. During summer when the weather is hot and dry, mite infestations are common in dusty areas of an orchard. Water-stressed orchards are particularly susceptible to mite problems. Mite populations also rise at this time when mite predators are killed off by certain insecticides, or when predators cannot control them adequately in warmer than average years.

Webspinning mites reproduce rapidly in hot weather and commonly become numerous in June through September. If temperature and food supply are favorable, a generation can be completed in 7 days or less, and 8 to 15 generations can occur during a single growing season. As foliage quality declines in heavily infested trees, mites leave the trees and are dispersed by wind currents to new locations. High mite populations generally undergo a rapid decline in late August or September when predation overtakes them, when host plant conditions deteriorate, and as the daylight hours decrease and the weather turns somewhat cooler. With the change in the quality of the host plant, day length, and weather, females enter a period of diapause for the winter.

Damage

Mites cause damage by sucking cell contents from leaves. Initially, they cause a stippling effect on the leaves; as feeding continues, stippling becomes more severe and the leaves turn light yellow and fall. When spider mites are abundant, they often cover tree terminals with copious amounts of webbing.

Webspinning mites have the greatest potential for damage from late June through August; high populations at this time can result in partial to complete defoliation. Leaves that remain on the trees will not recover from mite damage after mite control has been achieved.

Webspinning mites can produce copious amounts of webbing when they are abundant.

Defoliation by webspinning mites not only affects the following year's yield but also creates a trash problem at harvest.

No effect can be detected on tree growth and productivity during the season of infestation and defoliation. This is because annual growth and nut sizing are complete by the time mite infestations become serious. The following season, however, nut set, leaf size, extension of terminals, and total crop may be significantly lowered; crop reduction can approach 20%. Infestations on young trees affect tree size and trunk girth.

Trees under water stress will show mite damage more readily than well-irrigated trees. Transpiration and photosynthesis rates are depressed more under the combined stress of water shortage and mite leaf feeding than when either factor is operating alone. On the other hand, mite populations may be higher on trees that receive excess nitrogen fertilizer. Even though the potential for mite damage is great, this potential is reached only when pest mite populations get out of hand. Low pest mite populations are necessary to maintain predator populations in the orchard. Research indicates that until webspinning mites exceed a threshold of an average of 100 to 120 mite-days per leaf over the entire tree (one mite-day = one mite feeding on one leaf for one day) on well-irrigated trees, there is little effect on the tree. Therefore, effects from an average of four to five mites per leaf for 20 to 25 days may not be detectable. Because well-irrigated trees can tolerate feeding by more mites than water-stressed trees, the threshold for mite feeding in water-stressed trees will be lower.

Management Guidelines

Prevention is a key part of an IPM program for mites because damaging populations are often encouraged by orchard conditions that favor them or insecticide applications that kill their natural enemies. The need for acaricide applications is determined by brush and count or presence/absence sampling. Selective acaricides are available when treatment thresholds are reached. Release of predatory mites is an option in some situations.

Prevention. Good orchard management practices can reduce mite problems. For instance, dusty conditions often lead to mite outbreaks. Minimize dust by oiling orchard roads or watering roadways with a tank truck at regular intervals throughout the growing season. Maintaining a ground cover in the orchard also helps reduce dust and provides habitat for general predators. In problem areas of the orchard, limit the amount of traffic and always drive slowly on orchard roads. Do not overfertilize trees, because excess nitrogen can increase mite populations.

A key aspect of mite management is adequate irrigation. Because mites cause greater damage to water-stressed trees, keep all trees in the orchard well irrigated and monitor mite populations carefully. Unless mites are under satisfactory control, withdrawing irrigation water and drying orchards in preparation for harvest will intensify defoliation related to mite damage.

Biological Control. Several predators play an important role in regulating mite populations. Orchards that do not receive in-season sprays may have several species of predatory mites including the western predatory mite, *Galendromus* (= *Metaseiulus*, *Typhlodromus*) *occidentalis*, as well as *Euseius* (*Amblyseius*) *hibisci*, *Euseius tularensis*, *Typhlodromus caudiglans*, and *Typhlodromus citri*. In orchards that receive moderate applications of organophosphate insecticides, the predominant predatory mite is the western predatory mite, because it has developed resistance to these materials and commercially reared western predatory mites that are organophosphate-resistant have been released into orchards.

The western predatory mite is one of the most effective mite predators. Although it prefers the egg stage, it feeds on all stages of the pest mites. It is about the same size as the webspinning mite but lacks markings on its body. The color of its body may range from cream to amber to red, depending on what it has been eating, but it never attains the deep red color of the European red mite. Western predatory mite eggs are larger than those of webspinning mites and are oval in shape, whereas webspinning mite eggs are spherical.

Western predatory mite can complete a generation in 7 days under optimal conditions and build up rapidly enough to regulate webspinning mites. To be effective, however, western predatory mite must be present in the orchard by May and early June so that sufficient numbers are available to respond to the rapid increases in pest mite populations associated with the onset of hot weather. If pest mites greatly outnumber the predatory mites (more than 10 pest mites per predatory mite), a better balance can often be achieved by applying low rates of selective miticides to reduce the pest mite population while leaving the predator population intact. Also, augmentative releases of predatory mites, which are available from commercial insectaries, may be helpful. Selective miticides that control spider mites but have little impact on the western predatory mite at lower-than-label to low label rates include propargite (Omite) and fenbutatin-oxide (Vendex).

Insecticides applied to control navel orangeworm and other insect pests often kill the natural enemies of mites. Carbamate and pyrethroid insecticides are especially destructive. Pyrethroid residues remaining on leaves and bark can affect predatory mites for months following their application. Some native populations of western predatory mite in commercial orchards are resistant to organophosphate insecticides if the orchards have been frequently sprayed, but no native populations are known to be resistant to carbamates or pyrethroids.

Strains of western predatory mites that are resistant to some organophosphate insecticides and the carbamate insecticide carbaryl have been developed in the laboratory and are now commercially available. Successful establishment of these predatory mites, or conservation and enhancement of native populations of western predatory mites, may reduce the number of applications and rate of miticides needed.

Special management practices are required in orchards to establish organophosphate-resistant predatory mites. Predatory mites need a period of at least 6 months, and in some cases as long as 2 years, after they are released to become established. Low population levels of pest mites must be maintained in the orchard to serve as a food source for predatory mites. Maintaining ratios of 1 predatory mite to 5 to 10 pest mites (1:10 in early season, 1:5 in midsummer), or the number of leaves with predators equal to or greater than half the number of leaves with pest mites if using presence/absence sampling, will keep the pest mites under biological control by the predatory mites. If predatory mites are present but the

The most dependable predator of webspinning mites is the western predatory mite. This photo shows how the predatory mite adult (top and bottom) differs in appearance from the webspinning mite (center).

Sixspotted thrips adults are characterized by three dark spots on each forewing. They are efficient mite predators, with both nymphs and adults feeding on mites and mite eggs, but they move into orchards only when mite populations are high, often after damage to trees already has occurred.

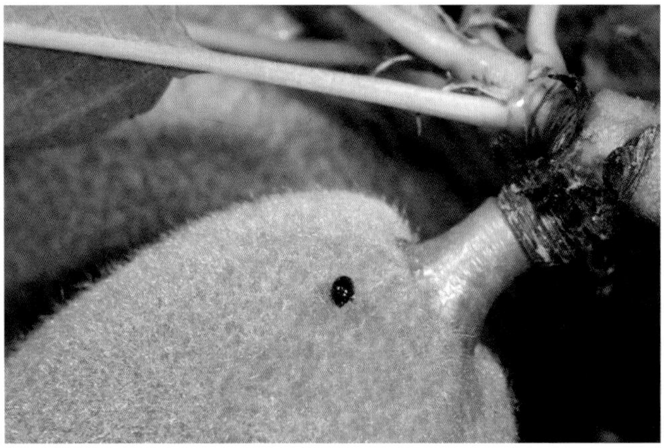

This small black beetle is an adult spider mite destroyer (*Stethorus picipes*). It preys on mites but becomes numerous only when mite populations are heavy.

SIX-SPOTTED THRIPS

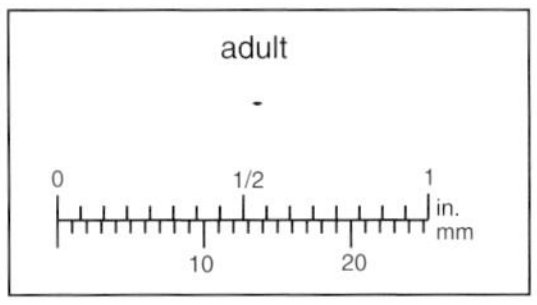

STETHORUS

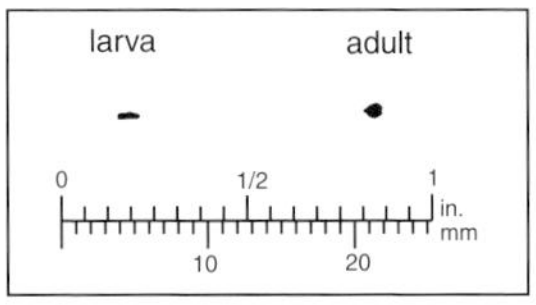

LACEWING

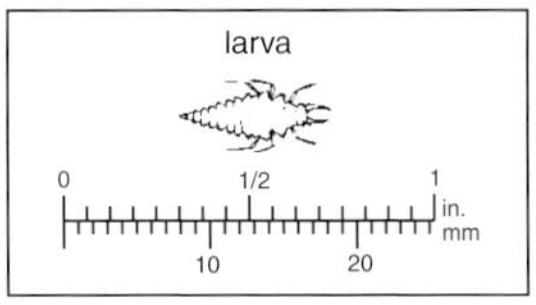

The larva of the spider mite destroyer also preys on mites.

This adult green lacewing is devouring a mite.

predator:pest ratio is out of balance, use lower-than-label rates of a selective miticide to reduce the pest mite population. Using selective miticides at rates lower than normal ensures that some pest mites remain and predatory mites do not starve.

Carefully monitor predatory mites and spider mites during the period of establishment to ensure that an adequate level of the pest population is present to serve as a food source, the predators have become established, and the pest mite populations are not reaching damaging levels. Insecticide resistance in the predatory mites developed because of the repeated use of the pesticide to which the predators have become resistant. If the insecticide is not applied occasionally, providing selection pressure that favors pesticide-resistant mites in the population, the predatory mites may become more susceptible to the insecticide when it is again applied.

Sixspotted thrips, *Scolothrips sexmaculatus*, is a mite predator frequently encountered in almond orchards, but it usually does not move into the orchards until mite populations are already high. Both adults and larvae prey on all stages of the pest mites and can rapidly reduce high mite populations. Monitoring is critical if sixspotted thrips are present in the orchard, especially if pest mite populations are near control action thresholds. If sixspotted thrips are present in samples and pest mite populations are near or below the thresholds, the sixspotted thrips may control the mites. If the pest mite populations exceed the control action threshold, a miticide application will be needed.

Spider mite destroyer, *Stethorus picipes*, is a small lady beetle that feeds on mites, particularly in unsprayed orchards. The larvae are dull brown or black and have a velvety appearance. Like the sixspotted thrips, however, it usually does not become numerous until mite populations are heavy. It is an active feeder and can reduce mite populations quickly when it is abundant. Brown lacewings (*Hemerobius* spp.), green lacewings (*Chrysopa* and *Chrysoperla* spp.), and predatory larvae of a cecidomyiid fly, *Feltiella* sp., also feed on all stages of spider mites.

Monitoring. Check orchards for webspinning mites once every 2 weeks from March to the first part of May. From May though August, monitor the entire orchard weekly and monitor problem areas, such as trees along roads or water-stressed trees, every few days.

For general monitoring in the early part of the season, observe individual leaves in the lower interior portions of trees throughout the orchard for signs of mite feeding. Use a hand lens to examine leaves for mite eggs, mobile mites, and predatory mites. As a rule of thumb, if there are more eggs than mobile mites, the pest mite population is increasing. Continue watching the western predatory mites to determine predatory mite:pest mite ratios, because these predators often control mites in the early season before the populations can build up. If a ratio of 1 predatory mite to 10

Brown lacewings feed on many species of pest mites.

The green lacewing larva (shown here) looks similar to the brown lacewing larva.

Cecidomyiid larva feeding on mite eggs.

or fewer pest mites is present or the number of leaves with predators is at least half the number of leaves with pest mites using a presence/absence sample, webspinning mites will probably not be a problem. Miticides in the early part of the season are seldom required if predators are already present.

If you are managing a pesticide-resistant predator population and pest mite populations increase with the onset of hot weather, use lower-than-label rates of a selective miticide or augmentative releases of predatory mites to adjust predator:prey ratios without removing all food sources for predators. If predator augmentation is used, it is best to do so while pest populations are still low, because the number of predators necessary to control high pest mite densities would be prohibitively expensive and may not work at all. Consult *Pest Management Guidelines: Almond*, listed in the suggested reading, for the latest information on miticides and rates.

Two techniques (described below) for sampling mites in almond orchards from mid to late season are brush and count sampling and presence/absence sequential sampling. Brush and count sampling can be used earlier in the season than the presence/absence sequential scheme to assess population levels of webspinning mites, European red mites, and western predatory mites, but it is more time-consuming than presence/absence sampling. The presence/absence sampling method has not yet been developed to assess European red mite populations. You must be able to correctly identify active stages and eggs of pest mites and western predatory mites to use these sampling methods. Photographs in this manual will help you to identify them.

Brush and Count Sampling of Mites

Brushing and counting mites using a mite brushing machine is an efficient alternative to actually counting the number of mites on individual leaves using a hand lens. References to control action thresholds and other guidelines provided in this manual that mention brushing and counting also apply to directly counting the number of mites on leaves.

Begin sampling for mites with the brush and count method in early May by establishing sampling sites in each orchard. If webspinning mites are the only pest mites present in the orchard, you can either continue to brush and count or switch to the presence/absence method in mid-May. If European red mites are consistently present in the majority of samples, you may want to use the brush and count method throughout the season.

To make the sample, a specified number of leaf samples are taken from selected trees, put in paper bags or closed containers, refrigerated, and taken or shipped to a laboratory for counting. The leaves should be kept refrigerated until brushed.

Establish sampling sites that consist of a cluster of five trees. Flag the trees so the same trees can be sampled on each sampling date. Unless the orchard is very small, use a minimum of four sampling sites in orchards of 200 acres (80 ha) or less, with each site located 10 rows and 10 trees in from a corner of the orchard. Use one sampling site per 50 acres (20 ha) for orchards larger than 200 acres (80 ha). Sampling sites can be established 10 rows in from orchard edges in addition to the sites near the corners. Take separate samples from orchard edges and hot spots where mites have been a problem in the past. These locations may require separate treatment even if predators are keeping pest mites under control in the rest of the orchard.

To take a sample, randomly select 10 leaves from each of the five trees at the sampling site. Select leaves from above the sprinkler line or at head height from inside and outside the canopy and from all four quadrants of the tree. Put the 50 leaves from each sampling site in a labelled paper bag and keep the samples cool until they can be brushed.

Use a mite brushing machine to brush mites from the leaves. Each leaf is run through the mite-brushing machine, which removes the mites onto a glass plate that is coated with a thin layer of vegetable oil. Place the glass plates over a paper grid and count the mites under a dissecting microscope (Fig. 46). On a sampling form, record the total number of twospotted and Pacific spider mites, the total number of European red mites, and the total number of predatory mites. Record active stages only. Divide these totals by the number of leaves to get the average number for each. By keeping the counts for each site separate, each portion of the orchard can be evaluated separately and treatments applied to only those areas that need them, reducing costs and potential harm to natural enemy populations.

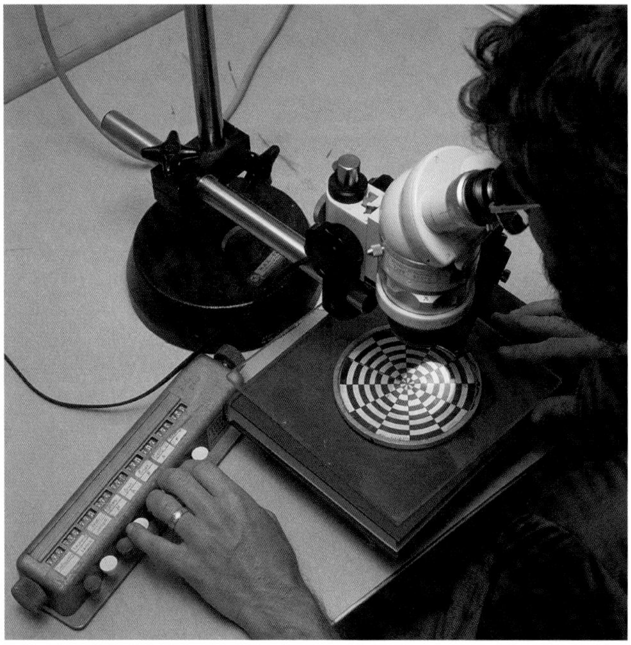

Figure 46. For brush and count sampling of mites, mites are counted under a microscope after they are brushed off the leaves onto a glass plate.

Determine the ratio of predatory mites to spider mites (P:S) by adding the average number of predatory mites per leaf and dividing by the total of the average number of webspinning spider mites and the average number of European red mites per leaf.

Calculate the number of spider mite days (SMD) to estimate the damage being done by spider mites using the following formulas:

weekly SMD = days since last sample × (current week average mites per leaf + last week's average mite's per leaf) ÷ 2

seasonal SMD = weekly SMD of current sample + total SMD from previous samples

Compute SMD for webspinning spider mites and European red mites separately and divide the SMD for European red mites in half before adding to the SMD for webspinning mites to get weekly and seasonal totals.

Use the calculated values for spider mites per leaf, P:S ratio, and SMD with the flow chart in Table 12 to reach a treatment decision for each sampling area.

Presence/Absence Sequential Sampling for Webspinning Mites

Presence/absence sequential sampling is presently applicable to webspinning mites only. Begin sampling in mid-May and continue to sample through August to determine if a miticide spray is needed. To use this method, you do not need to count the numbers of mites per leaf; you need only to determine if they are present or absent.

1. Before July 1, you only need to monitor hot spots—that is, areas that usually develop mites first. These include roadside trees and trees in dry or dusty areas or areas with sandy streaks in which the trees are water-stressed. Once a pest mite threshold is reached in these areas, sample the remainder of the orchard to determine if a spot treatment is sufficient or if the entire orchard needs to be treated. After July 1, monitor the entire orchard by dividing it into sampling areas. A sampling area is the minimum portion of the orchard to which you're willing to make a treatment. Spot-treating or treating only specific areas of an orchard where mites are a problem can reduce the total amount of miticide applied in an orchard while allowing beneficial mite predators to survive in untreated areas. If you plan to treat the whole orchard anyway if an application is made, it is not necessary to divide the orchard into different sampling areas.
2. Within each hot spot or sampling area, randomly select trees to sample. Do not choose trees that are obviously different from others in the area, such as replants.

Table 12. Using Mite Brushing Data to Reach a Treatment Decision for Spider Mites. Use calculated values for ratio of predatory mites to spider mites (P:S), weekly and total spider mite days (SMD), and average numbers of spider mites per leaf (SM/L) to reach a treatment decision based on the time of year as described below.

MAY	
1. If no predators present	
and total SMD for season is less than 10	DO NOT TREAT
and total SMD for season is 10 or more	
and SM/L has increased each of previous 3 weeks	TREAT
and SM/L has not increased each of previous 3 weeks	Go to step 2
2. If P:S is lower than 1:20	
and total SMD for season is less than 25	
and SMD for last week is less than 10	DO NOT TREAT
and SMD for last week is 10 or more	TREAT
and total SMD for season is 25 or more	TREAT
3. If P:S is between 1:10 and 1:20	
and total SMD for season is less than 25	DO NOT TREAT
and total SMD for season is 25 or more	TREAT
4. If P:S is 1:10 or higher	DO NOT TREAT
JUNE	
1. If no predators present	
and no predators were present last week	
and SM/L has increased for two straight counts in June	TREAT
and SM/L has not increased for two straight counts in June	Go to step 2
and predators were present last week	Go to step 2
2. If P:S is less than 1:5	
and SMD for week is 20 or less	
and SM/L has increased each of previous 4 weeks	
and P:S is less than 1:20	TREAT
and P:S is 1:20 or higher	
and P:S is less than 1:10	TREAT
and P:S is 1:10 or higher	
and SMD for week is less than 15	DO NOT TREAT
and SMD for week is 15 or more	TREAT
and SMD for week more than 20	TREAT
3. If P:S is 1:5 or higher	
and SMD for week is 20 or less	DO NOT TREAT
and SMD for week is more than 20	TREAT
JULY	
1. If hull split spray is scheduled	
and P:S is 1:10 or lower	Add acaricide to hull split spray.
and P:S is higher than 1:10	No acaricide needed unless carbaryl used.
2. If no hull split spray is scheduled	
and SMD for week is less than 20	DO NOT TREAT
and SMD for week is 20 or more	TREAT
3. If no hull split spray is scheduled	
and P:S is lower than 1:10	
and SM/L has increased each of the previous 4 weeks	TREAT
and SM/L has not increased each of previous 4 weeks	DO NOT TREAT
and P:S is higher than 1:10	DO NOT TREAT
AUGUST	

If total SMD for season will approach or exceed 110–130 within the month, treat with material and rate allowed depending on the preharvest interval.

3. As you walk around the tree, randomly select leaves both inside and outside the tree canopy from all four quadrants of the tree. In orchards with sprinklers, pick leaves from above the line where water from the sprinkler hits the tree.
4. Sample a minimum of five trees within each sampling area. Select 15 leaves from each tree and use a hand lens to examine both upper and lower leaf surfaces carefully for the presence or absence of webspinning mites and eggs, as well as predators such as western predatory mite and sixspotted thrips. Be sure to examine the entire leaf carefully because there may be only one or two mites on the leaf, but once a pest mite and/or mite predator is identified you do not need to examine the leaf anymore. Use a sampling form such as the one in Figure 47 to record your results. Don't count the mites; just record the number of leaves with webspinning mites or eggs in the first column of the sampling form and the number of leaves containing mite predators or eggs of predatory mites in the last column. Add together the total number of leaves with webspinning mites for all trees sampled and record this number in the second column. If western predatory mite or sixspotted thrips is present in your orchard, use the "Don't' Treat" or "Treat" figures given on the form under the heading "Predators Present." If no predators are present, however, use the treatment figures in the columns under the heading "Predators Absent." If the total number of leaves with mites falls between the "Treat" and "Don't Treat" figures given on the form, continue to sample until a definite decision can be reached.

For example, if predators are present in the orchard and you find mites on 36 leaves out of 75 in the first five trees you sample, continue to sample until either you can make a treatment decision or you sample 20 trees. If, on the other hand, 15 out of 75 leaves in the first five trees contain mites, don't treat your orchard.

WEBSPINNING MITES IN ALMOND TREES: A SAMPLING FORM

Date: ______________ Orchard: ______________

				Predators Present		Predators Absent		
Tree number	Total no. leaves sampled	Number leaves with mites	Total leaves with mites	Don't treat if accum. total less than or equal to:	Treat if accum. total more than or equal to:	Don't treat it accum. total less than or equal to:	Treat it accum. total more than or equal to:	Number of leaves with western predatory mite and/or sixspotted thrips
1	15	____	____					____
2	30	____	____					____
3	45	____	____					____
4	60	____	____					____
5	75	____	____	27	40	12	24	____
6	90	____	____	33	48	15	28	____
7	105	____	____	39	55	18	31	____
8	120	____	____	45	62	21	35	____
9	135	____	____	51	69	23	39	____
10	150	____	____	57	76	26	43	____
11	165	____	____	63	83	29	46	____
12	180	____	____	70	90	32	50	____
13	195	____	____	76	97	35	54	____
14	210	____	____	82	104	38	57	____
15	225	____	____	88	111	41	61	____
16	240	____	____	94	118	45	65	____
17	255	____	____	101	125	48	68	____
18	270	____	____	107	132	51	72	____
19	285	____	____	113	139	54	75	____
20	300	____	____	119	146	57	79	____

Stop Sampling

Figure 47. Use a sampling form such as this one for presence-absence sequential sampling of webspinning mites.

5. If a decision to treat is reached, consider the predator population. If the number of leaves with webspinning mites is about equal to or less than the number of leaves with western predatory mites, delay treatment for a few days; the predator population is likely to cause the pest mite population to decline. If you do treat and predators are present in any abundance, use low rates of miticides to reduce the webspinning mite populations without eliminating them as food for the predatory mites. If you are combining the miticide with a carbamate or pyrethroid hull split spray, both of which are very disruptive and likely will result in mite outbreaks in the absence of miticide treatment, use the full miticide rate.
6. If after sampling 20 tress you could not reach a decision, sample again 2 to 3 days later. If you did reach a decision to not treat at the present, sample again in 1 week.

Eggs of the European red mite are red, somewhat flattened, and have a characteristic stalk (stipe). The stipe may be absent following rains.

European Red Mite

Panonychus ulmi (Tetranychidae)

European red mites can be found on nuts, pome and stone fruits, berries, and some ornamental shrubs. High populations may occur in any almond-growing region of the Central Valley.

Description and Seasonal Development

European red mites overwinter as eggs. Look for eggs in roughened bark, at bases of buds and spurs on smaller branches and twigs, and in wounds. Eggs are globular in shape and are red with a slender stalk (stipe) arising from the top center and many grooves extending from top to bottom. During the growing season, eggs are laid on the leaves. Overwintering eggs are laid as early as August and, depending upon the weather, through October and November.

Eggs hatch when the trees bloom, and immature mites migrate to leaf buds and feed on emerging leaves. Three instars occur before the adult stage. Immature mites are bright red, except just after molting when they appear bright green. The green color turns to red after mites resume feeding. Adults are dark red and have six to eight white spots at the base of hairs on the back. European red mites are found mostly on the upper surface of leaves. They spin little or no webbing. As the population increases, the mites colonize both surfaces of the leaf.

Depending upon temperature, European red mites can have five to ten generations per year. Low spring temperatures can keep high populations from developing. If these mites build up in an orchard during the spring, the population may survive the hot, dry conditions of summer and become generally distributed throughout the orchard.

The adult female European red mite is dark red and has long, curved hairs growing from white spots on the back.

European red mites cause leaf stippling similar to that caused by webspinning mites. Dark spots on the surface of these leaves are adult European red mites.

Damage

Feeding by European red mites causes leaf stippling. Heavy or prolonged feeding eventually causes leaves to turn pale and appear bronzed and burned at the tip and margins; in severe cases or when trees are water-stressed it can even cause leaf drop. European red mite is not considered to be as damaging as webspinning mites, but repeated, heavy infestations may reduce yields. For example, it has been found that heavy infestations for 2 years in a row on walnut trees reduced yield the third year by 40%, and the trees did not completely recover the first year when subsequently protected. Damaging effects have also been recorded on apple and plum trees, but the effect of European red mite on almond trees has not been measured.

Management Guidelines

European red mite is often kept below damaging levels by natural enemies. Regular monitoring is recommended to detect when damaging population levels may be developing. The primary method of controlling European red mite in an IPM program is application of an oil spray to kill mite eggs during the dormant season.

Biological Control. Brown lacewing, *Hemerobius* sp., is an effective European red mite predator, but its occurrence is unreliable and not easy to predict. Western predatory mite feeds on immature and adult stages of European red mite but is unable to break through the egg shell. Therefore, it is not as effective against European red mite as it is against webspinning mites. Adult and immature stages of European red mite, however, serve as an early-season food source for predators, which may later feed on webspinning mites.

Monitoring and Control. The best way to control European red mite is with a dormant oil spray. In-season sprays usually are not necessary if oil is applied properly during the dormant season. Examine dormant spurs for red mite eggs. Keep records of the proportion of spur samples infested with eggs and relate this to red mite populations and damage during the season. A treatment threshold of 40% infested spurs for a dormant oil spray has been suggested.

Check orchards for European red mite once a week during the growing season as part of your regular monitoring program. You can check leaves for European red mite at the same time you are looking for webspinning mites. No treatment guidelines based on leaf sampling are available for European red mite. European red mite is not as damaging as webspinning mites, however, and it is thought that almond trees can tolerate a greater number of European red mites feeding per leaf than webspinning mites before economic loss occurs. Densities as high as 50 per leaf have not caused significant leaf drop in orchards that are not stressed for water or nutrients.

European red mite must be monitored more intensively in the spring than webspinning mites. Use a hand lens to examine leaves for mites and eggs. Also check the degree of leaf stippling that can be directly attributed to European red mite. European red mite may also be monitored with the brush and count method described above in the description of monitoring for webspinning mites. Do not treat low European red mite populations, because they help maintain predators for later control of webspinning mites. If extensive stippling is found in an area but predators are present, delay treatment a few days to see if the predators contain the red mite population. If high mite populations persist, treat the area of infestation.

An oil spray can also be used in-season to minimize impact on natural enemies. Use of high label rates and thorough coverage is essential for effective control with the oil. In-season insecticide sprays may disrupt biological control of European red mite and flareups may occur, so if you use an insecticide be sure to monitor carefully after treatment.

Brown Mite
Bryobia rubrioculus (Tetranychidae)

Brown mite, also called brown almond mite, is a pest of stone and pome fruits throughout the world. It is the largest of the mites that feed on almonds and the first to appear in spring. If not controlled with a dormant oil spray, brown mite can cause damage in the northern San Joaquin Valley and the Sacramento Valley.

Description and Seasonal Development

Like European red mite, brown mite overwinters in the egg stage. Eggs are red and spherical but do not have a stalk arising from the tops. They are laid in masses under bark scales and at the base of buds, and they hatch at the same time the leaf and flower buds open, usually 1 to 2 weeks before European red mite eggs hatch. Hatch is completed before full bloom.

Newly hatched brown mites have six legs and are bright red. After they molt for the first time, they turn brown, have eight legs, and resemble the adult but are smaller. Immature mites feed mainly on the undersides of leaves.

Adult brown mites are dark reddish brown, and the first pair of legs is longer than the other three pairs. During the warmest part of the day, brown mite rests and reproduces in woody parts of the tree. When temperatures cool off, it moves onto the leaves and feeds on both upper and lower leaf surfaces. Frequently, brown mite only infests a few trees within the orchard. Adult females live 2 to 3 weeks and reproduce without mating; there are no males. Brown mite is not active during hotter periods of the summer. Eggs laid during hot weather will not hatch until the following spring, and populations generally decline in midsummer. Usually two or three generations of brown mite occur between February and early June.

Damage

Feeding by brown mite causes whitish gray spots to appear on leaves. Brown mite does not produce webbing and rarely

Adult brown mites are flattened and dark green-brown, with very long front legs. Eggs are similar in appearance to European red mite eggs but lack a stipe when freshly laid.

Brown mite feeding causes leaves to turn whitish gray.

causes leaf drop. Affected leaves first become mottled, then bleached, but they do not turn bronze at the tips as do leaves damaged by European red mite. Badly damaged foliage is smaller than normal and covered with minute flecks of dried feces. Brown mite causes defoliation when population densities reach 50 per leaf.

Management Guidelines

Generally, brown mite is effectively controlled by a dormant oil spray. You can monitor spurs for eggs during the dormant season in the same way as for European red mite eggs, and the same treatment threshold can be used. Occasionally brown mite is a problem in the Sacramento Valley when there is a cool spring and inadequate amounts of oil were used in the dormant spray or spray coverage was poor. Brown mite has the same predators as European red mite, and moderate numbers of brown mite in the spring can be beneficial in providing a food supply for these predators. In-season sprays for brown mite are generally not needed. If you monitor brown mite populations during the season, take leaf samples early in the morning because the mites move off the foliage during the day. Alternatively, you can check spur scales for mites and eggs during the day.

Peach Silver Mite

Aculus cornutus (Eriophyidae)

Peach silver mite is sometimes found on almond foliage in the Central Valley. Although it may occur in large numbers (more than 100 per leaf), it causes damage only occasionally when populations build up on young trees up to 5 or 6 years old. Peach silver mite is usually considered a beneficial species because it serves as a food source for mite predators.

Peach silver mites are extremely small, pale yellow, and teardrop shaped with four short legs at the larger (anterior) end.

A leaf feeder, peach silver mite is white or cream colored and extremely small. Because it is much smaller than other mites on almonds, a microscope may be needed to see it clearly. Feeding by peach silver mite causes tiny, chlorotic spots that give the leaf a silvery appearance, especially along the midvein of the upper surface of the leaf. Symptoms resemble thrips or leafhopper damage. Brown, sunken spots sometimes develop along leaf margins following silver mite feeding early in the season. Unless silver mite numbers are high enough to cause defoliation, no treatment is necessary.

Pavement Ant
Tetramorium caespitum (Formicidae)
Southern Fire Ant
Solenopsis xyloni (Formicidae)

Pavement ants and southern fire ants can cause economic damage to almonds when the nuts are drying on the ground after harvest. Increased planting of almonds in the southern San Joaquin Valley on previously nonirrigated soils and changes in orchard floor and water management practices have led to increased ant populations and ant damage in almond orchards. These practices include change to nontillage orchard floor management, including vegetative ground covers, and greater reliance on cultural methods such as early harvest rather than hull split insecticides for control of navel orangeworm.

Description and Development

Ant hills in almonds orchards often appear as tiny mounds or small patches of loose soil. The ant colony consists of one to several queens and thousands of workers. Queens lay eggs and do not leave the nest. Workers care for queens and forage for food. At certain times of the year, winged males and winged females, or queens, are produced. They leave the nest in swarms and disperse to new locations. Males die shortly after mating, while females locate a new nest site, tear off their wings, excavate small caves in the soil, and lay eggs. The queen feeds and cares for the young maggotlike larvae in her cave until the larvae are full-grown. From that point on the adult workers take care of the queen, whose sole function is to lay eggs, and the developing young.

Although other ant species have been found in orchards and have caused occasional damage, the pavement ant and

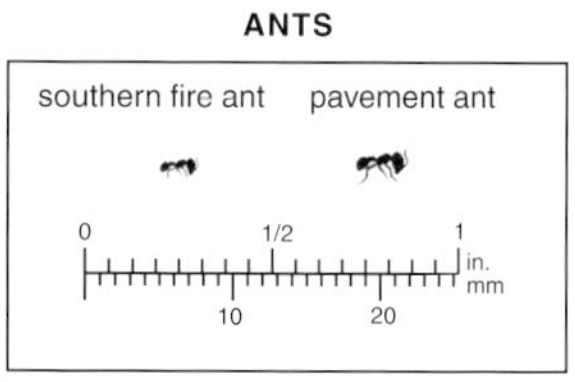

Ant hills in almond orchards often appear as tiny mounds or small patches of loose soil.

Red imported fire ants glue together soil as they construct nests, forming mounds that retain their shape when excavated. The southern fire ant does not form such hardened mounds.

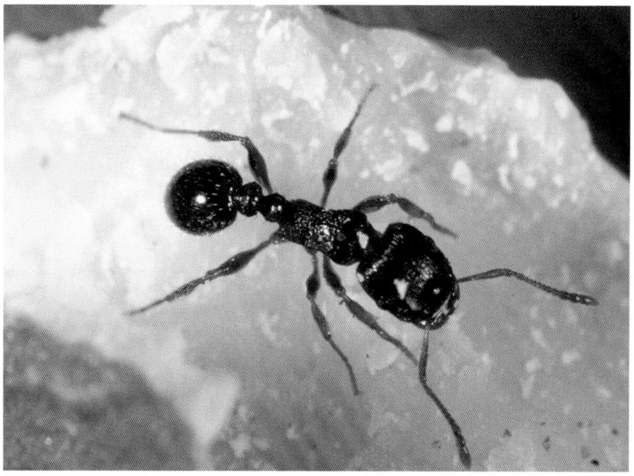

The pavement ant is dark brown and covered with coarse hairs.

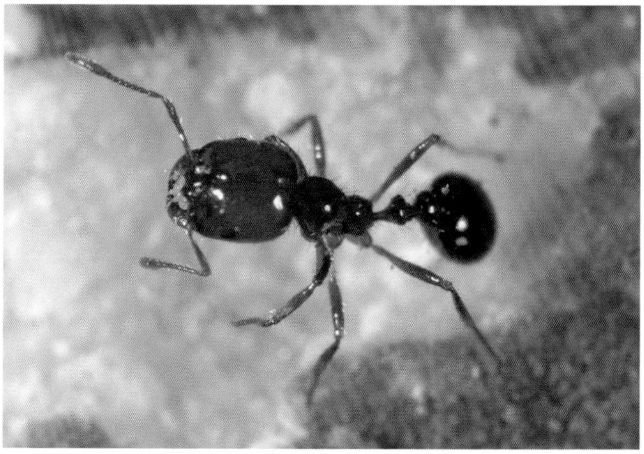

The southern fire ant has an amber head and thorax with a black abdomen.

the southern fire ant are the only ones that have been associated with major damage to almonds. Black harvester ants, *Veromessor andrei*, feed on almonds, but they are generally not very damaging because they are too large to enter the nuts. The red imported fire ant, *Solenopsis wagneri* (*S. invicta*), an exotic species to California, was accidentally introduced into a few orchard sites in 1998. This ant species poses a significant threat to California orchards should it become established and widespread, because of the harm it could cause to the crop as well as to workers.

The pavement ant worker is light to dark brown and covered with coarse hairs. Workers travel in single file in search of food and usually nest in sandy or loamy soils where the soil temperature is 58° to 82°F (14° to 28°C).

The southern fire ant has an amber head and thorax with a black abdomen. This ant has a painful sting that can cause visible swelling. It is found throughout the San Joaquin and Sacramento valleys and prefers soil temperatures of 70° to 95°F (21° to 35°C). The pyramid ant is a beneficial species that is similar to the southern fire ant in size but is active in midday when the fire ant is not. Pyramid ants do not swarm when their nest is disturbed as do southern fire ants. Southern fire ants prefer dry, undisturbed soil, so they are not as abundant as the pavement ant in orchards with cover crops.

The native gray ant, *Formica aerata*, which is an important predator of peach twig borer, nests in the soil but does not form hills of loose soil as do the pavement ant and southern fire ant. This ant has a gray or dark brown abdomen with dark red thorax and head.

Damage

The southern fire ant has a wider distribution and causes more damage in California almond orchards than does the pavement ant. Ants feed on a wide variety of plant material in the orchards. They usually do not feed on almonds until they are on the ground, and damage increases in proportion to the length of time the nuts are on the ground. Damage by both species of ants first appears as a scraping or peeling of the pellicle, or skin of the kernel. Eventually the meat is completely hollowed out, leaving only parts of the skin. Unlike navel orangeworm damage, there is no frass or webbing, although white, sawdustlike material from the ant's chewing may remain. In some orchards, ant-damaged nuts may account for the major portion of the nuts rejected at the processor.

Ants feed on almonds by hollowing out the nutmeats. Eventually all that is left is parts of the skin, as seen in the nut in the top of the photo.

Nuts with a tight shell seal or with shells split less than 0.03 inch (0.8 mm) are not likely to be attacked by ants. Shell seal can vary depending on cultivar, crop size, and cultural practices. Heavier crops have smaller nuts with tighter shell seals and therefore are less likely to be damaged by ants. Estimates of shell seal can be made by sampling nuts before harvest and using a spark plug gauge to measure shell seal.

Management Guidelines

If you think ant damage is a problem in your orchard, survey the orchard floor for colonies in the spring or shortly after an irrigation. The best time to monitor for ants is in April and May. If you plan to use bait to control ants, monitoring in late spring allows you to make a treatment decision in time for baiting to be effective.

Ant hills are easiest to see 2 to 3 days after an irrigation because the soil on the top of the mound will be freshly turned. Later in the season, especially when the orchard is being dried for harvest, ant hills are difficult to see. Also, when the soil cracks, ants may travel in the cracks rather than on the soil's surface and are harder to see. Examine the area from mid-alley to mid-alley along ten trees and count the number of active ant colonies. For each block evaluate a minimum of five such areas. It may be difficult to determine where one colony begins and another ends. Large colonies may have several entrance holes separated by more than 3 feet (90 cm). Active colony entrances more than 3 feet apart, however, should be counted separately. Ant activity appears to peak in the morning and again just before sunset; these are the best times to identify active colony entrances.

Once you have determined the number of colonies per sampling area, check Table 13 to find the corresponding estimate of damage. Note that the damage increases with the number of days that nuts remain on the ground. Picking up nuts quickly after shaking reduces ant damage. If your colony counts indicate that some areas may have a level of damage

Table 13. Percent Damage to Almonds Caused by Ants after Harvest.

PERCENT DAMAGED

Number of colonies	Days between shaking and pickup				
per 5,000 sq ft	4	7	10	14	21
15	0.9	1.6	2.1	3.1	4.9
45	1.4	2.3	3.2	4.7	7.0
185	2.0	3.6	5.0	7.0	11.1

that is not acceptable to you, measurements of shell seal indicate that the crop will be susceptible, and rapid pickup will not be possible, apply a treatment to those areas.

Ant baits are an effective and less disruptive alternative to organophosphate insecticides. Bait is taken back to the nest by workers and slowly kills the entire nest. Depending on the type of bait, 4 to 7 weeks is needed for baits to take full effect. A decision to apply ant bait must be made in June, so the material will have time to kill ant nests before nuts are shaken. Baits are most effective when applied in summer; results have been variable following spring applications. Baits are less effective in weedy orchards because weed seeds, especially spurge, tend to be more attractive to ants than the bait. Bait should not be applied within 48 hours before or 24 hours after an irrigation.

Organophosphate hull split sprays for navel orangeworm may incidentally suppress worker ants, but the colonies will eventually recover and persist from year to year. If ants are a problem, focus insecticide treatments on areas where ant colonies are concentrated to minimize impact on natural enemies or use baits, which are not harmful to beneficials. You may need to treat the entire orchard floor if ant colonies are widely distributed. Annual applications may not be required, but it is good practice to monitor your orchards each year to determine if a treatment is needed. The previous season's damage is the best indication of a need for treatment. Remove nuts from the orchard floor as soon as possible after shaking.

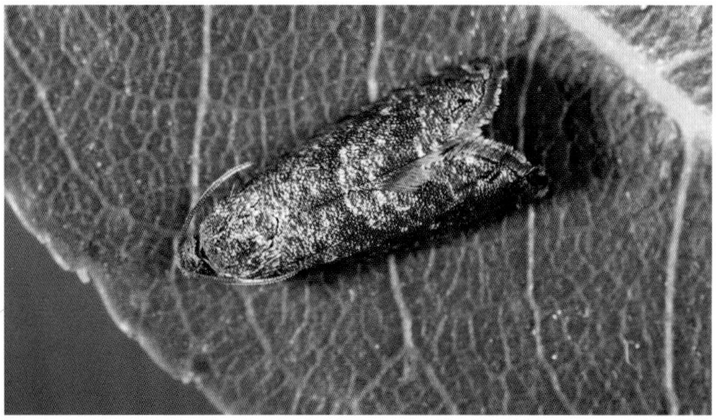

Oriental fruit moth adults are dark gray with alternating black and gray markings that form a chevron pattern on the folded wings. This moth is smaller than the peach twig borer moth and lacks the snoutlike projection at the front of the head.

The oriental fruit moth larva is white or pink with a brown head.

Oriental Fruit Moth

Grapholita molesta (Tortricidae)

Although it occurs commonly in almond orchards, oriental fruit moth only occasionally damages almond nutmeats. Early in the season, larvae bore into the tips of tender twigs and cause them to wilt and die. This damage can make it difficult to train first- and second-year trees. Oriental fruit moth shoot strikes are nearly identical to those caused by peach twig borer but are first seen later in the spring than peach twig borer shoot strikes. It is fairly easy to monitor shoot strikes for both species in the spring. Oriental fruit moth larvae mine deeper into a shoot than peach twig borer larvae, but identifying the cause of a shoot strike can be difficult unless the larva is still in the twig. The oriental fruit moth larva is white or pink with a brown head; the peach twig borer larva has dark brown bands and a black head.

ORIENTAL FRUIT MOTH

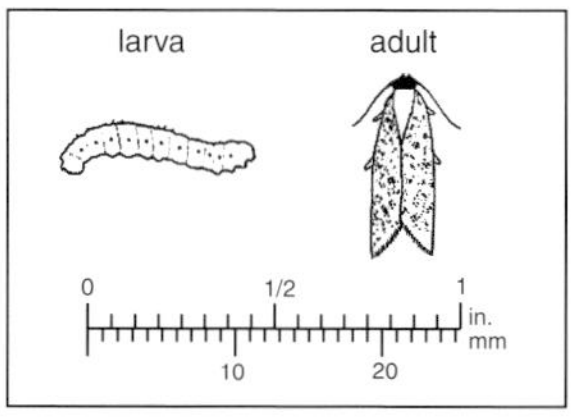

Oriental fruit moth larvae can be distinguished from other larvae, such as the navel orangeworm, by the presence of an anal comb, visible with a hand lens, that protrudes from beneath the sclerite of the last abdominal segment.

Oriental fruit moth larvae rarely attack almond fruit. If they do, they may be found feeding between the hull and shell after hull split when the hulls are still fairly green. In other crops, oriental fruit moth larvae prefer mature fruit, but in almonds they have been found boring into nuts in late May to early June. At hull split, the larvae enter the nuts primarily through splitting sutures and at the point of stem attachment.

Like navel orangeworm, oriental fruit moth may feed in groups of several larvae within a single nut. Oriental fruit moth larvae can be distinguished from navel orangeworm by the absence of the pair of crescent-shaped marks on the second segment behind the head and by the presence of an anal comb. Also, the oriental fruit moth larva does not produce webbing, and it produces reddish brown frass in the hull, in contrast to the brown and white frass of the navel orangeworm. Damage to the kernel by oriental fruit moth resembles that caused by peach twig borer. If you are using a program of Bt sprays to control peach twig borer, it is important to distinguish between peach twig borer and oriental fruit moth damage because Bt does not control oriental fruit moth. Also, oriental fruit moth is not controlled by dormant or bloomtime sprays as is peach twig borer.

A method for monitoring this pest with pheromone traps and for estimating when generations will occur, as well as the ideal timing for treatment using degree-days, is outlined in *Integrated Pest Management for Stone Fruits*, in the "General" suggested reading. A degree-day model for oriental fruit moth is available at the UC IPM Web site (www.ipm.ucdavis.edu).

Peachtree Borer
Synanthedon exitiosa (Sesiidae)

Peachtree borer, a clearwinged moth, is a sporadic pest of almond in the Central Valley. It is a more common pest of tree fruits in the Santa Clara Valley and Contra Costa County.

Peachtree borers overwinter as larvae in the tree trunk. Larvae have light-colored bodies and dark heads. Figure 48 illustrates how to distinguish peachtree borer larvae from American plum borer larvae, another pest that can be found boring in the crown area. Peachtree borers feed in the crown area and burrow up into the tree. In late spring they pupate

PEACHTREE BORER

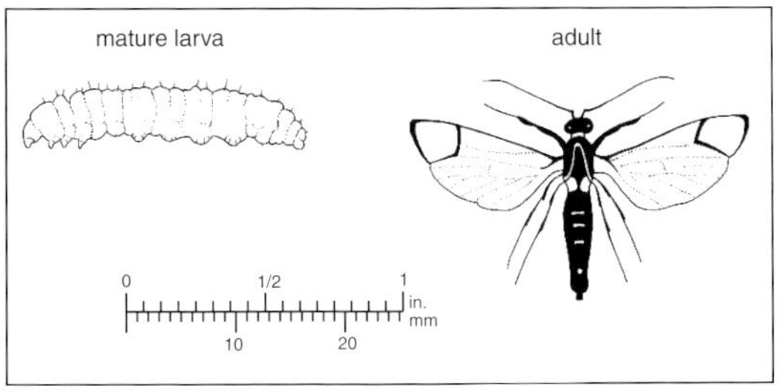

Nut damage caused by oriental fruit moth larvae is similar to that caused by peach twig borer larvae. Oriental fruit moth larvae do not bore deeply into the nutmeats and do not produce webbing.

Peachtree borer larvae feed in the crown area of trees.

When monitoring for peachtree borers, look for small accumulations of red-brown frass on the lower tree trunk, such as on the tree here. The empty pupal case lying at the base of the tree indicates that a peachtree borer moth has recently emerged.

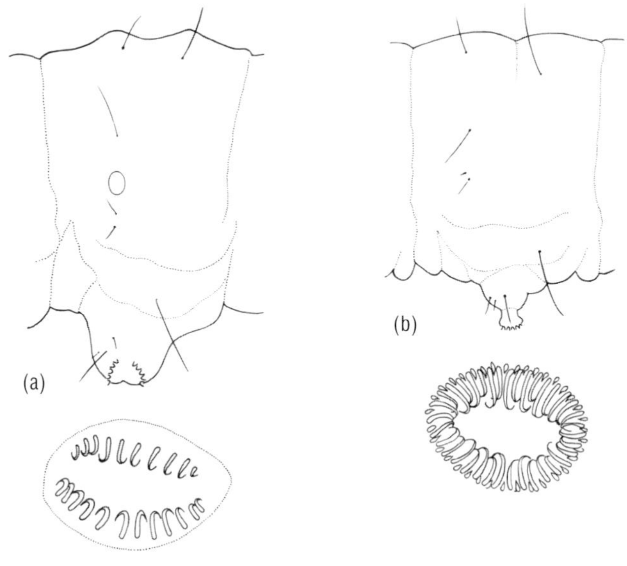

Figure 48. Distinguishing peachtree and American plum borers. The upper diagram is the lateral view of the fourth abdominal segment of (a) peachtree borer and (b) American plum borer. Note the size and shape of the proleg in relation to the abdominal segment. The lower diagram illustrates the crochet pattern that can be seen with a hand lens on the bottom of the proleg.

near the entrance of their burrows or in the soil. Steel blue to black moths emerge in 20 to 30 days, and the adult flight period extends from May through September. Eggs are laid during the summer on the bark of tree trunks. When the eggs hatch, the young larvae emerge and tunnel into the tree at or slightly below ground level. One generation occurs per year.

Peachtree borers can girdle and kill a young tree. Older trees are sometimes attacked but can sustain the damage unless there are many larvae or the tree is attacked several years in a row.

Management Guidelines

Almonds grown on peach rootstock are more susceptible to peachtree borer infestations than almonds grown on other rootstocks. If peachtree borers are a problem in your area, weigh potential damage by this pest along with other factors when choosing rootstock for new trees.

Damage to trees can occur above or below the soil line. If you live in an area where peachtree borers are a problem, examine your trees yearly. Pheromone traps are available to monitor populations of peachtree borer moths and can help you determine if they are in your orchard. Consult your UCCE farm advisor for more information on the use of these traps.

The best time to look for peachtree borer is in the fall. Look for small accumulations of frass on the lower tree trunk, especially on trees that appear weakened. After the first rains, sap will exude from the larval tunnels. Remove 6 to 8 inches (15–20 cm) of soil from around the trunk and peel back the bark to observe the burrows. If infestations are found, apply a chemical treatment the following season in mid-May and in mid-July to trunks of infested and adjacent trees to kill newly hatched larvae when they begin to bore into the trunks. If the infestation is extensive, treat the entire orchard. Remove soil from the base of the tree trunk and spray the trunk from the scaffold crotch to the ground so that some of the spray solution puddles and drenches the area below the soil surface.

American Plum Borer
Euzophera semifuneralis (Pyralidae)

Prune Limb Borer
Bondia comonana (Carposinidae)

American plum borer and prune limb borer are sporadic pests in young almond orchards from Tehama County in the north to Merced County in the south. They occur on all major almond cultivars, but first- and second-leaf Sonora, Carmel, and Price cultivars are attacked most frequently. They also attack other stone and pome fruits.

Plum borer and prune limb borer have about three to four generations per year, and each generation takes about 4 to 6 weeks for the insect to develop from egg to adult. They overwinter as mature larvae in cocoons in sheltered locations on

AMERICAN PLUM BORER

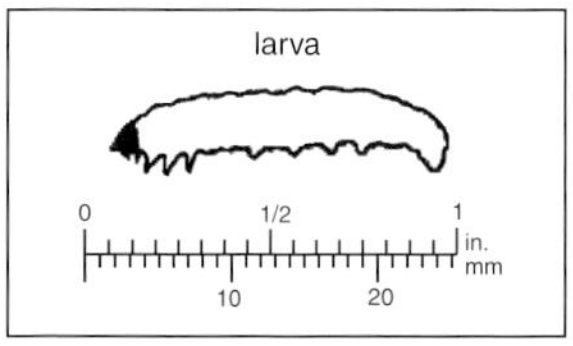

American plum borer larvae can be found boring in the crown area of the tree.

American plum borer larvae also bore in the scaffold crotches of young trees. Extensive gumming may occur around the feeding sites.

Young trees damaged at the scaffold crotch by plum borer larvae frequently lose the weakened limb during windy periods.

the tree. The majority of moths emerge in April and May. Adults are gray moths with a wingspan of about ¾ inch (19 mm) and brown and black markings on the wings. Following mating, females lay opaque, white eggs near pruning wounds, in scaffold crotches of young trees, near graft unions of young trees that have been recently topworked, or on crown galls of older trees. Young larvae are active and have white bodies with large, dark brown heads. When mature, the larvae are about 1 inch (25 mm) long and are dull white, pinkish, or dull green

Plum borer and prune limb borer larvae bore into the tree and tunnel in the cambium layer; reddish orange frass, webbing, and gum pockets indicate their presence. These larvae are most damaging when they tunnel in the scaffold crotches or graft unions of young trees. Vigorous trees usually heal over and continue to grow following a light infestation. Prolonged, heavy infestations, however, can weaken or girdle a scaffold, which then breaks off during windy periods or when the tree has its first heavy crop. Carmel and Sonora are the most severely affected cultivars. After the fifth year, trees are rarely attacked by plum borer larvae, possibly because the bark is thicker and tough enough to prevent penetration by this insect.

Look for signs of borer activity when monitoring young orchards in the spring. Examine scaffold crotches for small piles of reddish orange frass, webbing, and gum pockets. If larvae are present, treat trees with an approved insecticide. Use a hand-held sprayer to apply the material from 1 foot (30 cm) above to 1 foot below the scaffold crotch about two to three times during the growing season. Make two or three applications, the first application from mid to late April and subsequent applications at about 6-week intervals. This spray will repel plum borer invasions, as well as kill some of the larvae that are already present in the tree.

Pacific Flatheaded Borer

Chrysobothris mali (Bostrichidae)

Pacific flatheaded borers are attracted to diseased or injured limbs, such as those affected by sunburn, scale insects, bacterial canker, mallet wounds, or major pruning cuts. Unlike peachtree borers, which attack undamaged tree trunks, Pacific flatheaded borers attack aboveground portions of the tree that have been previously injured by

PACIFIC FLATHEADED BORER

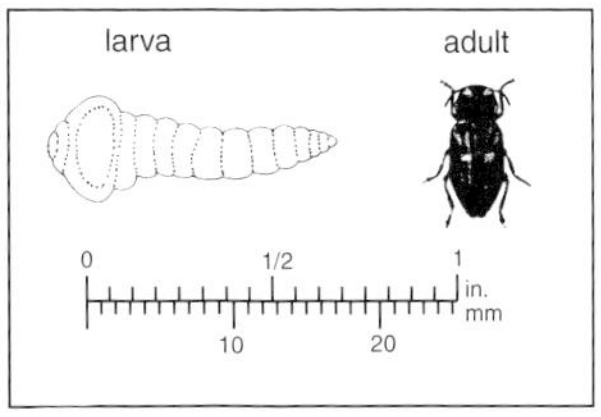

sunburn, water stress, or mechanical damage. These beetles lay eggs in the injured area, and the larvae excavate large caverns just beneath the bark and bore tunnels deep into the tree's cambium tissues. Excavations are usually filled with finely powdered sawdust. A full-grown larva is light colored, with a prominent, flat enlargement of the body just behind the head. Adult beetles are about ⅜ inch (10 mm) long with dark bronze bodies and have coppery spots on the wing covers.

Larvae of the Pacific flatheaded borer feed in previously injured aboveground portions of the tree.

The adult Pacific flatheaded borer is often found in moth pheromone traps.

Adult female shothole borers are tiny black beetles.

Injury by this borer causes sap to flow, and the affected area appears as a wet spot on the bark. Later, these areas may crack and expose the mines. Feeding by Pacific flatheaded borers may cause a portion of the bark to die, or it may girdle and kill young trees.

Management Guidelines

Flatheaded borers often invade sunburned areas on the trunk of newly planted first-year trees. Wrapping or painting the tree trunk above and 1 inch (2.5 cm) below the soil line with white, water-based paint or whitewash will protect the trunk from sunburn and flatheaded borer invasions.

In older trees the best way to avoid infestations is to keep trees sound and vigorous. Prune out all badly infested wood, and burn or remove it from the orchard before the growing season starts. Spraying for this insect is not recommended.

SHOTHOLE BORER

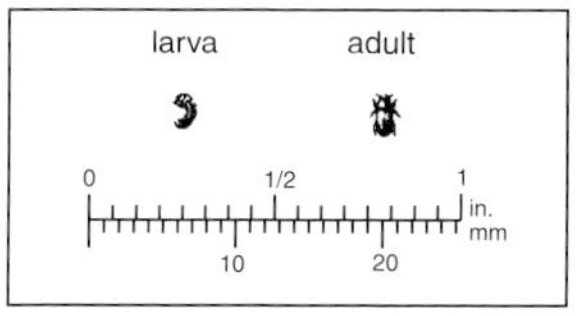

Shothole borers invade the bark of previously damaged trees and lay their eggs in tunnels than can be seen on this limb where the bark has been cut away. Externally, all that can be seen is resin exuding from small holes.

Shothole Borer

Scolytus rugulosus (Scolytidae)

The shothole borer is a small beetle that usually attacks trees already weakened by root diseases, insufficient irrigation, chronic mite infestations, sunburned limbs, Pacific flatheaded borer, or other causes of stress. Adult females bore tiny holes in the bark and lay eggs in the cambium layer of the tree. When the eggs hatch, young larvae feed and excavate secondary galleries at right angles to the egg gallery. The outline of the gallery system resembles a centipede. Normally, a number of shothole borer adults invade a tree at the same time. Healthy trees exude resin, which usually kills the insects. If the tree has injured or weakened areas, this resin buildup does not develop and the invasion will be successful. Ultimately the larvae may girdle part or all of the tree, killing branches or the entire tree.

Management Guidelines

Shothole borers invade trees that have been previously damaged. The nature of this damage dictates the course of preventive action. Maintaining trees in a sound and vigorous condition, with sufficient fertilizers and water, will prevent attack by this beetle and keep uninfested tree limbs from damage. Often, shothole borers occur in young trees that are already infested with American plum borer or Pacific flatheaded borer. If the tree is vigorous in all other respects, prune out the infested scaffold. Pruning can also be helpful in eliminating wounded areas in older trees infested with shothole borer. Severely infested trees should be removed. Burn or remove all infested wood from the orchard before the growing season starts. Do not leave pruned limbs or stumps (healthy or infested) near orchards (for example in woodpiles), as beetles can emerge from these materials before they dry out and migrate into orchards. Spraying for this insect is not recommended.

Boxelder Bug

Leptocoris trivittatus (Rhopalidae)

Boxelder bugs are a more serious problem in almond orchards in the Sacramento Valley than those in the San Joaquin Valley. The bugs' dark, oval body has many fine red lines on its back. The red undersides of the wings can be easily seen when it is in flight. These insects can invade almond orchards anytime after blossoms fade through the end of May, when their preferred hosts (maple, oak, elderberry, and boxelder) leaf out. Adults overwinter in grassy areas. Migration into almond orchards depends on how long the overwintering site remains suitable. In a cool, wet year, overwintering vegetation may remain succulent until the preferred hosts leaf out, and the bugs may not enter the almond orchards at all. On the other hand, mowing the overwintering area may trigger an invasion.

Boxelder bugs feed directly on young fruit. In the most severe cases, feeding in green fruit before shell hardening can cause the embryo to wither and abort. Larger nuts can sustain several attacks without withering, but the mature kernel will have small dark spots on its surface. Larger, damaged fruit gum internally, and during April some will drip gum in long, twisted threads. If you cut into the damaged fruit, the string of gum will lead down to a bump on the shell, where the bug's mouthparts entered the nut. A small, dark depression will be evident on the nutmeat just under this bump.

Once the shell hardens in late May or early June, the bug can no longer damage the crop because its mouthparts cannot pierce the shell.

Management Guidelines

No registered chemical treatments exist for this insect in almonds. It does not occur yearly in almonds, and usually by the time the damage is noticed, it has left the orchard.

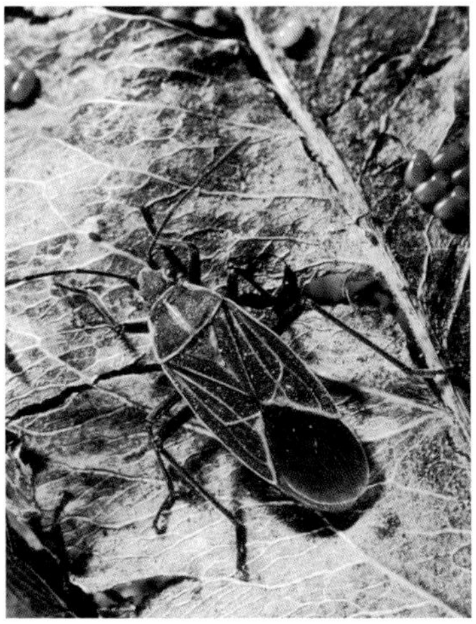

Boxelder bug adult and eggs.

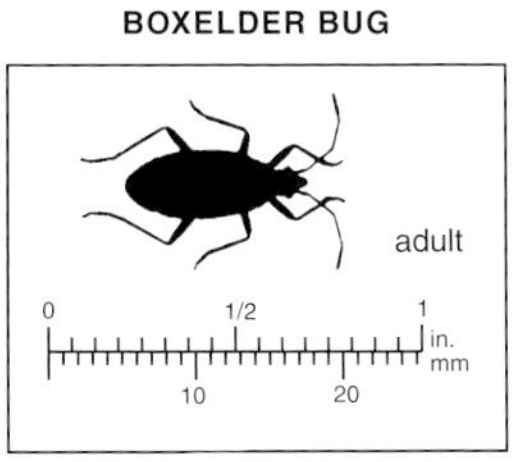

A leaffooted plant bug has tiny leaflike enlargements on its hind legs and a yellow zigzag line across its back.

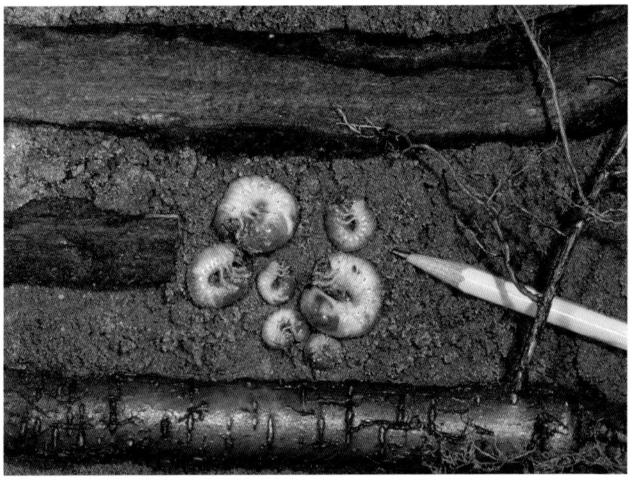

Larvae of June beetles are C-shaped grubs that live in the soil for two years and feed on the roots of almond trees. Both first- and second-year larvae are shown here, as well as a severely damaged almond root (above) and a healthy almond root (below).

Root feeding by June beetle larvae causes almond trees to slowly decline and die.

SUZANNE PAISLEY

June beetle adults are large brown beetles with white stripes on the back. In summer, males can be seen flying at dusk in the vicinity of infested sites.

Leaffooted Bug

Leptoglossus clypealis (Coreidae)

The leaffooted bug is an infrequent pest in almonds, causing damage when it feeds on nuts. It can be recognized by the tiny leaflike enlargements on its hind legs. The narrow, brown body has a yellow zigzag line across its flattened back. Adults are about 1 inch (2.5 cm) long.

Leaffooted bugs overwinter in or near orchards and begin damaging nuts in late spring. When bugs feed on young nuts before the shell hardens, embryos can wither and abort, or gum deposits may form inside the nut, resulting in a bump or gumming on the shell. Feeding after shells harden can cause black spots on the surface of nutmeats or wrinkled, misshapen nutmeats.

No treatment thresholds are available for leaffooted bugs. Treat with insecticide only if high populations are present. Available materials may cause outbreaks of mites, necessitating treatment with a miticide.

Tenlined June Beetle

Polyphylla decemlineata (Scarabaeidae)

Larvae of the tenlined June beetle feed on the roots of shrubs and trees, including fruit and nut trees. Infestations occur in almonds in certain locations of the San Joaquin Valley. Affected trees decline slowly over a period of years and eventually die. If not controlled, infestations spread slowly from tree to tree in a circular pattern.

Adult June beetles are about 1 inch (2.5 cm) long and are brown with white stripes running lengthwise on the back. They emerge from the soil in summer, from late June through August. Males fly at night and can readily be seen in infested orchards at dusk after they have emerged. Females usually do not fly but burrow into the soil after mating to lay eggs on or near living roots of host plants. Larvae are white or cream-colored grubs with brown heads. The tip of their abdomen is dark. They curl into a C shape when disturbed. Young larvae may be found near the soil surface until early summer. After the weather gets warm, older larvae may be found as deep as 3 feet (90 cm). Larvae take two years to mature and reach a length of about 1 inch (2.5 cm). A portion of the population reaches maturity each summer.

JUNE BEETLE

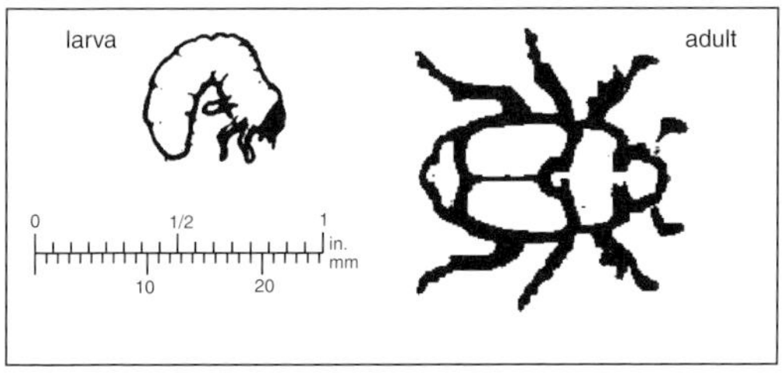

Nymphs of white apple leafhoppers do not have black spots on the back. Leaf feeding by both nymphs and adults of this leafhopper species causes damage similar to that caused by spider mites.

RICHARD L. COVIELLO

Feeding by apple leafhoppers causes leaf margins to turn brown and may result in severe defoliation.

Control requires removing affected trees and fumigating the soil before replanting. Dig up roots and soil around the base of trees to determine the extent of an infestation. Remove all trees in the infested area and one or two trees on all sides of the infested area. Fumigate the soil and replant following label directions.

Leafhoppers
Cicadellidae

Leafhoppers, primarily the apple leafhopper, *Empoasca maligna*, and the white apple leafhopper, *Typhlocyba pomaria*, sometimes reach high densities in almond orchards. They feed on the undersides of leaves and in some cases cause premature defoliation. Their effect on almond productivity has not been determined.

White apple leafhoppers overwinter as eggs laid in 3- to 4-year-old wood. Eggs hatch in April and the nymphs feed on the undersides of the newly emerging leaves. Nymphs are small, light-colored, and fast moving. Adults are white and fly readily when disturbed. Adults of the overwintering generation lay eggs of a second, summer generation in July and August. Feeding by white apple leafhoppers causes a pale yellow or whitish stippling on leaves and may result in premature defoliation.

Apple leafhoppers overwinter as eggs laid in 2- or 3-year-old wood. Eggs hatch in April. Both nymphs and adults are green. Feeding by apple leafhoppers causes a browning of leaf margins and premature defoliation that may be severe. In contrast to the white apple leafhopper, the apple leaf hopper has only one generation per year, with egg laying being completed by early summer.

No treatment threshold has been established for leafhoppers in almonds. In apples, populations are monitored weekly, beginning in April, by taking four leaves per tree from 25 trees selected randomly throughout the orchard and counting the number of nymphs or number of infested leaves. Treatment is recommended when leafhopper densities reach 50 nymphs or 30 infested leaves per 100-leaf sample.

LEAFHOPPER

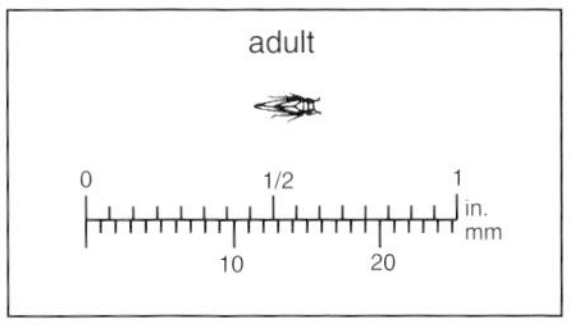

Stink Bugs
Pentatomidae

Stink bugs are shield-shaped and grayish brown to green in color. Like the boxelder and leaffooted bugs, stink bugs occasionally invade almond orchards and cause gumming and spotting of the kernel. High populations occur sporadically, usually in late May or early June. Control measures are not recommended.

STINK BUG

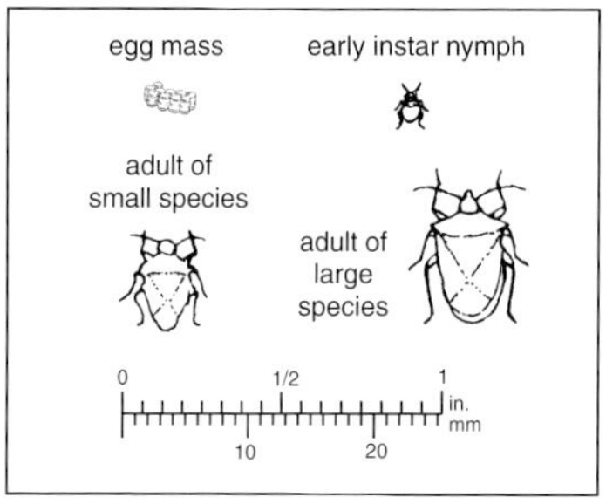

The consperse stink bug is shield shaped and gray-brown with small black spots. Other stink bug species are similar in shape but may be green or green with red or brown markings.

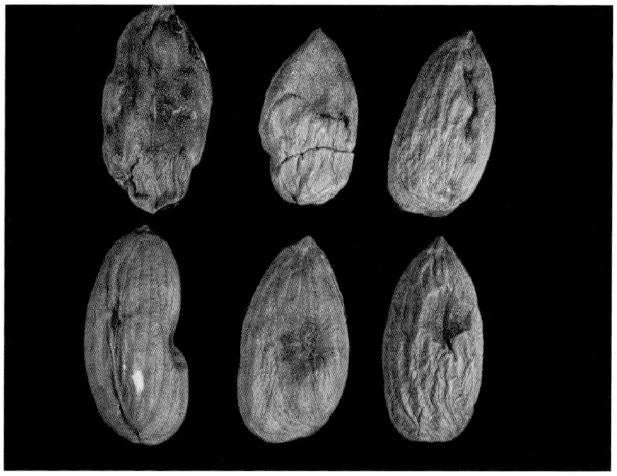

RICHARD E. RICE

Nuts damaged by bugs have small, dark spots on the surface and may be distorted in shape.

Lace Bugs
Tingidae

Lace bugs occasionally build up to high densities on almond foliage. Adults have a lacelike, convoluted thorax and wings. Nymphs are oval and wingless with long body spines. All stages of these bugs feed on the undersides of leaves and cause a yellow stippling similar to that caused by spider mites. Their damage can be distinguished by the presence of the bugs or their cast skins and dark spots of excrement. Lace bugs have several generations per year and overwinter as adults under fallen leaves or in concealed places on the bark. Treatment for lace bugs rarely is necessary. Damaging populations can be reduced with a summer oil spray that covers the underside of foliage. A second spray 7 to 10 days later may be needed if the population density is very high.

Fruittree Leafroller
Archips argyrospila (Tortricidae)
Obliquebanded Leafroller
Choristoneura rosaceana (Tortricidae)

Fruittree and obliquebanded leafrollers are green caterpillars that usually have dark heads. Fruittree leafroller overwinters as egg masses on branches or trunks. Eggs hatch in early spring and adults emerge in May or June. There is only one generation each year. Obliquebanded leafroller overwinters as larvae inside cocoons spun in protected locations in the orchard. These larvae pupate in early spring, and the first generation of adults emerges in March or April. There are two or three generations each year. The larvae of both species feed on tender new leaves that they roll and tie together with silken threads. When disturbed, they wiggle vigorously and often drop from a silken thread. Leafroller larvae also may enter young almond fruit and devour the kernel before it hardens. By May, damaged nuts are dry and collapsed with large slotlike holes into the empty kernel chamber. The number of nuts attacked is usually not significant. Leafrollers appear to occur most frequently in orchards where dormant-season insecticides are not used. Regular monitoring for leafrollers may be needed in locations where they have caused problems. Young larvae can be controlled with Bt sprays before they have rolled leaves.

LACE BUG

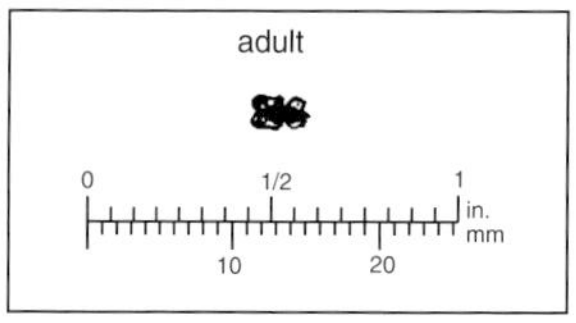

FRUITTREE LEAFROLLER

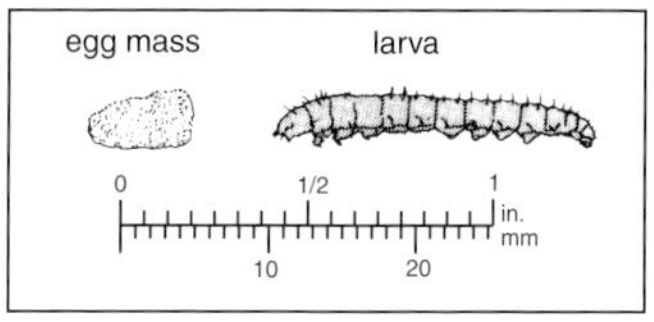

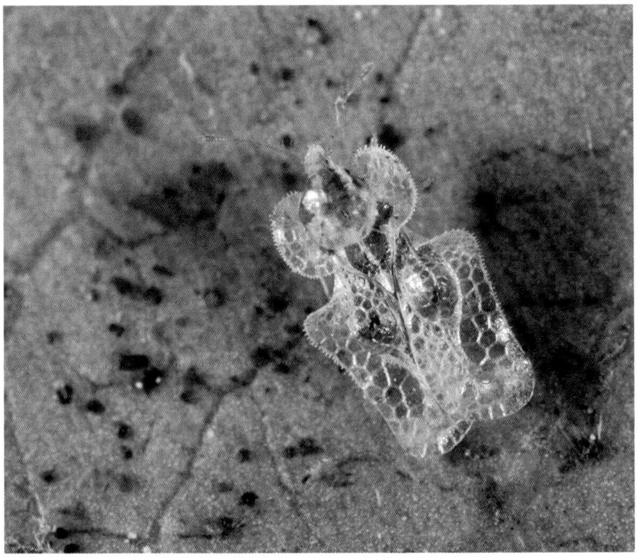

Lace bugs have a convoluted, lacelike thorax and wings. The presence of bugs or their cast skins and dark spots of excrement distinguishes lace bug damage from that caused by spider mites.

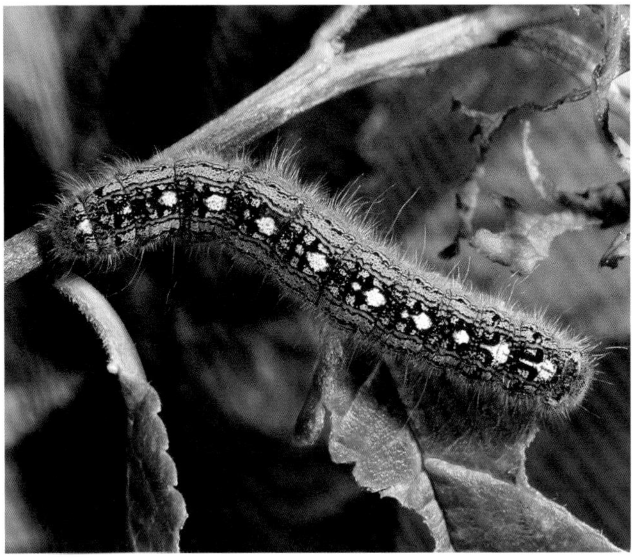

Forest tent caterpillar larvae have broad blue stripes and a row of keyhole-shaped spots on the back. Colonies of larvae build mats of webbing and move away from the webbing to feed on foliage.

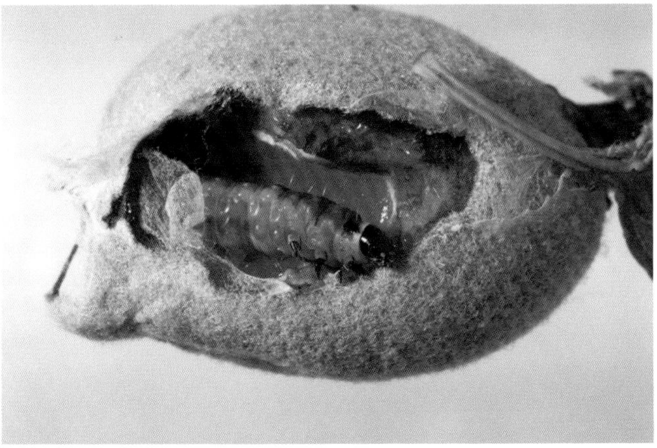

RICHARD E. RICE

The fruittree leafroller larva is a green caterpillar with a black head. Leafroller larvae occasionally feed in young nuts.

TENT CATERPILLAR

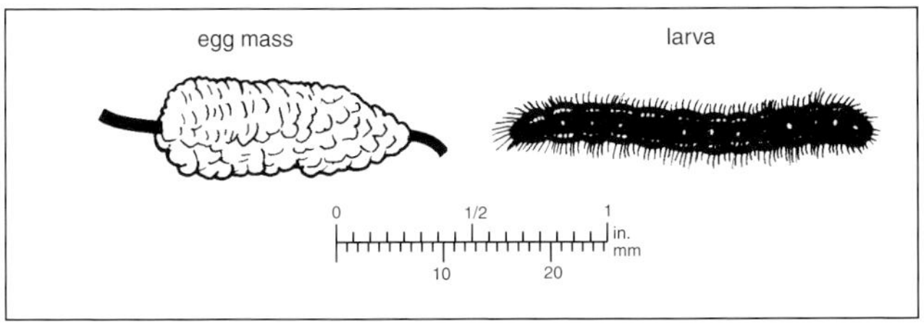

Tent Caterpillars

Malacosoma spp. (Lasiocampidae)

Infestations of tent caterpillars occur occasionally in almonds. They tend to be concentrated on individual trees scattered throughout the orchard. Tent caterpillars have one generation per year. They overwinter as eggs laid on host branches in masses that are covered with a hard brown material. Larvae hatch in spring and early summer, then colonies of larvae skeletonize leaves.

Forest tent caterpillar, *Malacosoma disstria*, is seen most frequently in almond orchards. Larvae are grayish with yellow stripes along the side, separated by a broad blue stripe, and they have a row of white, keyhole-shaped spots on the back. Colonies build a mat of webbing and move away from it to feed. Western tent caterpillar, *M. californicum*, is widespread in California but is seldom seen in almond orchards. Larvae are yellowish brown with long hairs and rows of blue and orange spots along the back. Colonies of this species build silk tents around branch crotches and feed outside the tents.

Watch for the appearance of tent caterpillars during routine monitoring in spring and summer. Damage by tent caterpillars may become serious, especially on young trees. On small trees, infested branches can be cut out and destroyed. On larger trees, spot treatment of infested branches with Bt usually is all that is needed. This material must be applied while larvae are still small to be most effective.

Diseased trees, such as this one with Phytophthora crown and root rot, frequently exhibit similar overall symptoms: pale foliage, weak growth flushes, leaf drop, and twig dieback.

Diseases

Almond diseases are caused by a number of microorganisms and environmental stresses. Most diseases do not affect all almond cultivars to the same degree, nor do they occur in all almond-producing districts. Many diseases, especially those affecting flowers, fruit, and leaves, occur only at a particular time of the year; other infections are present in the tree and produce symptoms or weaken the tree throughout the year.

The severity of a disease is influenced by virulence of the pathogen, the susceptibility and maturity of the tree, and environmental variables such as temperature, rain, humidity, and soil moisture and texture. The impact of the pathogen on the tree and the symptoms that develop also depend upon where infection occurs on the tree (Table 14).

Trees with root and crown rots usually exhibit similar aboveground symptoms. Damage to crown or roots disrupts uptake and transport of water and nutrients to and from roots. Affected trees variously show poor shoot growth, small chlorotic leaves, premature defoliation, decreased productivity, dieback of terminal shoots, and subsequent death.

Trunk and branch cankers damage bark and rapidly growing cambium tissues just beneath the bark. Cambium cells produce food-conducting tissue, or phloem, in the inner bark, and they produce water- and mineral-conducting tissue, or xylem, in the wood. Collapse of the cambium layer prevents production of new cells for these conducting tissues. Symptoms—including reduced growth and wilting or dying leaves—reflect partial or complete cutoff of food and water, depending on how much tissue is affected. If a canker kills the bark of the tree as deep as the cambium around the entire circumference of a trunk or branch, the portion of the tree above the canker will usually die within a few months; this is called *girdling*.

Vascular pathogens invade and plug the xylem vessels, halting movement of water and nutrients up from roots. In the case of Verticillium wilt, leaf and branch symptoms are similar to those caused by root and crown rots, while symptoms associated with almond leaf scorch are similar to those caused by an excess or deficiency of certain mineral elements.

Some pathogens, such as those responsible for brown rot, shot hole, and anthracnose, cause infections that are limited to branches, foliage, and fruit. They produce localized lesions or larger infections such as shoot strikes. Infections frequently occur after rains or sprinkler irrigation when water gets into the tree's canopy, spreading the pathogen and creating conditions favorable for infection. The tree also is susceptible to infection by some pathogens during bloom because flowers provide a good entry point.

Table 14. Almond Diseases and the Location of Infection.

Crown or roots	Trunk and branch	Vascular system	Branch, foliage and/or fruit
Armillaria root rot	bacterial canker	almond leaf scorch	anthracnose
crown gall	band canker	Verticillium wilt	Alternaria leaf blight
Phytophthora root and crown rot	brown line		bacterial blast
wood rotting fungi	Ceratocystis canker		brown rot blossom and twig blight
	foamy canker		calico
	Phytophthora canker		green fruit rot
	silver leaf		hull rot
	union mild etch		leaf blight
			rust
			scab
			shot hole
			yellow bud mosaic

Major almond diseases include Phytophthora root and crown rot, Ceratocystis canker, bacterial blast and canker, brown rot, anthracnose, and shot hole. Phytophthora root and crown rot is a major problem because it can kill the tree, and few controls, aside from careful water management and use of resistant rootstocks, are available. Serious losses can occur during extended wet periods caused by heavy rainfall or overirrigation. Ceratocystis canker is associated with bark injuries created during harvesting operations. The canker eventually kills the affected limb of the tree, or the entire tree if it is located on the main trunk.

Bacterial blast and canker can be serious in young orchards growing in coarse-textured soils. Bacterial canker frequently kills 2- to 5-year-old trees, but losses can be minimized by using soil treatments to reduce harmful nematode populations and by following cultural practices that maintain tree vigor. Brown rot affects blossoms and fruiting twigs and can reduce fruit set substantially. Anthracnose attacks flower, fruit, and fruiting wood and can debilitate trees over time. Shot hole, a disease of leaves and fruit, can defoliate the tree and cause crop reduction in many almond-growing regions if not controlled annually with fungicides.

Monitoring and Diagnosis of Almond Diseases

Familiarize yourself with symptoms of common almond diseases and watch for signs of tree stress as part of regular orchard monitoring. Watch for weak trees and irregular growth patterns. Check low-lying sites and areas with fine-textured or shallow soil; trees growing in these locations tend to have a higher incidence of diseases. Monitor more frequently during and after periods of rainfall. To diagnose a problem, examine the trunk for gumming, discoloration, or sunken areas. Also examine roots, branches, leaves, and fruit. You may have to scrape off bark, dig up some feeder roots, or remove soil from the crown or main lateral roots. A shovel, hatchet, chisel, knife, and hand lens are useful tools for field inspection. The text and illustrations of specific diseases in this manual will also help you to identify disease symptoms.

Time of year influences what symptoms you may see: fungal and bacterial diseases of fruit and leaves cause most damage during rainy or foggy periods, or in sprinkler-irrigated orchards where the water hits the foliage and fruit. Pathogens attacking roots and trunks in fall, winter, or early spring may not produce symptoms until sometime later in spring or in summer, when the tree is unable to get sufficient quantities of water and nutrients from the soil after active growth begins.

Certain cultivars are prone to disorders that are not caused by pathogens and may not be evident until long after the trees are in production, or unless the trees are grown in certain environments. Less commonly, a mutation may occur in part of a tree, affecting fruit shape or color, or the growth habit of leaves and twigs.

Many diseases ultimately have the same effect: a declining tree with light green foliage, weak growth flushes, leaf drop, twig dieback, and reduced crop. Primary symptoms may appear on roots and trunk, fruit, leaves, and twigs. They can affect overall growth habit, size, and crop yield. These symptoms help identify a disorder. In this manual, individual diseases are organized according to where the pathogen infects the tree. Keep in mind that other pests (such as nematodes), nutrient deficiencies, poor water management practices, or the environment may also be the cause of decline.

Compare your field observations with the descriptions and photographs in this chapter. A combination of symptoms is usually needed to identify a disease. Some diseases, however, are difficult to identify by field symptoms, and positive identification will require laboratory diagnosis. If you find symptoms not described here, see the suggested reading for publications on diseases, or consult your UCCE farm advisor.

Keep accurate records of your observations and note infected trees, or those suspected of being diseased, on a map of your orchard. Positive diagnosis, especially of slowly developing viral diseases, is often not possible at first inspection. Repeated monitoring and mapping will help you confirm your diagnosis, follow development of a disease, and make proper management decisions.

Prevention and Management

The pathogen, the almond tree host, and the environment must all be considered in a disease management program. The pathogen is usually difficult to control; thus, emphasis lies on manipulating the tree and its environment to prevent disease outbreaks or at least lessen the impact of a disease. Prevention is especially important in disease management because many pathogens establish themselves slowly over years. Once disease symptoms are apparent, treatment options may be limited.

Preventive measures include preplant soil preparation, resistant rootstocks and cultivars when available, careful water management, sanitation, and fungicidal sprays. In addition, adequate fertilization, pruning, and control of other pests (insect and mites, weeds, vertebrates, and nematodes) also helps maintain vigorous trees that are more resistant to infection than stressed, weakened trees.

Disease management in almond orchards starts before planting the trees. Fewer problems occur if the site has deep, properly graded soil with good drainage. Low areas and finely textured and shallow soils have the potential for waterlogging. Where hardpans obstruct water penetration, ripping or backhoeing completely through the hardpan before planting helps improve water penetration.

Good water management is essential for maintaining healthy roots and vigorous trees throughout the life of the orchard. A carefully designed irrigation system and irrigations scheduled according to physical and climatic conditions of the site and needs of the trees also contribute to disease management (see the discussion of irrigation in the chapter "Managing Pests in Almonds"). As a general rule, do not irrigate during bloom; use low-angle sprinklers in sprinkler-irrigated orchards so that leaves and fruit do not get wet. In poorly drained soils, use sprinklers or drip irrigation systems. Controlling weeds, especially around the base of trees where cool temperatures and moisture favor development of crown rot, may also help prevent disease problems. Also, during periods of water stress, competition from weeds can further weaken trees, making them more susceptible to certain diseases.

To prevent or reduce the impact of diseases, select resistant rootstocks and cultivars (if available), fumigate planting sites where almond or related fruit trees have been pulled out, and pretreat the rootstock to protect against crown gall. Tolerant cultivars are available for some common diseases, although no cultivar-rootstock combination is resistant to all important diseases and environmental stresses. The choice, therefore, depends on prevailing local disease problems, on soil conditions and climate, and on cultural and marketing considerations. For more information on choosing rootstocks and cultivars, applying pesticides, irrigating, and preplant fumigating, see the chapter "Managing Pests in Almonds."

ROOT AND CROWN ROTS

Root and crown rot diseases disrupt uptake and transport of water and nutrients to and from roots. Sometimes apparently healthy trees collapse and die quickly in early summer. In other cases, aboveground symptoms of affected trees are poor shoot growth, small chlorotic leaves, premature defoliation, decreased productivity, dieback of terminal shoots, and subsequent death. Many of these symptoms could be caused by factors other than root and crown rot pathogens. If your trees exhibit the above symptoms, remove soil from around the crown and main lateral roots to look for symptoms outlined in the following sections.

Phytophthora Root and Crown Rot
Phytophthora spp.

Phytophthora root and crown rot kills more almond, apple, cherry, nectarine, peach, and walnut trees in California than any other disease. Root and crown rot is caused by several species of *Phytophthora*, which are widespread and found in many orchards. Historically, *Phytophthora* spp. have been classified as water mold fungi, but they are now placed in the group Stramenopila because of distinctive characteristics including a mobile spore stage. However, for practical purposes they are still referred to as fungi. *Phytophthora* spp. do not require wounds to penetrate the host: if a favorable environment exists, they infect directly. Phytophthora infections occur most frequently throughout fall, winter, or spring during periods of cool to moderate soil temperatures and abundant or excessive water.

JOHN M. MIRCETICH

Phytophthora crown and root rot can infect large numbers of trees in an orchard. It may be difficult to discern any pattern in the distribution of the disease; infected trees may occur singly or in groups.

Symptoms and Damage. At least 11 species of *Phytophthora* infect rootstocks on which almond trees are grown. There are differences among these species with respect to the portion of the tree primarily attacked and temperature requirements (Table 15). The pathogen may infect roots, crowns, trunks, or branches. Primary symptoms include root rot, crown rot, and trunk or branch cankers. Copious quantities of gum are produced on the surface of trunk and branch cankers. A secondary but more obvious symptom is overall tree decline.

A tree infected by *Phytophthora* spp. can either undergo a period of slow decline that may last a few years or suddenly collapse and die with the advent of warm weather in spring or early summer. The severity and rate of disease development depends in part upon the combination of rootstock and species of *Phytophthora* involved.

Slow decline is a result of root infection by certain *Phytophthora* species that primarily kill feeder roots. In the early stages of root infection, diseased trees may be difficult to distinguish from healthy ones. The aboveground symptoms that develop are essentially the same for other root rots, especially Armillaria root rot. With the progression of root rot, infected trees develop open canopies as foliage becomes pale and sparse, and there is a lack of terminal shoot growth. The rotted root system cannot adequately supply the aboveground portion of the tree with water and minerals. Eventually leaves drop, terminal shoots die back, and trees die. Often a grower mistakenly identifies these symptoms as water stress and irrigates the orchard more frequently; this, in turn, favors further disease development and increased incidence of infected trees. The bark and outer wood of infected roots are reddish brown. The discoloration extends into the wood, and an active canker has a zonate margin. Infected roots are firm and brittle.

Although aboveground symptoms are easily confused with those caused by *Armillaria mellea* (oak root fungus),

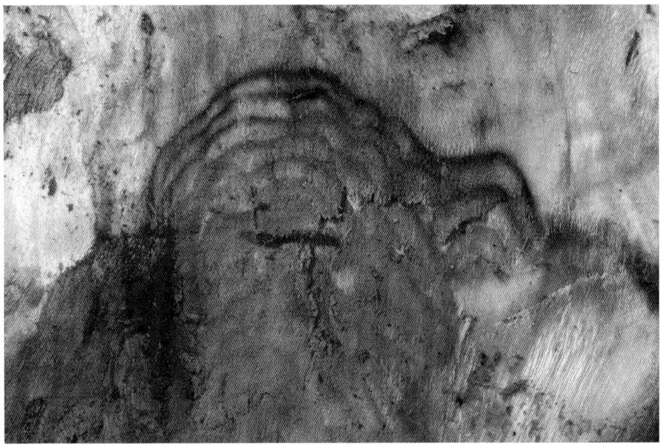

Margins of Phytophthora cankers often show a zonate pattern.

Discolored wood in the crown area of this tree indicates a Phytophthora crown rot infection.

Phytophthora has directly penetrated the trunk of this tree, resulting in a trunk canker. Suckers at the base of the tree indicate that the roots have not been infected and remain alive.

Table 15. Species of *Phytophthora* that Attack Almond, Part of Tree Affected, and Optimal Temperature for Pathogen Growth.

Phytophthora species	Part of tree affected	Optimal temperature °F (°C)
P. cactorum	roots, crown	77 (25)
P. cambivora	roots (Lovell peach), crown	72–75 (22–24)
P. cinnamomi	roots	75–82 (24–28)
P. citricola	roots (peach), crown	77–82 (25–28)
P. citrophthora	roots (peach), crown	75–82 (24–28)
P. cryptogea	roots (peach), crown	72–77 (22–25)
P. drechsleri	roots (peach, plum), crown	82–88 (28–31)
P. megasperma	roots, crown	77 (25)
P. nicotianae (= *P. parasitica*)	crown	81–90 (27–32)
P. syringae	trunk, branches	59–68 (15–20)

Phytophthora spp. do not produce mycelial plaques under the bark, rhizomorphs on the surface of affected roots and crowns, or mushrooms at the base of infected trees. Oxyporus wood rot sometimes occurs as a secondary disease on trees with Phytophthora root rot. White fungus growing on the crown area indicates the presence of *Oxyporus* (see the discussion of wood rots later in this chapter.)

Infections in both crown and trunk can damage trees substantially in a short time if conditions are favorable. Crown infections, however, are usually more destructive than trunk cankers because trunks rapidly warm up in spring to temperatures that no longer favor disease development. Most Phytophthora trunk cankers are chronic, but the fungus does not grow during periods of hot or cold temperatures. Trees with Phytophthora crown rot infections usually decline rapidly. Trees infected in fall may be completely girdled by crown rot and killed before new growth starts in spring. These trees have enough stored energy to blossom and leaf out, but when the weather warms up and water is required from the roots, they quickly die. The bark at and below ground level is brown, dead, and often gum soaked; the underlying wood is discolored.

Phytophthora trunk cankers are irregularly shaped; their margins often exhibit a zonate pattern of light and dark lines. They are frequently confused with Ceratocystis cankers. *Ceratocystis* can enter the tree only through bruise-type wounds on the trunk and scaffold branches. Association of bark damage with Phytophthora cankers is purely coincidental in most cases, although *Phytophthora syringae* primarily invades pruning wounds in fall, winter, or early spring during periods of heavy rainfall. Phytophthora trunk cankers grow rapidly, often 2 to 4 feet (0.6–1.2 m) per year, while the growth of a Ceratocystis canker is slow. Also, Ceratocystis cankers do not extend below the soil line, whereas Phytophthora cankers frequently extend into the crown and main lateral roots. Profuse gumming usually occurs on Phytophthora trunk and branch cankers, but not on Ceratocystis cankers. However, both *Phytophthora* and *Ceratocystis* can be present in the same tree.

Branch cankers, often called aerial Phytophthora, become evident when numerous gum balls appear in groups on a scaffold or at pruning wounds. Gumming occurs on the outer surface of the cankered area. Under the bark, the margin of an active canker consists of reddish concentric rings.

Occasionally, nursery trees are girdled by a canker on the trunk. This disease is known as "bundle rot" and is caused by *Phytophthora.* When almond trees are dug up, tied together in bundles, and heeled-in before planting, *Phytophthora* infections can occur and can spread to a large number of trees if conditions are wet and heeling beds are infested with *Phytophthora.* Often these infections may not become apparent until after the tree is planted in the orchard. The symptoms, including gumming along the trunk, are sometimes confused with bacterial canker infections, but bacterial canker is not known to occur in first-leaf trees.

Phytophthora branch cankers may occur anywhere along a scaffold.

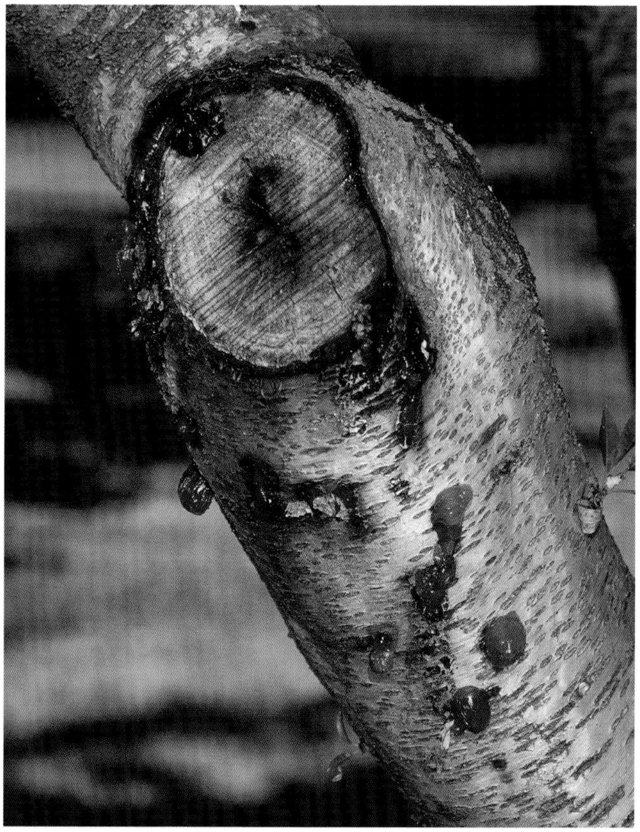

Phytophthora branch cankers frequently occur at pruning wounds.

Seasonal Development. All *Phytophthora* species that attack almond trees are soilborne and have a wide geographical distribution, although not every orchard is infested nor do all species occur in all soils. *Phytophthora* spp. can persist in soil without a host for a long time but require free water to reproduce and infect the host plant. Thus, root and crown rots frequently occur in orchards with fine-textured soils that are poorly drained or have low areas where water collects. Epidemics of the disease occur in years when rainfall is plentiful in late fall, winter, and early spring, and the soil remains flooded or saturated with water for extended periods.

Phytophthora species most often infect the tree at sites where infested, waterlogged soil or floodwater touches the roots, crown, or trunk. Once in the root or crown, the infection may extend into the crown, trunk, or branches, depending on the species of *Phytophthora* involved. If *Phytophthora* spores are present in water that is used to sprinkler-irrigate the orchard, the pathogen may also enter the trunk or branches at sites where water collects. Other sources of inoculum for *Phytophthora* infections include decaying, infected roots and active cankers. Spores produced on these infected tissues can move short distances through saturated soil or can be transported by water movement, splashing rain, or on contaminated field equipment.

The cool to moderate temperatures and moist conditions that favor Phytophthora root and crown rot occur during fall, winter, and spring, although certain species of the pathogen may be most active during more limited time spans. When conditions are less favorable for the pathogen, trees infected with *Phytophthora* spp. that primarily cause root rot often are able to regenerate roots and recover from the damage. Recovery depends on good soil water management and rootstock tolerance to the disease.

Phytophthora syringae, which causes branch and trunk cankers, is active only when temperatures are below about 75°F (about 24°C). It may be spread by windblown rain and enters primarily through pruning wounds that have not healed. With the onset of hot weather, the fungus dies within the canker and the affected area can heal over.

Management Guidelines. Proper water management is the most important aspect in controlling root and crown rot. Water should not be allowed to accumulate around crowns of trees. Low spots in the orchard and areas that flood frequently or have poor water penetration should be provided with adequate drainage or left unplanted. In some cases, amendments added to the irrigation water or applied directly to the soil may help improve soil permeability; consult your local UCCE farm advisor for more information. Once *Phytophthora* is present in orchard soil, the pathogen will remain: eradication is impossible. Be careful not to introduce the pathogen into uncontaminated soil by infected plant material, infested irrigation water, or infested soil on farm equipment.

When replanting an area where *Phytophthora* is present, plant trees on small mounds as shallowly as possible or on broad ridges with the upper roots near the soil level. Establish berms before planting; the ridges should be 8 to 10 inches high (20–25 cm). The planting depth after settling should be no deeper than in the nursery, and the graft union should be well above the soil line. Almond scions generally are more susceptible to *Phytophthora* infection than the rootstocks on which they are grown.

Use a rootstock that has some resistance to *Phytophthora*. Of the rootstocks used in almond orchards, almond is the most susceptible to Phytophthora root and crown rot. Limited evidence suggests that some peach × almond hybrids are more susceptible than either Lovell or Nemaguard peach rootstocks. The plum rootstock Marianna 2624 is the most resistant of the rootstocks used for almonds, but it is not compatible with some almond cultivars such as Nonpareil.

To protect the trunk and scaffolds from infection, use low-angle sprinklers and water guards on sprinklers to keep from wetting the trees, and remove weeds around the base of the tree. Weeds may maintain an environment favorable for disease development.

A limited number of chemical control options are available for Phytophthora diseases of almond. For further information, consult your UCCE farm advisor and the *UC IPM Pest Management Guidelines: Almond*, listed in the suggested reading.

Crown Gall

Agrobacterium tumefaciens

The bacterium that causes crown gall, *Agrobacterium tumefaciens*, is widespread and occurs in most agricultural soils throughout the world. Although this pathogen can cause galls in established orchard trees, it is most damaging to young trees, especially nursery stock.

Symptoms and Damage. Although almond trees do develop galls visible at ground level, most galls develop on the root system and cannot be detected unless the soil is removed. Aboveground symptoms of trees infected with crown gall can include sparse canopies, lack of vigor, reduced yield, poor terminal growth, and stunted leaves. Although young trees are usually not killed by this disease, they never become very productive.

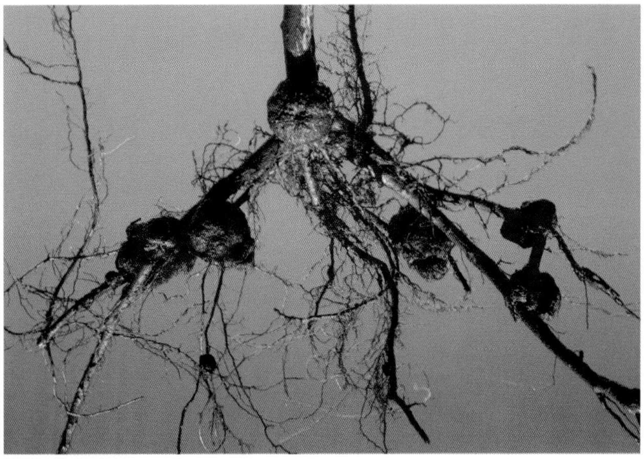

JOSEPH M. OGAWA

Crown and root galls are most damaging to the rootstock of young almond trees. Galls disrupt root functions and severely debilitate the trees.

Galls are commonly the size of a baseball but can vary from microscopic to more than 12 inches (30 cm) in diameter. Wood inside a gall lacks growth rings and is softer than healthy wood, but the color is the same as healthy wood. The outside color of the gall is the same as normal bark. Young galls have a smooth surface that becomes rough and irregular as they enlarge and outer layers slough off. Portions of the gall may provide entry points for wood-rotting fungi, and the gall may eventually decay.

The severity of damage caused by crown gall depends upon several factors, including the number and location of infections, size of galls, age of the tree, and establishment of secondary infections. Infections that develop during the first 3 years after planting are much more debilitating than those beginning later in the tree's life. Older trees can usually tolerate these galls if they are not on primary roots. However, if wood-rotting fungi invade the tree through gall tissue, the tree may eventually be damaged seriously or lost.

Seasonal Development. When galls are broken, sloughed off, or moistened, crown gall bacteria can enter the soil, where they will survive for a short period. *Agrobacterium tumefaciens* can be spread in contaminated soil on equipment or with infected plant material. The pathogen infects trees only through wounds, such as those caused by cultivation, pruning, frost injury, or growth cracks. Each gall that forms is caused by a separate infection. Nursery trees are particularly prone to infection because of the many potential injuries during handling.

Management Guidelines. Treat your trees for crown gall just before planting them in the orchard. Spray roots for 30 seconds in a protective solution of commercially available biocontrol agent, being sure to thoroughly cover and wet the root system. The agent, a nonpathogenic strain of *A. tumefaciens*, provides a protective shield against the pathogenic strain; it does not eradicate existing galls. A few strains of *A. tumefaciens* are occasionally encountered that are more difficult to control with the biocontrol agent. Research is currently under way to develop better control measures for these strains. Trees that are heeled-in and root-pruned may require treatment before each operation. If only one treatment with biocontrol agent is economically feasible, use it before planting the trees in the orchard.

The following measures will also help to prevent crown galls: purchase young trees from a nursery that follows recommended procedures for gall control; plant trees with a minimum of handling, and avoid injury to trees at planting and during cultural practices involving machinery; and remove soil from around the roots before planting, and cut off damaged or torn roots before treating the root system as described above.

If a tree is infected with crown gall, you can paint the gall tissue while the trees are dormant with a bactericide registered for this purpose. Expose the gall by carefully removing soil with a stream of water or compressed air. If you use shovels to remove the soil, use considerable care to avoid further injury to the tree. Do not treat more than 50% of the tree's circumference with the bactericide at one time. Keep soil away from the treated area for several weeks.

Most rootstocks used for almond are susceptible to the crown gall bacterium, but Marianna 2624 is more resistant to invasion by secondary wood-rotting fungi than the other rootstocks. This may provide some protection from the damage that wood rots can cause to older trees.

Armillaria Root Rot (Oak Root Fungus)

Armillaria mellea

Armillaria root rot, also known as oak root fungus and shoestring root rot, affects a wide range of woody plants. The disease occurs more frequently in the Sacramento Valley and is seen most often in orchards planted where forest trees or oak woodlands were recently cleared. Affected trees decline and die over a period of one or a few years. Although not a widespread, major problem, the disease is very serious where it occurs.

Symptoms and Damage. The first signs of Armillaria root rot are poor shoot growth in the spring or a premature yellowing of foliage during the season, often on one side of the

Armillaria mycelial plaques grow on the surface of the wood, usually in a fan-shaped pattern. Presence of mycelial plaques helps distinguish Armillaria root rot from Phytophthora crown and root rot.

Brittle, black rhizomorphs of *Armillaria mellea* can be difficult to find because they tend to blend in with the surface of the root or crown on which they are growing. They are easily broken if care is not taken when removing soil from these areas.

White fungal growth at the base of a tree is usually the first indication of Oxyporus wood rot.

Clusters of light brown *Armillaria mellea* mushrooms may appear at the base of infected trees following rain in fall or winter.

tree first. Trees may suddenly wilt in hot weather. Similar aboveground symptoms develop on trees damaged by Phytophthora root and crown rot or rodents. Armillaria root rot tends to be a greater problem on coarse-textured, well-drained soils, whereas Phytophthora root rot is more serious on fine-textured, poorly drained soils. Over time, Armillaria root rot symptoms will spread outward to adjacent trees, usually in a circular pattern.

Armillaria root rot is distinguished by the presence of mycelial fans—thin, white or yellowish layers of fungus mycelium that can be seen by peeling bark away from diseased wood on the trunk where it meets the soil. The mycelium has a distinctive, mushroom odor. Dark brown or black filaments of mycelium called *rhizomorphs*, which resemble thin shoestrings or small roots, are sometimes present on the surface of diseased wood and, when present, extend into the soil. The fungus causes a "white rot" type of wood rot in which the wood is bleached white or lighter than healthy wood and is soft and punky. Damage to the wood may spread into the crown and lower trunk.

After periods of rain in fall or winter, clusters of light brown mushrooms may form at the base of trees infected by *Armillaria mellea*. The stipes (stems) of these mushrooms are lighter brown than the cap and they have an annulus, a ring of tissue that remains clinging to the stipe after the cap opens. The gills on the underside of the cap are white, and a white spore print often can be seen on the upper surface of a cap that is beneath another cap in a cluster of mushrooms.

Seasonal Development. *Armillaria mellea* can survive for many years on dead roots or other fragments of woody host tissue in the soil. Rhizomorphs are closely associated with infected tree roots. The fungus grows along roots or through the soil for short distances as rhizomorphs, which infect new hosts by penetrating the root surface. The disease is spread over longer distances by the movement of contaminated soil and infected root fragments. *Armillaria* also can spread into the roots of an uninfected tree that are in contact with the roots of a diseased tree. The pathogen moves through the root system and into the crown, killing host tissue as it progresses. Scion wood is eventually colonized by the fungus. Trees usually die within 1 to 3 years, depending on the susceptibility of the rootstock.

Management Guidelines. Avoid planting almond trees where forest or oak woodland has recently been cleared or where there is a history of Armillaria root rot. All rootstocks used for almonds can be attacked by *Armillaria mellea*, but Marianna 2624 is less affected than others. By maintaining the trees' vigor, you can help the trees resist attack. Follow the irrigation management recommendations given for Phytophthora root and crown rot.

Infested sites can be fumigated, but often this procedure does not prevent recurrence of the disease. Remove diseased trees and adjacent symptomless trees, which are likely to be infected with *Armillaria*. Remove the stumps, backhoe the site, and remove all root fragments that are 1 inch (2.5 cm) or larger in diameter. As with Phytophthora root and crown rot, when you remove host material from the orchard, be careful not to spread the fungus to new locations. Methyl bromide is most effective against *Armillaria* if the soil is dried out before treatment. Dry out the infested site's soil by withholding water and planting a deep-rooted cover crop such as sudangrass or safflower for one summer. After drying the soil, work it thoroughly in the fall and fumigate it before replanting. If methyl bromide is being used to control other pest problems as well, such as root knot nematodes, or if sodium tetrathiocarbonate is being used, the material should be applied to moist soil.

It is a good idea to replant the site with Marianna 2624 rootstock, although some cultivars, including Nonpareil, are not compatible with it. Avoid ripping the soil or deep plowing in locations known to be infested with *Armillaria*, because this will spread the infestation.

Wood Rots

Of the several species of wood-rotting fungi that can affect almond trees, the most common are *Oxyporus latemarginatus* (formerly known as *Poria ambigua*), *Ganoderma* spp., and *Trametes* spp. *Oxyporus latemarginatus* also is found on cherry; species of *Ganoderma* and *Trametes* are found on cherry and peach. These fungi cause white rot decay. Another fungus, *Laetiporus sulphureus*, causes a brown rot of wood. Both white and brown rots lead to limb breakage or uprooting of trees during windstorms or mechanical harvesting.

Oxyporus wood rot appears first as a white fungal growth around the base of an affected tree in late summer. The growth, which is made up of fungal fruiting structures, extends for a short distance up the trunk and out into the soil. *Ganoderma* and *Trametes* produce fruiting structures that appear as shelflike brackets on the trunk or branches. By the time fruiting structures appear, the wood-rotting fungus is well established in the inner structural tissues of the tree. Although the tree is still productive and may appear perfectly healthy otherwise, it will probably fall over during a windstorm because the interior wood of the tree is weak, soft, and decayed.

Most wood rots are secondary diseases that invade only injured or dead tree tissue. The best way to protect a tree from wood rot fungi is to follow recommended cultural practices to maintain vigorous trees, use careful soil and water management to avoid crown and root problems, and take steps to avoid mechanical injuries and sunburn. The incidence of wood rots is higher in orchards irrigated with sprinklers; if you use sprinklers, avoid wetting tree trunks as much as possible. No chemical treatments are recommended for wood-rotting fungi, and destroying the conk, or the white fungal growth at the base of the tree, is useless—the fruiting bodies are only an indication of extensive inner rot.

Remove and destroy diseased wood. When a tree falls over and is removed, no treatment is necessary for wood rots before planting another tree in the same spot because the fungi are not a threat to healthy young trees. If other disease organisms or nematodes are present in the soil, however, preplant fumigation may be necessary.

TRUNK AND BRANCH CANKERS

A canker is an area of necrotic or dead bark tissue. It can be any size. Most cankers are found on large branches, scaffolds, or trunks, but they can also occur on small, young branches. On the outside of the tree, a canker usually appears as a depressed area of bark that is somewhat darker in color than the surrounding healthy bark tissue. All cankers have definite margins where healthy and necrotic tissue meet. Usually, gum balls are produced at the margin or over the surface of cankered areas.

Cankers may be caused by physical injuries such as sunburn, mechanical injuries such as shaker or mallet wounds, or by plant pathogens such as bacteria and fungi. Cankers described in this section are caused by plant pathogens. Distinguishing characteristics of each canker are outlined in Table 16 and described in the following sections. In addition, several species of *Phytophthora* that cause trunk and branch cankers are described above in the section on Phytophthora root and crown rot.

Ceratocystis Canker
Ceratocystis fimbriata

Ceratocystis or mallet wound canker is a widespread disease of almond. Cankers develop in areas of the trunk or branches that have been damaged by cultural equipment or by rubber-covered mallets or mechanical shakers during harvesting. They may also develop at pruning wounds.

Cultivars most susceptible to the disease are Mission, Ne Plus Ultra, and Nonpareil.

Symptoms and Damage. Ceratocystis cankers first appear as elongated water-soaked areas with amber-colored gum at the canker margins. Infected tissue turns dark brown, and

Table 16. General Characteristics of Almond Cankers.

	Bacterial canker (*Pseudomonas syringae*)	Ceratocystis canker (*Ceratocystis fimbriata*)	Dothiorella (band) canker (*Botryosphaeria dothidea*)	Foamy canker (unidentified)	Phytophthora branch canker (aerial Phytophthora) (*Phytophthora syringae*)	Phytophthora trunk canker (*Phytophthora* spp.)
Causal agent	bacterium	fungus	fungus	bacterium	fungus	fungus
Period of activity	winter	summer	summer	summer	fall, winter, spring	summer
Annual or chronic	usually annual	usually chronic	not strictly annual	annual	annual	usually chronic
Type of margin	flecking	zonate	zonate	diffuse	zonate	zonate
Gumming	occasional gum balls in cankered area	gum balls around edge of canker	gum balls in band around limb or trunk	oozing reddish gum and white foam	clustered gum balls, usually around a pruning wound	numerous random gum balls throughout cankered area

A shaker wound at the base of the scaffold on the left side of this tree provided an entry point for *Ceratocystis*. The resulting canker has girdled and killed the entire scaffold.

RICHARD M. BOSTOCK

Ceratocystis cankers are always associated with bark injuries. Generally, gumming occurs at the margins of the canker.

eventually the affected area becomes sunken. Dark stains may permeate the heartwood and extend longitudinally 20 inches (50 cm) or more past the margin of the canker on the bark. Cankers can girdle and kill infected branches, scaffold limbs, or entire trees. A limb 4 to 6 inches (10–15 cm) in diameter can be girdled in 3 to 4 years. However, some cankers become inactive and heal after two or three growing seasons. Trees consistently damaged by harvesters year after year are most severely affected by this canker.

Cankers caused by *Ceratocystis fimbriata* may be confused with branch cankers caused by *Phytophthora* spp. Gumming occurs with both types of cankers but usually in a different pattern. Gum balls generally occur at the margins of a Ceratocystis canker, whereas in Phytophthora cankers, they generally occur throughout the diseased area. Ceratocystis cankers grow slowly from year to year, whereas those caused by *Phytophthora* advance rapidly but stop when the pathogen dies out in hot weather and do not resume. Unlike Phytophthora cankers, which can develop on uninjured portions of the tree, cankers caused by *Ceratocystis* are always associated with bark injuries.

Obvious wounds such as those caused by harvesting equipment or pruning are the most common entry sites for *Ceratocystis*, but inconspicuous wounds on small twigs or branches also are susceptible. When pruned branches are pulled from the tree, small twigs are broken. If these or minor bark abrasions are infected by *Ceratocystis*, small cankers form. A branch with several such cankers may be girdled and die.

Seasonal Development. *Ceratocystis* is spread by several species of sap-feeding beetles and a fruit fly. These insects feed on the fungi in diseased trees, including other stone fruit trees, and either ingest fungal spores and later excrete them or come into contact with the spores and transport them on their bodies to new locations. Insects that have fed on spores as larvae can retain the fungus through pupation and emerge with it as adults. Adult insects can also be contaminated by the fungus in winter months if they inhabit old bark wounds.

Spores deposited on bark by insects may be washed into fresh pruning wounds or other injury sites by winter rains, where they are able to infect exposed cambium tissue. Following a bark injury, most almond trees are susceptible to *Ceratocystis* infections for 8 to 14 days. Once it infects cambium, the fungus can invade healthy bark tissue and young xylem tissues. Dark stains permeate the sapwood and heartwood, but the fungus seldom penetrates further than the xylem of the previous year's growth. The fungus grows more rapidly in smaller branches, and they are killed sooner.

Management Guidelines. The most effective way to prevent Ceratocystis canker is to avoid shaker injury to trunks and scaffolds. If the bark is injured, shave the rough portions to promote callus formation. Whenever possible, avoid pruning immediately before rainy weather.

To control established cankers, you can perform tree surgery. Cankers are not easy to remove, however, and if you miss some of the infected tissue, the fungus can continue to grow and the canker will return. More often than not, tree surgery must be done year after year until you finally remove all infected tissue, and it may not be economically feasible. The best time to carry out tree surgery is from December to February; both the pathogen and the insect vectors are less active at this time. (However, surgery at this time of year may provide an opening for infection by aerial *Phytophthora* fungi.) Remove infected bark and ¼ to ½ inch (6–12 mm) of the woody tissue underneath the bark. Extend the cut at least 1 inch (2.5 cm) beyond the visible canker margin. Recheck the area the following year. If the canker has returned, repeat the process.

If surgery is done in the winter months when few insects are active, leave the wound undressed. The value of putting dressings on tree wounds following surgery has not been established, and it slows the healing process. Avoid shaking areas of the tree that have had surgery, as this can reopen wounds.

Bacterial Canker

Pseudomonas syringae pv. *syringae*

Bacterial canker, also known as gummosis or sour sap, occurs on all commercially grown stone fruit trees. In almond, the disease is more common in areas with coarse-textured or sandy soils than in areas with fine-textured or heavy clay soils. The bacterium, *Pseudomonas syringae* pv. *syringae*, is commonly present on the surfaces of many plants and causes both bacterial canker and bacterial blast (discussed below under "Branch, Foliage, and Fruit Diseases"). No consistent correlation exists between the presence of bacterial canker and blast; orchards with one disease may or may not have the other. However, evidence suggests that temperatures at or near freezing, in addition to favoring blast, may also enhance the development of cankers.

Symptoms and Damage. On almond trees, bacterial canker can develop into either a limited canker or a more severe form of the disease known as "sour sap." Sour sap generally occurs in younger trees, whereas limited cankers occur in older trees or trees of any age that are infected in late dormancy. In epidemic years heavy tree losses have occurred, especially in orchards where the trees are 2 to 7 years old. Older trees are usually not seriously damaged, and trees less than 2 years old are rarely attacked by this pathogen. First-leaf trees showing similar girdling and gumming symptoms may have bundle rot, a disease caused by *Phytophthora* spp. (discussed earlier in this chapter).

Bacterial cankers can occur on trunks, scaffolds, or small branches. Active bacterial cankers are usually not visible externally unless there is some gumming. Internally, the bark and outer layers of wood are dark reddish brown in the canker area. Often, small brown flecks in the bark beyond the canker margin occur in active cankers. Those dark spots represent new infections caused by the bacteria as they move out to infect healthy wood. The irregular shape of the bacterial canker distinguishes it from most fungal cankers, which usually are symmetrical in shape and have contoured margins. Although bacterial cankers are active during dormancy, they are not apparent until early spring, when they gum or produce amber-colored sap; the name *gummosis* refers to this symptom. Branches that are girdled by a canker may die and "sour out," producing a brownish watery ooze on the surface of the bark. Inactive, healed cankers are rough, elongated, and sunken into the bark.

Sour sap occurs when a tree is invaded by *Pseudomonas syringae* early enough in dormancy to allow the bacteria sufficient time to kill a large branch or the entire aboveground portion of the tree. Generally, the younger and smaller a tree is, the more likely it is to be completely invaded by the bacteria. The dead wood is brown, moist, and sour smelling. In the spring, when other trees bloom and leaf out, diseased trees produce a watery sap that runs down the tree and has a vinegarlike odor. Invasion by the pathogen can cross the graft union but is restricted to aboveground portions of the tree. The tree may produce suckers from the base of the trunk, but the entire aboveground portion of the tree is dead.

Seasonal Development. Bacteria survive the summer on leaves and bark of trees. In fall and winter the pathogen is moved by rain and enters dormant trees through buds or natural openings such as leaf scars. Rain appears to play a major role in the dispersal of the bacteria to natural openings in the tree.

The disease progresses throughout the dormant season, forming cankers or souring out. When the tree resumes active growth in spring, it forms callus tissue around the canker and the activity of the bacterial canker pathogen is suppressed. In rare instances, the canker may not be completely encircled by callus tissue. In this case, the bacterium may become active the following winter and the canker will continue to grow. Also, bacteria may occasionally remain alive throughout the summer in small areas of bark outside the margins of a canker and resume growth in the fall.

Many factors appear to influence a tree's susceptibility to the canker phase of the disease. In general, bacterial canker is less damaging on vigorous trees. Freezing or near-freezing temperatures appear to enhance development of bacterial cankers, and trees from 2 to 7 years old have a higher incidence of those cankers than do older trees. Trees are more susceptible if grown on coarse-textured, shallow, or nitrogen-deficient soils, or on soils containing a high population of ring nematodes, *Mesocriconema xenoplax*.

Trees infected early in dormancy are more seriously damaged than trees infected as spring approaches. The earlier the invasion occurs during the dormant season, the more time the bacteria have to extend the infection before active tree growth resumes. If active growth of the tree in the spring is impaired by other diseases or environmental stresses, the development of bacterial canker will be enhanced.

Management Guidelines. Research in other stone fruits indicates that preplant fumigation can reduce the incidence and severity of bacterial canker, especially in coarse-textured soils and shallow soils above a hard pan. Preplant soil fumigation controls ring nematodes, which may predispose trees to bacterial canker. Preplant fumigation will protect young trees from ring nematodes for 2 to 3 years, the period when they usually are most susceptible to the disease.

More information is needed on rootstock selection for areas where bacterial canker is a problem. Research on prune trees indicates that trees grown on Lovell and Nemaguard peach roots suffer less damage from bacterial canker than do those grown on plum rootstocks.

Generally, vigorous trees have less trouble with bacterial canker. Practice measures, such as adequate fertilization

This young tree was killed during the winter by bacterial canker. Because only the aboveground portion of the tree is affected, roots put out suckers in the spring.

Although this bacterial canker has crossed the graft union, it is restricted to the aboveground portion of the tree. The red-brown color of the diseased inner bark is characteristic of a *Pseudomonas syringae* infection.

The dead spur on this branch served as the entry point for the developing bacterial canker, shown by the gumming just to the left of the spur.

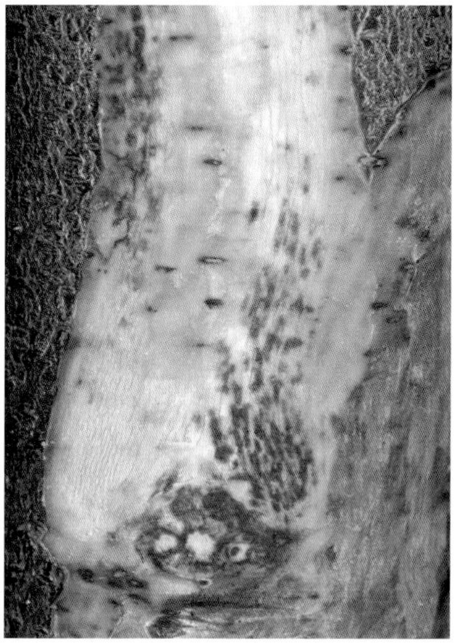

Active bacterial canker infections have red-brown flecks in the phloem just beyond the margin of the main canker. The flecks signify that the canker is expanding and also aid in distinguishing bacterial canker from most fungal cankers.

Removing bark from an infected branch reveals the extent of a bacterial canker infection.

Foamy canker produces copious amounts of red gum that drain down the scaffold and trunk of the tree to form a puddle on the ground.

and irrigation, that encourage good plant growth. Delaying pruning until late spring may help reduce problems with bacterial canker.

Prune severely cankered or dead scaffold limbs to discourage secondary invasion by wood rot organisms. Experiments on the effectiveness of copper sprays have yielded no data that suggest this approach works.

Foamy Canker
Unidentified causal agent

Foamy canker was first reported in 1974. The pathogen is thought to be a bacterium, but this has not been confirmed. Although it occurs throughout the San Joaquin and Sacramento valleys, foamy canker generally affects only a few scattered trees within an orchard. The disease has been found primarily on second- and third-leaf trees of the Carmel cultivar, but it also occurs on healthy or weakened trees of various cultivars and ages. No particular pattern or cultural practice has been associated with the disease, nor has this disease been reported in areas outside the Central Valley.

Symptoms and Damage. When the canker is active, copious amounts of watery, reddish gum drain down the scaffolds and trunk to form a puddle on the ground. In addition, white foam cascades down the tree surface. The foam bubbles forth from the bark and resembles the foam of beer. It has an alcoholic odor that suggests fermentation is taking place. Under the bark the cambium, or outermost layer of wood, is rotted, white, and mushy. Ultimately, bark and wood die and turn dark brown.

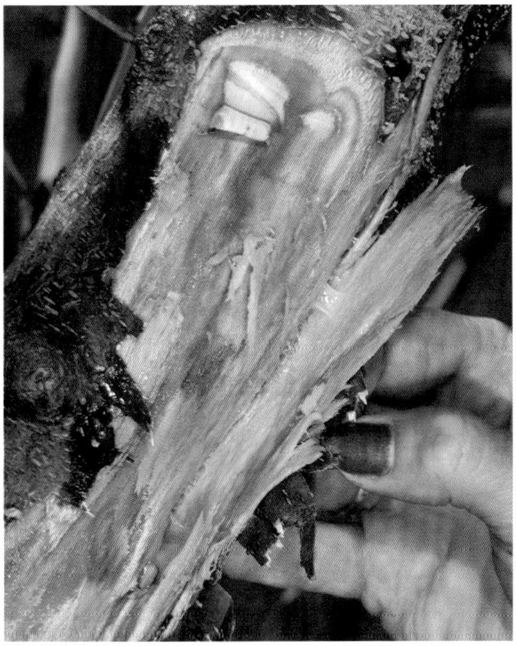

Foamy canker foam is produced under the bark in the rotted and mushy cambium layer.

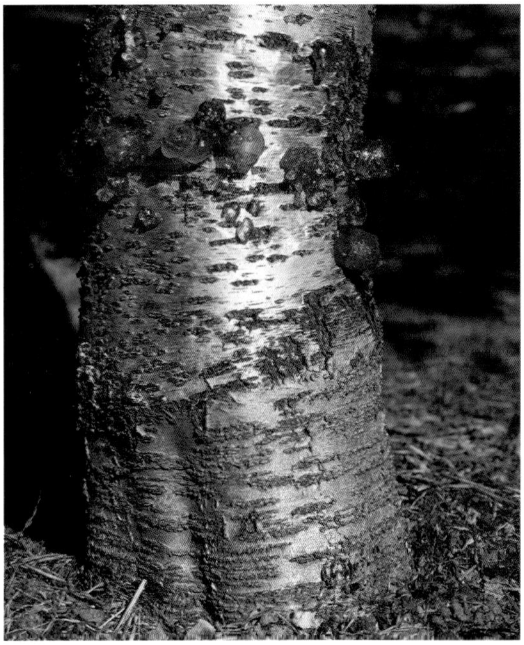

Band canker produces a band of gum balls around the trunk or affected limb.

Leaves remain throughout the growing season on limbs killed by foamy canker. From a distance the disease may resemble Verticillium wilt.

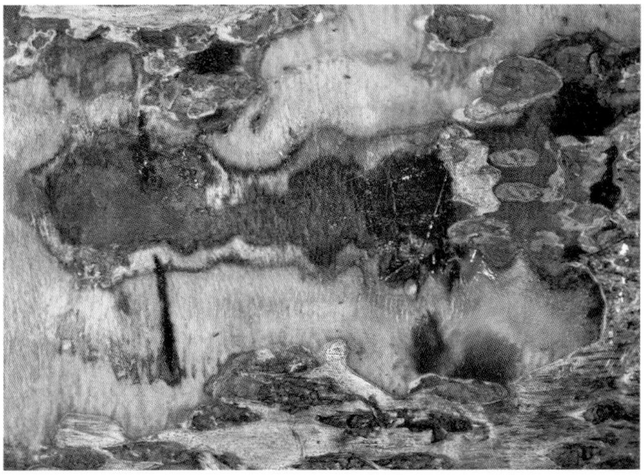

Underneath the bark, band cankers grow around the trunk instead of longitudinally, as do cankers caused by other pathogens.

Like bacterial canker, foamy canker produces very irregularly shaped lesions. Diseased tissue may completely surround a healthy area, leaving islands of tissue that are still capable of producing shoots. These shoots do not live very long, however. The disease is active only in the growing season, and cankers heal over in the winter. If the tree trunk or limbs are girdled by the disease, leaves die and remain on the tree. From a distance, this symptom resembles Verticillium wilt, but unlike Verticillium, the lower trunk and rootstock remain unaffected and produce vigorously growing suckers. Foamy canker might be confused with bacterial canker, but bacterial canker is active in the dormant season, not in the summer.

Seasonal Development. Foamy canker usually begins at the scaffold crotch of the tree and advances up into primary and secondary limbs and down toward the bud union. It does not progress past the bud union, and the roots are not affected. Symptoms first appear in late July. Disease activity usually tapers off in fall, and the canker becomes inactive. Most cankers do not reactivate the following spring. Warmer-than-average temperatures may be connected to the occurrence of this disease. The identity of the pathogen is uncertain, and nothing is known about how the disease is spread or how infection occurs.

Management Guidelines. Young trees that aren't killed by the canker can be pruned and retrained, or removed. If only a portion of an older tree has collapsed, prune out the dead wood. Make your cut below the dead tissue, in healthy wood. No chemical controls are available.

Band Canker
Botryosphaeria dothidea

Band or Dothiorella canker occurs sporadically throughout the Central Valley. It primarily attacks trees in their third to fifth year. It has most commonly been reported on vigorous Nonpareil trees and also on Mission, Ne Plus Ultra, Davey, Drake, Carmel, and Price.

Band canker is unusual because unlike other cankers it extends around the branch or trunk instead of longitudinally along the affected part. Cankers occur on the trunk or the lower portion of scaffold branches, with infection taking place through growth cracks in the bark. The fungus kills the bark and cambium layer, and the affected area becomes sunken and frequently girdles the limb. During the growing season copious amounts of amber-colored gum exude from the cankered area, forming a necklace or band of gumballs around the affected part of the tree. Band cankers occur infrequently, are active only during warm summer weather, and usually do not reactivate the following year. Some cankers become chronic and sometimes trees are killed. Band canker is of minor importance in almond orchards and no controls are recommended.

Foliage on trees infected with almond leaf scorch will gradually turn golden in midsummer. Leaves may drop prematurely, but many remain on the tree until the end of the growing season.

The lower leaf in this photo is just beginning to show symptoms of almond leaf scorch. In the upper leaf, the zonate pattern typical of this disease is just beginning to appear.

VASCULAR SYSTEM DISEASES

Pathogens that attack the vascular system of a tree plug the xylem vessels and halt the movement of water and nutrients up from the roots. As a result, leaves yellow and die on one or more branches. Usually the dead leaves remain on the tree throughout the growing season.

Almond Leaf Scorch
Xylella fastidiosa

Almond leaf scorch, or golden death, was first discovered in Mendocino County in the 1940s; soon after, the disease was found in the Lancaster area of Los Angeles County and the Brentwood-Antioch area in Contra Costa County. The bacterium that is responsible for almond leaf scorch also causes alfalfa dwarf and Pierce's disease of grapevines. It is spread from plant to plant by xylem-feeding insects such as spittlebugs and sharpshooters. Almond leaf scorch is also found in a few scattered locations in the San Joaquin and Sacramento valleys. The disease may become more important if the glassy-winged sharpshooter, which can spread the disease, becomes established in these areas. Almond leaf scorch remains a serious threat to almond trees grown in the Contra Costa County area.

Symptoms and Damage. Leaves on infected trees appear normal in spring but develop symptoms in early to mid-July. Tips and margins of leaves turn yellow, then brown, and a golden yellow band forms between the green and brown leaf areas; the intensity of this yellow band varies with cultivar. This leaf symptom frequently is confused with that caused by salt injury. Affected leaves do not wilt, but eventually the entire leaf blade turns brown and necrotic. Some leaves may drop prematurely, but many remain on the tree until the end of the growing season.

Newly infected trees typically have one terminal branch that is affected. But if you examine the trees closely, several developing infections on the same or nearby scaffolds will be apparent. As the disease progresses from year to year, trees have reduced terminal growth, large numbers of dead spurs and small branches, and dieback of terminal shoots. Within 3

As leaves on trees with almond leaf scorch become necrotic, the zonate pattern in the scorched area becomes more prominent.

The most common sharpshooter vector of almond leaf scorch is the green sharpshooter.

to 8 years, the disease spreads over the entire tree, yields are greatly reduced on affected branches, and eventually the tree dies. No obvious effects of the disease can be seen in the wood without a microscope.

Seasonal Development. Almond leaf scorch is caused by a bacterium that invades xylem tissues of almond trees and causes leaf burn symptoms by plugging the xylem and possibly by producing toxins. The bacterium resides in a wide array of host plants, including barnyardgrass, bermudagrass, fescues, rye, blackberry, cocklebur, elderberry, and nettle. Most of these hosts do not exhibit symptoms but act as reservoirs for the bacterium. Strains of *Xylella fastidiosa* that cause Pierce's disease of grapevines can cause almond leaf scorch. The same strains have been shown to multiply within most plant species tested without causing disease symptoms but at levels high enough for insect vectors to acquire the bacterium.

Spittlebugs and several species of leafhoppers called sharpshooters transmit the pathogen that causes almond leaf scorch. The most likely vectors in almond are the redheaded sharpshooter, *Carneocephala fulgida*, and grass (green) sharpshooter, *Draeculacephala minerva*. These insects become carriers of the bacterium when they insert their mouthparts to feed in the xylem of an infected plant. An almond tree is infected when an insect that carries the bacterium feeds on it. Occurrence of almond leaf scorch depends upon the amount and type of vegetation near the orchard, susceptibility of the cultivar, and environmental conditions that are favorable to high populations and dispersal of sharpshooters and spittlebugs. The most common habitats for sharpshooter vectors in the Central Valley are pastures, alfalfa infested with grass weeds, and permanent covers.

The glassy-winged sharpshooter, *Homalodisca coagulata*, is expanding its range into central and coastal California. It is now abundant in numerous citrus and avocado orchards and

Color patterns of spittlebugs, another vector of almond leaf scorch, can be highly variable, as these three different patterns of the meadow spittlebug illustrate. Spittlebugs can be recognized by their general shape and a circlet of spines on the outer end of each hind leg's largest segment.

The glassy-winged sharpshooter is much larger than other sharpshooters that are vectors of almond leaf scorch.

on some woody ornamental species such as crape myrtle and sumac in southern California. It has established large populations on citrus in the southern Central Valley and is expected to move northward to other citrus areas. The glassy-winged sharpshooter flies farther and more frequently than other vectors of *Xylella fastidiosa* and has been observed to feed more frequently on stone fruit trees. It is now rare to find vectors on almond in major almond-producing areas. The glassy-winged sharpshooter feeds on dormant trees and vines during winter and feeds on larger stems than do other vectors. All of these characteristics may increase the incidence and severity of almond leaf scorch in California. Almonds within 1 mile (1.6 km) of citrus orchards are at the greatest risk for developing almond leaf scorch problems caused by glassy-winged sharpshooter, but this sharpshooter may find other habitats suitable for developing damaging populations as it expands into new areas within California. For the latest information on glassy-winged sharpshooter and almond leaf scorch, check with your local UCCE farm advisor and the *Xylella* web site (www.cnr.berkeley.edu/xylella).

A Verticillium wilt infection usually becomes apparent when leaves on one or more branches suddenly wilt, turn light tan, and die. Dead leaves generally remain on the tree throughout the growing season.

Management Guidelines. In areas where almond leaf scorch is prevalent, annually inspect orchards in mid to late July. Prune out all infected limbs in newly infected trees 2 to 3 feet (60–90 cm) below symptoms. Do this as soon as infected shoots are detected. In heavily infected areas, replant with cultivars such as Harvey, Ruby, Padre, Solano, Carmel, Butte, and Fritz, which have shown resistance in field and laboratory tests. Susceptible cultivars include Nonpareil, Ne Plus Ultra, Jordanolo, Milow, Peerless, and Long IXL. Sharpshooter vectors require year-round access to host plants, therefore clean cultivation for a 6-week period, such as during harvest, will prevent establishment of vector populations within the orchard.

Verticillium Wilt

Verticillium dahliae

Verticillium wilt, or blackheart, is caused by the soil-inhabiting fungus *Verticillium dahliae* that is widespread in soils throughout temperate regions of the world. It attacks a wide variety of fruit trees and field crops such as cotton, tomato, and potato. Verticillium wilt often becomes a serious problem in cultivated soils where the presence of suitable host plants allows the fungus to multiply. Young trees, especially those of the Ne Plus Ultra and Carmel cultivars, are highly susceptible to Verticillium wilt. All of the rootstocks used in almonds are highly susceptible except Marianna 2624. Cool spring weather and high soil moisture favor disease development.

Symptoms and Damage. The first evidence of Verticillium wilt in almond trees is sudden wilting of foliage on one or more branches in late spring or early summer. Leaves turn light tan and die, although they often remain on the tree. Sometimes the lowest leaves on a branch will wither and fall. The pathogen gradually kills other leaves as it moves up into the branch. Internally, the xylem of the wood turns dark, and cross-sections of the branch show a full or partial ring of discolored tissue. This darkened tissue extends down the affected scaffold through the trunk and to its point of ori-

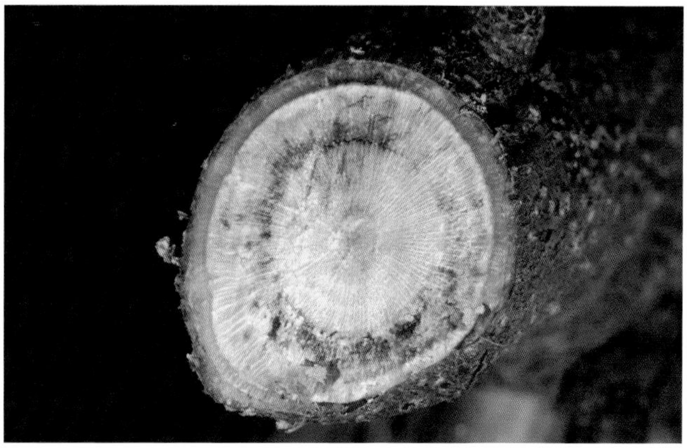

A cross section of a branch infected with *Verticillium dahliae* shows a full or partial ring of discolored tissue.

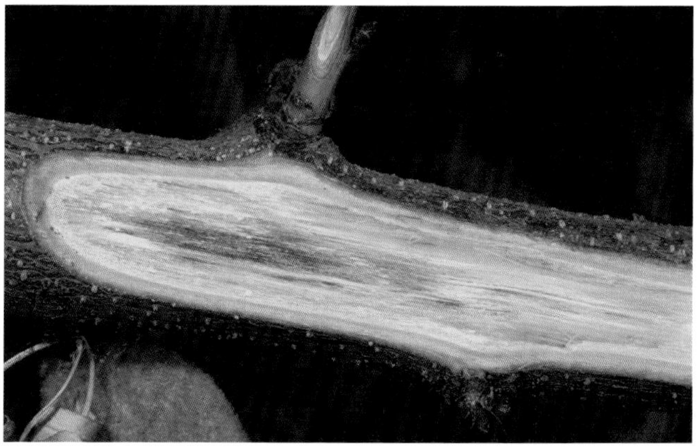

Vascular tissue discolored by *Verticillium dahliae* also can be seen by removing bark from an infected branch.

gin in the root system. Symptoms typically affect one side of the tree, depending upon which portion of the root system is infected. Several scaffolds on a tree may be involved, however, if the pathogen infects a large portion of the roots.

Two- or three-year-old trees usually are affected more severely than older trees. Verticillium wilt causes poor growth and low productivity but does not usually kill almond trees. Terminal growth on infected and previously infected branches is poor, and these branches are most likely to be reinfected by the fungus the following year, resulting in further stunting. Once trees are 4 to 6 years of age, they recover from the disease. By this time, almond tree roots have penetrated below the top 3 feet (90 cm) of soil, where most of the inoculum resides, and trees may have developed some tolerance to the disease. However, nut yields may be reduced even when foliar symptoms of Verticillium wilt are not apparent.

Seasonal Development. The fungus can survive for years in the soil. It enters the tree through young roots, establishing itself in the water-conducting tissues of the plant, and renders the xylem vessels nonfunctional. Cool spring weather and abundant soil moisture are most favorable for disease development. High summer temperatures may actually kill the fungus in the tree. In such cases, if new infection through the roots does not occur, trees recover and grow normally the next year. Although the pathogen may reside in roots for a period of time, old infections eventually die. Once roots have penetrated below the level of inoculum in the soil, new infections do not occur and symptoms will virtually disappear from a previously infected young orchard.

Management Guidelines. When establishing a new orchard, try to select a site that has never been used to grow cotton, tomatoes, melons, potatoes, or other highly susceptible crops or weed species (Table 17). When replanting an old orchard site where susceptible perennial crops were grown, try to remove as many of the old roots as possible because they may harbor the fungus. Thorough sampling of soil for populations of *V. dahliae* is helpful to determine the level of risk prior

Table 17. Partial List of Crops that Are Susceptible or Resistant to Verticillium Wilt.

SUSCEPTIBLE CROPS		RESISTANT CROPS[1]	
Common name	**Scientific name**	**Common Name**	**Scientific name**
apricot	*Prunus armeniaca*	asparagus	*Asparagus officinalis*
cantaloupe	*Cucumis* sp.	barley	*Hordeum vulgare*
Casaba melon	*Cucumis* sp.	bean	*Phaseolus* spp.
castor bean	*Ricinus communis*	carrot	*Daucus carota*
cherry	*Prunus avium*	celery	*Apium graveolens*
cotton	*Gossypium* sp.	corn	*Zea mays*
Crenshaw melon	*Cucumis* sp.	milo	*Sorghum vulgare* var. *Durra*
eggplant	*Solanum melongena*	oat	*Avena sativa*
honeydew melon	*Cucumis* sp.	pea	*Pisum sativum*
mint	*Mentha* spp.	sorghum	*Sorghum vulgare*
muskmelon	*Cucumis* sp.	sweet potato	*Ipomoea batatas*
nectarine	*Prunus persica*	wheat	*Triticum aestivum*
okra	*Hibiscus esculentus*		
olive	*Olea europae*		
peach	*Prunus persica*		
peanut	*Arachis hypogaea*		
Persian melon	*Cucumis* sp.		
pistachio	*Pistacia vera*		
plum	*Prunus domestica, P. salicina*		
potato	*Solanum tuberosum*		
prune	*Prunus domestica*		
safflower	*Carthamus tinctorius*		
strawberry[2]	*Fragaria X ananassa*		
tomato[3]	*Lycopersicon esculentum*		
watermelon	*Citrullus vulgaris*		

1. Although certain other crops are resistant to *Verticillium dahliae*, they are not listed because they are good hosts for another almond pest, the root knot nematode.
2. Resistant cultivars of strawberry are Blakemore, Catskill, Sierra, Siletz, Surecrop, Vermilion, and Wiltguard.
3. The VF-numbered cultivars of tomato are resistant to Verticillium wilt and Fusarium wilt.

to planting. Intercropping with susceptible crops in young orchards is not recommended.

Inoculum levels can be reduced prior to planting by fumigating the soil, flooding the fields during summer, solarizing the soil, growing several seasons of grass cover crops (especially rye or sudangrass), or a combination of these treatments. If the fungus has infected your trees, proper fertilization and irrigation will help young trees to recover from the attack. Prune out all weakened and dead limbs in midsummer when the fungus is no longer active.

Silver Leaf

Chondrostereum purpureum

Silver leaf or silver blight occurs rarely on pome and stone fruits and until recently had never been reported on almonds in California. The disease is now being found in almond orchards of the northern San Joaquin Valley. Silver leaf is caused by a fungus that infects wood and the water-conducting xylem through fresh wounds. A toxin produced by the pathogen is carried through the xylem to leaves, causing them to turn a silvery gray. Affected leaves curl upward at the edges and may turn brown. As the disease progresses over a few years, limbs, scaffolds, and eventually the whole tree will die.

ROGER A. DUNCAN

A toxin produced by the silver leaf pathogen causes leaves on infected branches to turn a pale, silvery color.

ROGER A. DUNCAN

The heartwood of limbs infected by the silver leaf pathogen turns a dark brown.

The pathogen, *Chondrostereum purpureum,* attacks a wide range of woody plants, including many indigenous to riparian habitats such as willow, poplar, birch, and oak. It invades the sapwood and may kill branches or entire trees. Dark brown discoloration of the heartwood in dead or dying limbs is a characteristic symptom of the disease. Spore-forming basidiocarps develop on the surface of trunks and branches that have been killed by the fungus. These are small, leathery structures that are often shelflike in shape and frequently form on the north side of affected trees. Their upper surface is grayish white and indistinctly zoned, and their lower surface is smooth and purplish. They may appear at any time of the year, but most often they are formed in the fall. Spores are ejected from the basidiocarps' lower surface during rainy or moist weather and are spread by wind. A basidiocarp can produce spores for 2 years. Sapwood-exposing wounds that have not healed over are susceptible to infection. Spores infect exposed xylem, and the pathogen remains confined to the xylem tissue until the infected branch dies.

Certain cultural practices help reduce the spread of silver leaf. Avoid excessive and improper pruning, including pruning of large branches that may require long periods for wound healing. Basidiocarps may form on infected wood after it is dead, so be sure to remove and burn or bury any prunings, branches, or stumps of diseased trees. Prune young trees in late spring and bearing trees immediately after harvest to reduce the likelihood of infection during rainy weather. Control of silver leaf in almond is being studied. Consult your UCCE farm advisor for the latest information.

BRANCH, FOLIAGE, AND FRUIT DISEASES

Several pathogens cause infections that are limited to branches, foliage, and fruit. They produce localized lesions or larger infections such as shoot strikes. Almond trees are susceptible to infection by many pathogens during bloom because the flowers provide a good entry point. Infections also occur following rain or sprinkler irrigation when water gets into the tree's canopy, spreading the pathogen and creating conditions favorable for infection.

Brown Rot Blossom and Twig Blight

Monilinia laxa

Brown rot occurs in most fruit-growing regions of the world, where it affects all stone fruit crops including apricot, almond, cherry, plum, nectarine, peach, and prune. On almond and apricot, brown rot is caused primarily by the fungus *Monilinia laxa*. On other stone fruits, *M. fructicola* also causes blossom and twig blight. Brown rot first appeared on California almonds in the late 1800s and is currently common in the northern San Joaquin Valley and the Sacramento Valley. Disease development is favored by frequent rains or fogs during bloom. Every almond cultivar is susceptible to brown rot, but Butte, Carmel, Drake, Jordanolo, Ne Plus Ultra, IXL, and Ruby are the most severely affected. Severe blossom infection is less common in Peerless and Davey, and the susceptibility of the Mission cultivar is variable.

Symptoms and Damage. The first symptom of brown rot is the withering of almond blossoms. Withered blossoms remain on the tree and are covered with light brown, powdery masses of fungal spores during humid weather. The fungus moves from the blossoms into adjacent twigs and kills them as well. Gumming may occur at the base of dead flowers and on the bark of twigs at the junction of dead and live tissue. Dead twigs eventually turn a light yellow-brown color. Leaf clusters or shoots that are not adjacent to infected flowers will not be infected, although they may collapse if a blossom that is lower on the branch becomes infected and girdles the twig. In contrast, another disease known as bacterial blast affects both leaf and blossom clusters. Because flowers are destroyed, the net effect of brown rot blossom and twig blight is a reduction in fruit set.

Infection of green fruit by *Monilinia laxa* can occur but is uncommon in almond; green fruit is more likely to be infected by two other species of fungi, *Botrytis cinerea* and *Sclerotinia sclerotiorum*. Both *Monilinia laxa* and *M. fructicola* do infect splitting hulls, however, and cause hull rot (which is discussed separately later in this section).

Seasonal Development. Brown rot fungi survive from one season to the next in infected blossoms, twigs, and fruit. In late winter the fungus begins to develop light gray tufts of spores (sporodochia) on the infected surface. These spores are spread by wind or by rain, which splashes them to new infection sites, and they infect blossoms. The critical period for flower infection is from the time unopened flowers emerge from winter buds (pink bud stage) until petals are shed (petal fall). Flowers are most susceptible when fully open. Optimal conditions for infection occur in foggy or

The tip of this twig is being girdled by a brown rot infection. Leaves at the terminal have wilted and will eventually die.

Brown areas on these petals signal the beginning of a brown rot infection.

Gumming on this branch infected with brown rot marks the junction of dead and live tissue. Note that nearby leaf clusters have not been infected, showing that the problem is brown rot, not bacterial blast.

Brown rot fungi can overwinter in twig cankers that were infected the previous year. Reproductive structures of the fungus appear as gray pustules on margins of the canker.

Brown rot fungi also can overwinter on dead blossoms.

rainy weather when temperatures are around 77°F (25°C) or higher. Disease will develop at temperatures as low as 50° to 60°F (10° to 15.5°C), however, if there is sufficient rain.

Management Guidelines. On highly susceptible cultivars (see Fig. 11), fungicides are the recommended means of control. In most cases, a single application at bloom gives adequate control. Make your first treatment at the pink bud stage, when petals start to emerge beyond the green sepals, using a fungicide having some systemic properties. Ideally you should apply the fungicide to each cultivar when it reaches the pink bud stage. If this is not possible, time your spray to the pink bud stage of the most seriously affected cultivar in your orchard. If you spray with a contact fungicide, you will need to spray at both pink bud and full bloom. In orchards where brown rot has been severe or when bloom is prolonged and weather conditions remain favorable for disease, repeat treatment every 14 days until bloom is completed. Strains of M. *laxa* resistant to benzimidazole fungicides have been found in almond orchards, but they do not occur widely.

Various cultural practices can be helpful in managing brown rot. If you are planting an orchard in areas where brown rot is prevalent, consider using less-susceptible cultivars, and space your trees so that there will be adequate air movement between them. Never sprinkler-irrigate trees during bloom, and use low-angle sprinklers in sprinkler-irrigated orchards to prevent water from splashing the spores to new infection sites. Follow recommendations for maintaining adequate but not excessive nitrogen; leaf nitrogen levels above 2.2 to 2.5% exacerbate brown rot.

Bacterial Blast

Pseudomonas syringae pv. *syringae*

Bacterial blast is caused by the same pathogen as bacterial canker. It affects all stone fruits but is particularly severe on almond, apricot, plum, and peach. Occurrence of bacterial blast is sporadic, depending upon the climatic conditions of a given year. Optimal conditions for this disease are cold or freezing temperatures and rain during the bloom period; the pathogen is spread by water. Cultivars that bloom early, such as Ne Plus Ultra, are more susceptible to bacterial blast because the buds and blossoms are more likely to be damaged by frost or experience frosty conditions.

Symptoms and Damage. The bacterium attacks just before, during, or shortly after bloom and kills buds, blossoms, young leaves, and shoots. Blossoms or opening buds turn black and remain attached to the tree. Leaves may develop dark spots that fall out, leaving a hole. When the petiole is infected, the leaf often becomes distorted. Entire spurs and young shoots may be killed, but the bacteria usually do not move very far into older twigs. Dead spurs with dark brown to black leaves and flowers remain in the tree, and these "flags" are an indication of a previous blast infection. Blast can be devastating, and under certain conditions it can cause almost complete crop loss.

Seasonal Development. Wet weather activates the bacterium, which is commonly present on leaves and bark of the tree. In addition, the pathogen can induce formation of ice crystals at slightly higher temperatures than normal, resulting in increased plant injury. Because it is associated with frost injury, bacterial blast most commonly occurs in low-lying, cold pockets of an orchard and in lower parts of trees.

Freezing temperatures at bud break or during the bloom period injure flower tissue and provide a point of entry for the bacteria. In the absence of freezing temperatures, blast usually is of little consequence. If freezing temperatures occur and the bacteria invade the tree during bud break, bacterial blast occurs in flower and leaf buds, which die before opening. If infection occurs during bloom, blast symptoms occur on flower blossoms and young leaves. In both cases, twig and spur dieback can also occur. Less commonly, young green shoots are attacked directly and killed. The bacteria cease to be active in late March, when the tree begins active growth and develops resistance to the pathogen.

Management Guidelines. Because occurrence of bacterial blast is sporadic and unpredictable and the only material available, copper, can be very phytotoxic, chemical control is not recommended. Protecting blossoms from frost may provide control because of the connection between freezing temperatures and incidence of infection. More research is needed before specific management recommendations can be made.

Shot Hole
Wilsonomyces carpophilus (Stigmina carpophila)

Shot hole is prevalent in the Pacific coast states, as well as other parts of the world, and is a major disease of peach, nectarine, apricot, and almond. It used to occur most frequently on almonds growing in the Sacramento and northern San Joaquin valleys, until sprinkler irrigation was widely adopted in the central and southern San Joaquin Valley. The disease is most severe during prolonged wet spring weather.

Ne Plus Ultra, Drake, Peerless, Thompson, Jordanolo, and Nonpareil cultivars are more susceptible than Mission.

Symptoms and Damage. *Wilsonomyces carpophilus* causes spots, or lesions, on blossoms, leaves, fruit, and twigs. Only the current season's growth is susceptible to shot hole infection. Early in the season, small purplish areas develop on the young leaves. These areas expand into spots that range from ⅛ to ⅜ inch (3–10 mm) in diameter. In the early stages these lesions are surrounded by a narrow zone of light green to yellow tissue. Later on, leaf tissue in the lesions dies, turns light brown, and is surrounded by a dark margin. The dead tissue may fall out, leaving a "shot hole." Otherwise, the tissue may remain in the leaf and one or more small dark bumps or tufts form in the center of the lesion; these are the fruiting structures (sporodochia) of the shot hole fungus, and they produce spores. Some other agents, such as viruses, fungi, bacteria, and herbicides, may cause spots or shot holes, but none produce a fruiting structure in the lesion. Infections that produce shot holes tend to occur on young leaves because young leaves expand as they grow, causing nongrowing dead tissue in lesions to break away and eventually fall out. Infected young leaves also drop more readily than mature diseased leaves, which may remain on the tree for the remainder of the growing season. Because mature leaves are

BETH L. TEVIOTDALE

The tip of this shoot was killed by a bacterial blast infection while the tree was leafing out. The dead leaves will remain on the tree throughout the season.

Shot hole lesions on fruit begin as small, dark spots that eventually become rough and corky.

Shot hole leaf lesions begin as small red freckles. As lesions expand, the center turns brown and is surrounded by a darker margin. Several lesions in the leaf on the lower right are breaking away from the surrounding leaf tissue and will eventually fall out, leaving a shot hole.

The small, dark fruiting structures (sporodochia) in the center of these lesions are characteristic of shot hole infection.

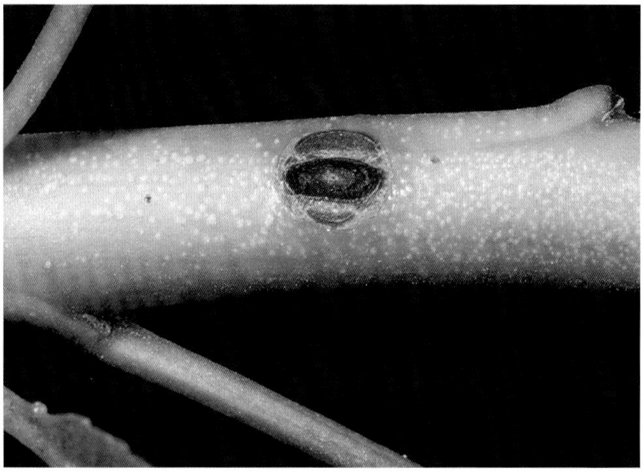

Shot hole twig lesions begin as small, purple spots.

Twig lesions serve as overwintering sites for the shot hole fungus.

no longer expanding, the dead lesion tissue is less likely to fall out. Lesions on mature leaves, therefore, are more likely to show fruiting structures and produce spores. Shot hole can cause defoliation of trees, and, if left unchecked for several years, may debilitate trees and result in yield reductions.

On young fruit, the disease first appears as small, round, purplish spots that eventually develop a lighter center. These spots usually form on the upper side of the fruit, possibly because the hairs on the hull trap the waterborne spores. Many spots may run together. Fruit lesions become rough and corky as they grow older. Severe fruit infections can cause young fruit to drop.

Small purple spots may also appear on almond twigs. Examine these spots with a hand lens. If a black spore mass, or fruiting structure, is visible in the center of the lesion, it is shot hole. Shot hole twig lesions occur rarely on almond and are not considered harmful.

Seasonal Development. During winter, the fungus survives within infected buds and twigs. Cool to moderate temperatures and free moisture are necessary for shot hole spores to be produced, germinate, and infect plant tissues. Spores are disseminated to leaves, twigs, and fruit in droplets of water. Persistent, warm rains during spring or sprinkler irrigation enhances the spread and intensity of the disease, and lesions can appear 5 to 6 days after infection. The disease is found more frequently on lower parts of almond trees where sprinkler water reaches into the canopy. Lower limbs are gradually weakened by repeated loss of leaves and become unproductive.

Management Guidelines. Monitor orchards in fall and spring for shot hole lesions and the presence of sporodochia in these lesions. If sporodochia are present in leaf lesions in the fall, apply a fungicide for shot hole at petal fall the following spring. If you do not find sporodochia in the fall, a petal fall treatment for shot hole will not be needed in the spring. Whether or not you treat at petal fall for shot hole, monitor leaves in the spring for shot hole lesions with sporodochia. Apply a fungicide if sporodochia are found and continue treating, following label directions, as long as conditions are favorable for shot hole development. If no sporodochia are found, continue monitoring until conditions are no longer favorable for shot hole development.

Sprinkler irrigation intensifies disease incidence. In sprinkler-irrigated orchards where water hits the trees, you may reduce infection by shortening the irrigation period to reduce the time that trees are wet.

Dormant application of a fungicide in November or early December should be considered only if the disease is so severe that twig infections are occurring. This is extremely unusual.

Anthracnose infections on fruit appear as rusty orange lesions on the hull. Lesions on older fruit often exude amber gum.

Anthracnose

Colletotrichum acutatum

Anthracnose occurs sporadically on almond throughout the Central Valley. Following a number of unusually wet springs in the late 1980s and early 1990s, anthracnose has become a serious problem in almond-growing areas of the northern Sacramento Valley and northern San Joaquin Valley. Symptoms include blossom blight, fruit infections, and dieback of foliage and fruiting wood.

Anthracnose is most damaging to Thompson, Merced, Price, Monterey, and Butte. Harvey, Carmel, Ne Plus Ultra, Fritz, Peerless, Padre, and Mission are moderately susceptible. Nonpareil is less susceptible.

Young fruit infected by anthracnose become shriveled. The zonate pattern of rusty orange anthracnose lesions can be seen on these fruit.

Symptoms and Damage. Blossom blight symptoms of anthracnose are similar to brown rot blossom blight. Affected blossoms wither and turn dark, but tufts of spores are not produced during damp weather as they are in the case of brown rot.

Fruit infections appear as circular, sunken, rusty orange lesions on the hull. Symptoms usually are seen first on young fruit 2 to 3 weeks after petal fall. Infected young fruit quickly become shriveled and turn rusty orange, and they look like "blanks," or aborted fruit. The developing kernels of these young fruit are killed by the disease. Anthracnose infections on older fruit are usually accompanied by the production of amber gum. The kernels of these fruit may be destroyed or may be partly invaded by the pathogen, in which case they become shriveled and bluish brown. Diseased fruit remain on the tree as dried, shriveled mummies.

Anthracnose fruit infections can lead to shoot dieback, apparently caused by a toxin produced by the pathogen. Symptoms resulting from the infection of young and older fruit can be seen in this photo.

Symptoms may develop on twigs and branches adjacent to infected fruit. Shoots are apparently killed by a toxin produced by the pathogen in infected fruit. Leaves on affected shoots turn brown, curl up, and may remain attached to the shoot. If leaves are attacked directly by the pathogen, they develop watersoaked patches or spots along margins and at the tip. These patches turn yellow. As the disease progresses, branches die back and trees become unproductive.

Seasonal Development. The fungus that causes anthracnose survives from one season to the next in mummified fruit and may also survive in infected peduncles. During warm, wet weather in the spring, orange spore masses are produced on infected fruit, giving the disease lesions a rusty orange color. Spores are splashed by rain and spread to young developing fruit and leaves. The cycle of spore production and infection continues as long as wet weather persists. Death of fruiting wood may be caused by a toxin produced by the pathogen rather than direct colonization of the wood.

Anthracnose becomes most severe in years when rainy weather persists through the spring.

Management Guidelines. Fungicide treatments are recommended in orchards with a history of anthracnose. Make the first application at the pink bud stage with a systemic fungicide. Make a second application at full bloom, using a mixture of contact fungicide and systemic of a different type than used at pink bud. Repeat the treatment every 10 to 14 days as long as rains persist. Be sure to alternate the types of chemicals to reduce the likelihood of resistance buildup. For the latest information on available materials and treatment recommendations, see *UC IPM Pest Management Guidelines: Almond,* listed in the suggested reading. Prune out and destroy diseased wood to reduce inoculum levels.

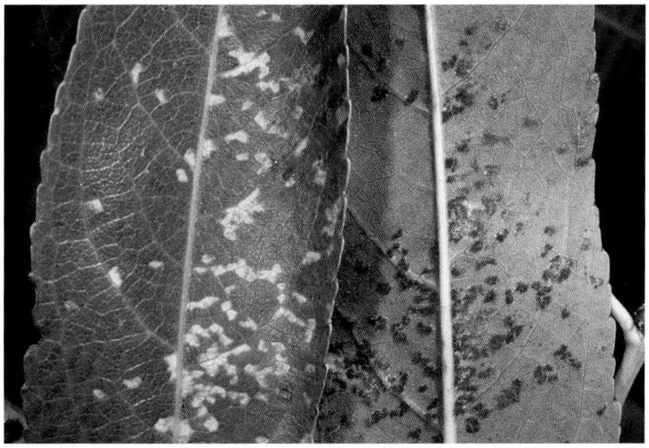

Rust pustules on the underside of the leaf (right) are red-orange from the production of urediniospores. The lesions appear yellow on the upper side of the leaf (left).

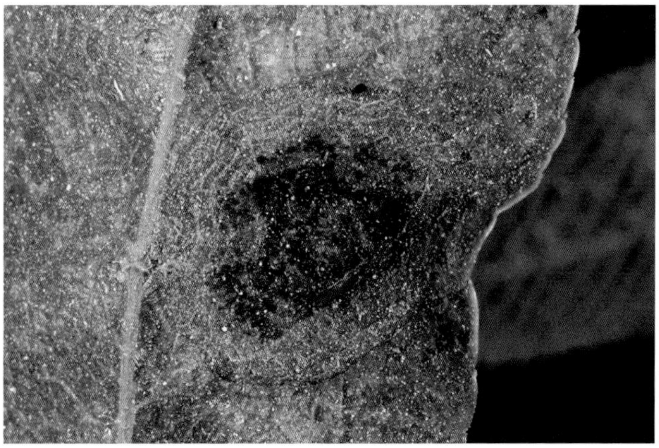

Alternaria leaf spot appears as large, brown spots that have a zonate pattern of lighter and darker brown. The lesions turn black as the fungus produces spores.

Rust

Tranzschelia discolor

Rust occurs sporadically throughout almond-growing areas in California. It is most likely to become serious in orchards near rivers or streams or other locations where humidity is relatively high in spring and summer. The disease causes leaves to fall prematurely and will weaken trees if not controlled.

Rust appears as small yellow spots on the upper surface of leaves. On the lower surface of the leaf these spots take on a rusty red appearance when the rust-colored spores produced in the lesions erupt through the surface. These spores are spread by air movement and infect other leaves to continue the disease cycle. Young twigs may be infected, but twig lesions are seldom seen on almond. Rust development is favored by humid conditions, and the disease becomes worse when rain occurs in late spring and summer. Trees can be defoliated quickly when rust becomes severe. The rust fungus survives from one season to the next in infected leaves and possibly also in infected twigs.

In orchards with a history of rust, apply fungicide in late spring and summer to control leaf infections. Two or three applications may be needed in orchards that have had severe rust problems. To be effective, fungicide must be applied before rust symptoms are visible.

Alternaria Leaf Spot

Alternaria alternata

Alternaria leaf spot occurs on almond trees in the San Joaquin Valley and northern Sacramento Valley. It has been most serious in the southern San Joaquin Valley.

Alternaria leaf spot appears as fairly large brown spots on leaves, about ½ to ¾ inch (12–18 mm) in diameter. The spots turn black as the fungus produces spores. Leaf spot develops most rapidly in June and July, and trees can be almost completely defoliated by early summer when the disease is severe. The disease appears to be most severe where dews form, humidity is high, and air is stagnant.

Changes in cultural practices have not reduced Alternaria leaf spot. However, the disease occurs first and is most severe on exposed leaves. Trees trained to an open and spreading canopy usually have more severe Alternaria leaf spot.

Alternaria alternata is difficult to control with fungicides. Treatment programs are being developed; contact your local UCCE farm advisor or pest control professionals for the latest information.

Corky Spot

Causal Agent Unknown

Corky spot has been observed primarily in Contra Costa and Colusa counties. The cause of this disease is unknown, but it seems to be associated with local conditions because the disease is not perpetuated by propagation. Affected cultivars have been Nonpareil, Ne Plus Ultra, Long IXL, and Jordanolo.

Small, yellow necrotic spots appear on leaves, particularly around the margins. As the spots turn brown, they fall out, giving the leaves a shot hole appearance. Premature leaf fall occurs, followed by new leaf formation. This pattern of shot holing, leaf drop, and new leaf formation may repeat itself several times during the season and can weaken the tree.

Corky spot lesions resemble those of shot hole, but they are smaller than shot hole lesions and do not contain small fruiting bodies in the center of the lesion.

Corky spot has been associated with almond trees grown on almond rootstock rather than on peach rootstock. No other control measures are known.

Corky spot lesions may resemble those caused by shot hole, but lesions are generally smaller, concentrated around the leaf margin, and do not develop black spore structures in the center.

Scab infections first appear on leaves as small, yellow spots.

Scab lesions eventually darken and look greasy gray when the fungus begins to produce spores.

Almond Scab

Fusicladium (Cladosporium) carpophilum

Almond scab has been in California since the early 1900s. Previously it was most damaging in the northern Sacramento Valley, where it occasionally caused trees to defoliate. Scab seldom caused extensive damage until the 1950s, when use of sprinkler irrigation became widespread. Though all California almond cultivars are believed to be susceptible to scab, Ne Plus Ultra and Drake are most severely affected.

Scab infects leaves, fruit, and twigs in almonds, causing dark spots to form. The major concern with almond scab is partial or complete defoliation of the tree. If early defoliation is severe, fruit drop will also occur. Scab infections left uncontrolled for several years will weaken trees and reduce crop yield.

On twigs, lesions start as indistinct, water-soaked spots that gradually turn brown in the center and have a lighter-colored margin. In spring, when the fungus resumes growth, it produces spores at the margin of the lesions, causing the margin to develop a distinct dark color. Lesions are superficial and do not girdle the shoots. Twig lesions are present throughout the year and can be used to confirm the presence of the disease in an orchard. Twig lesions should not be confused with the natural patchy darkening of maturing wood.

The first visible evidence of a scab infection on leaves occurs in late spring as small, indistinct, yellowish spots. These enlarge to about ⅕ inch (5 mm) in diameter and become a greasy gray when the fungus begins to produce spores. Later, the lesions become brown and necrotic. On leaf petioles and midribs, lesions are the same color but are more elongate.

Small circular spots can also develop on the hulls. These spots are more common on the upper side; eventually they coalesce into large, irregular dark blotches. The kernel is not affected, however. Fruit symptoms caused by the fungus *Alternaria*, sometimes referred to as black rot, may be confused with scab lesions because both appear on the upper fruit surfaces and often coalesce.

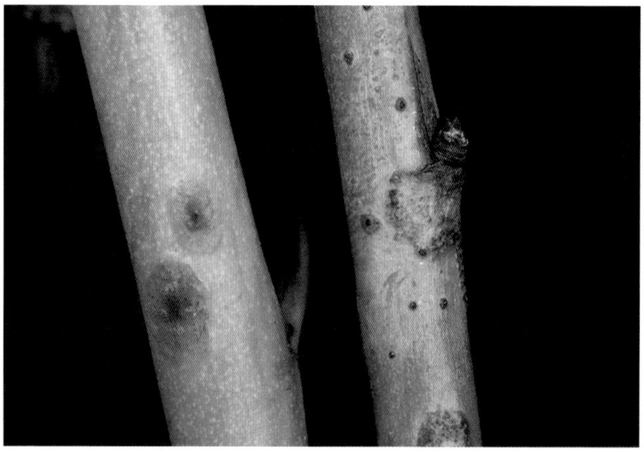

On twigs, scab lesions begin as water-soaked spots that eventually turn brown in the center and develop distinctly dark margins when the fungus produces spores. Twig lesions serve as overwintering sites for the fungus.

Scab lesions on fruit tend to concentrate on the upper side of the hull and appear as small, dark, circular spots. Numerous lesions may coalesce to form large irregular, dark blotches, as seen on the fruit in the left side of this photo.

The scab fungus overwinters in lesions on infected twigs and resumes growth in March. It is spread by spring rains and sprinkler irrigation, but leaf and fruit lesions usually do not appear until June or July. Treatment with fungicide 2 weeks after petal fall for shot hole will help control scab. Adequate control of scab requires two treatments with materials effective against scab. One of the two may be a dormant treatment with liquid lime sulfur, or both treatments may be applied during spring.

Leaf Blight

Seimatosporium lichenicola

Leaf blight of almond was first identified in 1950, when it developed in a few Sacramento Valley orchards. Areas of high spring rainfall, such as Contra Costa County and the Sacramento Valley, have the highest incidence of leaf blight.

Dead leaves at the base of this leaf cluster are typical of leaf blight infection. Next to the dead leaves, on the previous year's growth, are infected petioles remaining from leaves infected the previous season. These petioles serve as an overwintering site for the fungus.

Almond cultivars most susceptible to leaf blight are Drake, Ne Plus Ultra, and Peerless; moderately susceptible cultivars are Nonpareil, Mission, and IXL.

The most characteristic symptom of leaf blight is sudden withering of leaves. Leaves at the base of shoots and spurs begin to show symptoms in June; first they become yellow, then wither and die. Some affected leaves fall off, but most remain on the tree. Other leaves on the same spurs may wither and die throughout the summer. No sign of leaf blight fungus is present on affected leaves, although they may be covered with the dark growth of other fungi.

The fungus overwinters in the base of infected leaf petioles that remain on twigs throughout the winter. During late autumn and winter, the fungus may grow into the twig from the petiole and kill axillary buds. After the first fall rains, the lower end of the petiole turns light tan and becomes covered with dark fruiting bodies. Spores are spread by winter and spring rains, and the fungus infects petioles of new leaves in spring. Once inside the petiole, the fungus interferes with water conduction and the leaf withers. Flowers also may be destroyed in spring as they emerge from buds adjacent to infected leaf petioles.

The disease is not a serious problem unless it destroys leaves several years in a row and weakens the tree. Death of leaf and flower buds may contribute to yield loss. In orchards where the disease is a problem, apply fungicide between full bloom and petal fall to protect new leaf growth. Treatment for shot hole at this time will also control leaf blight.

Green Fruit Rot

Botrytis cinerea, Sclerotinia sclerotiorum, Monilinia laxa

Green fruit rot, also called jacket rot, calyx rot, and blossom rot, occurs on almond, apricot, prune, plum, and occasionally cherry. The disease is of little importance in most years, but it sometimes causes severe crop losses in all almond-growing regions. Ideal conditions for green fruit rot are cool temperatures and high levels of moisture during bloom. The disease is most damaging to cultivars such as Ne Plus Ultra and Carmel that have tight clusters of blossoms.

In addition to *Botrytis cinerea* and *Sclerotinia sclerotiorum*, the brown rot pathogen *Monilinia laxa* also attacks green fruit.

Symptoms and Damage. Green fruit rot begins during the latter part of the bloom period, when the fungus infects senescing petals and anthers. Infected petals develop water-soaked brown spots. Usually, diseased petals just fall to the ground, but some falling petals may land on leaves and cause secondary infections. Anther infections, on the other hand, spread to senescing floral tubes or flower jackets. Diseased jackets wither and stick to the developing fruit, especially if the jacket becomes wet. As the fruit sets and starts to grow, a brown spot develops where the jacket contacts the fruit. The infected spot expands rapidly until the entire fruit rots. Frequently, fruit will fall off the tree, and the net result of the damage is a reduction in yield.

Seasonal Development. *Botrytis cinerea* overwinters in decaying plant material and possibly in broadleaf winter annual weeds such as filaree, shepherd's-purse, and redmaids. It produces and releases spores in cool, wet weather. Spore release is greatest after a rapid rise or fall in relative humidity, but spore germination requires moisture. Spores are spread by air currents and deposited on blossom parts; they can survive 1 to 3 months if the weather remains cool and humid. Once an infection is established, moisture level and temperature are less important for survival of the fungus.

Initially, blossom parts are somewhat resistant to infection by *Botrytis* spores, but as soon as they begin to wither, they become highly susceptible. Decaying flowers provide the necessary nutrients for spores to germinate and the fungus to grow. *Botrytis* spores easily penetrate withered flower parts. Infection of leaves or fruit requires contact with an infected flower part. The fungus can be identified by observing with a hand lens the typical clusters of spores produced on branching sporophores.

Sclerotinia sclerotiorum overwinters in the soil. In the spring, during bloom, it produces spores that are carried to the blossoms by wind. If adequate moisture is present, the spores germinate and infect flower parts. The white fungal growth of *Sclerotinia* is readily distinguishable from the gray fungal growth of *Botrytis*.

Botrytis spores easily penetrate withered flower parts. Small fruit are susceptible to green fruit rot until the withered flower jacket falls off.

Botrytis cinerea produces clusters of spores on diseased flower parts.

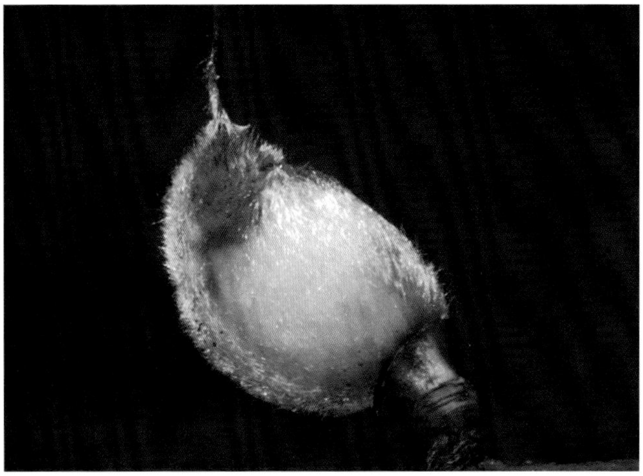

The brown area at the top of this young fruit indicates a green fruit rot infection. Eventually the infected spot will expand and the entire fruit will rot.

Management Guidelines. Green fruit rot does not occur yearly. When bloom is extended by cool temperatures and moisture is abundant, apply a fungicide at full bloom to prevent green fruit rot. Be sure the fungicide you use is effective against the two green fruit rot pathogens (*Botrytis cinerea* and *Sclerotinia sclerotiorum*), as well as against *Monilinia* spp., which are primarily responsible for diseases such as brown rot and hull rot but also may be involved in the green fruit rot complex.

Yellow Bud Mosaic

Tomato Ringspot Virus

Yellow bud mosaic, caused by a strain of tomato ringspot virus, was first found in California in 1936 on peaches and almonds in Solano and Yolo counties. It can affect most stone fruits, but in almond, yellow bud mosaic seriously damages only the Mission cultivar. Many native plants provide natural reservoirs of tomato ringspot virus, which is transmitted from plant to plant by the dagger nematode.

Trees with yellow bud mosaic may have rosettes of stunted leaves occurring on branches with normal-appearing leaves. The fruit on Mission trees appears wrinkled, rough, and larger than normal, because the hulls are abnormally thick.

Trees infected with yellow bud mosaic for several years have foliage that is concentrated on the terminals of branches that lack lateral growth.

Symptoms and Damage. Mission trees with yellow bud mosaic appear open and stunted; foliage is crinkled and distorted. Occasionally, chlorotic spots may develop that eventually die and drop out of the leaf, causing holes or tattering. Leaves may also be stunted and form small rosettes or short tufts. They stand out in sharp contrast to normal leaves on the same branch or in other parts of the tree. In severe cases the leaves may develop a mottled appearance.

Most of the leaves are concentrated at the branch terminals. Affected branches lack lateral growth and have little terminal growth from one year to the next. Frequently, one or more normal shoots are produced on a branch that is otherwise completely diseased. Fruit on diseased trees have wrinkled, rough hulls and appear larger than healthy fruit because the hulls are abnormally thick. Trees with yellow bud mosaic will live for many years but do not produce well; yield reduction is directly proportional to the severity of the symptoms.

The yellow bud mosaic strain of tomato ringspot virus can infect most rootstocks used for almond, but in scion cultivars other than Mission it rarely spreads up from the roots into the main portion of the tree, where most of the damage caused by this virus occurs. Occasional symptoms may be seen on water sprouts in the lower scaffold limbs of Nonpareil and Ne Plus Ultra, and a form of the disease has occurred in the Patterson area of Stanislaus County that caused Nonpareil trees to grow in a spiral form, similar to the characteristic spiraling of the IXL cultivar.

Seasonal Development. Tomato ringspot virus is indigenous to California and is widely distributed in the coastal areas and the Sacramento Valley; it also occurs in a few scattered areas of the San Joaquin Valley. Tomato ringspot virus is spread by budding and grafting and by dagger nematodes, *Xiphinema* spp., in the orchard soil. The virus is seedborne in dandelion, and infects a number of other broadleaf weeds (see Table 18) as well as grapevines, apples, and cane berries. The nematode vector acquires the virus by feeding on the roots of infected weed hosts. *Xiphinema* juveniles remain infective until they molt. Adults remain infective for 3 to 8 months.

Susceptible rootstocks become infected with tomato ringspot virus when infected dagger nematodes feed on their roots. Gradually, over the course of several years, the virus spreads to the upper shoot terminals. If nematodes carrying the virus feed only on one root, the disease will initially develop only on one side of the tree. If the nematodes are distributed throughout the soil and feed on all the roots, the disease will progress uniformly throughout the tree.

Table 18. A Partial List of Plants that Host Tomato Ringspot Virus.

Common name	Scientific Name
apricot	*Prunus armeniaca*
bristly oxtongue	*Picris echioides*
bull thistle	*Cirsium vulgare*
cheeseweed (little mallow)	*Malva parviflora*
common chickweed	*Stellaria media*
common currant	*Ribes sativum*
common lambsquarters	*Chenopodium album*
common mullein	*Verbascum thapsus*
dandelion	*Taraxacum officinale*
Fullers teasel	*Disacus fullonum*
Himalaya blackberry	*Rubus procerus*
oxeye daisy	*Chrysanthemum leucanthemum*
peach	*Prunus persica*
plantain	*Plantago* spp.
red clover	*Tifolium pratense*
red raspberry	*Rubus idaeus*
red sorrel	*Rumex acetosella*
sand strawberry	*Fragaria chiloensis*
seaside daisy	*Erigeron glaucus*
spurge	*Chamaesyce* spp.
white clover	*Trifolium repens*
wild carrot	*Daucus carota*
wild mustard	*Brassica kaber*
wild strawberry	*Fragaria vesca*

Tomato ringspot virus is spread to new locations when floods or cultural operations move infected nematodes. Trees in areas adjacent to east-west streams flowing from the coastal mountains or Sierra foothills have a higher incidence of this disease. The disease also can spread when wind-disseminated seeds of virus-infected wild plants such as dandelion germinate and grow in or near an orchard and dagger nematodes are present in the soil. Within an orchard, the disease spreads slowly to adjacent trees, gradually enlarging the original area of infection. Frequently, symptoms first develop in border trees and subsequently spread to adjacent trees. Although it is possible for infection to spread from tree to tree through root grafts, this is not common. The disease spreads mainly by the movement of dagger nematodes that carry the virus through the soil.

Mission trees planted in infested soil will exhibit symptoms in 2 years and will probably never be economically productive. When older Mission trees become infected as a result of natural spread of nematodes in the soil, they generally become uneconomical producers 3 to 5 years after the symptoms first appear.

Management Guidelines. Once yellow bud mosaic symptoms appear in an orchard, it is extremely difficult to eliminate the virus. Deep soil fumigation at rates high enough to kill the nematode is expensive and often not practical, especially if other factors may also be affecting the growth of the tree.

If you have an area that contains tomato ringspot virus, try to limit its spread to other parts of the orchard. Do not perform any cultural operations, including flood irrigation, that may move soil from this area to other locations. If the affected area involves only a small part of the orchard, remove infected trees and trees in at least two rows beyond. If several areas of diseased trees are present, you may need to remove the entire block. Before removing the trees, treat them with an herbicide or girdle them to aid in killing the roots. Remove as many roots as possible. For best results, fallow the ground for 2 years and then fumigate the soil to kill the nematodes.

Do not plant Mission trees on peach or almond rootstock in infested areas. Marianna 2624 plum rootstock is resistant to the virus. Replant with cultivars that are compatible with Marianna 2624, or with nonhosts of the virus, such as walnut, pear, and plum or prune on Marianna 2624 rootstock.

Hull Rot

Rhizopus stolonifer and *Monilinia* spp.

Hull rot can be caused by either the common bread mold, *Rhizopus stolonifer*, or by the brown rot fungi *Monilinia fructicola* and *M. laxa*. Of the two Monilinia species associated with hull rot, *M. fructicola* is responsible for most hull rot in almonds. Although it is occasionally associated with hull rot, *M. laxa* is primarily responsible for causing brown rot blossom and twig blight in the spring.

Hull rot was first noticed causing severe twig blight in the 1950s and now occurs in most almond-growing districts. It is most severe, however, in well-managed orchards with large vigorous trees.

Symptoms and Damage. The first indication of hull rot usually comes several weeks before harvest, when leaves on a shoot wither and die. Closely examine fruit on this shoot for a brown area on the outside of the hull and either tan fungal growth in the brown area on the inside or outside of the hull (*Monilinia* spp.), or black fungal growth on the inside of the hull (*Rhizopus* spp.). Fungi invade almond hulls after the hulls begin to split in early July. Once in the hull, fungi apparently produce a toxin that causes the shoot and leaves to die from the point where the fruit is attached to the tip of the spur. Fungi can also move in the other direction on the branch. Because the shoot is killed, other green fruit on the shoot won't mature, and they remain on the tree after harvest.

In sprinkler-irrigated orchards, hull rot frequently occurs on lower tree branches that are hit directly by water. Such infections can result in a loss of lower limbs on many of the trees throughout the orchard.

Dieback of shoots and fruiting wood due to hull rot fungi is the most damaging aspect of this disease. Fruit killed by hull rot remains on the tree after harvest.

Hull rot caused by *Rhizopus* is evidenced by black fungal spore structures growing inside the hull.

Tan fungal growth on the outside of the hull indicates hull rot caused by *Monilinia* sp.

The most damaging aspect of this disease is dieback of shoots and fruiting wood and the resulting yield reduction. When more than 30% of the fruiting wood is killed, a significant reduction in yield results. If less than 20% of the wood is killed, economic losses are minimized by the fact that most trees can offset the loss with new growth. Lower branches, however, do not usually generate new growth and will be lost.

Seasonal Development. Almond hulls are susceptible to hull rot fungi from the beginning of hull split until the hulls dry, a period that can range from 10 days to 2 months. Several factors affect the length of time it takes hulls to dry, including the amount of fertilization and irrigation an orchard receives, the amount of space between trees, and the openness of the tree's canopy. Rainfall or sprinkler irrigation between hull split and harvest will also slow drying and increase the length of time that hulls are susceptible to infection.

Rhizopus spores are commonly found in the soil. *Monilinia laxa* spores are found on blighted blossoms and sticktights on the tree. It is not known where spores of M. *fructicola* reside in the orchard. Spores of both *Rhizopus* and *Monilinia* fungi are readily airborne and can be spread by winds and possibly by dried fruit beetles that are attracted to almond hulls when the hulls split. The first fruit on a tree to split are usually blanks, or fruit without kernels. Spores carried by air currents and in the dust created when the orchard floor is prepared for harvest infect the blanks, and infected blanks subsequently provide a ready source of inoculum for the remainder of the fruit on the tree.

Management Guidelines. When planting an orchard, consider selecting resistant cultivars. The most susceptible cultivars are Nonpareil, Kapareil, Sonora, Jordanolo, and IXL. Merced, Thompson, and Ne Plus Ultra are frequently affected by the disease, but the damage is minimal. Hard-shelled cultivars such as Mission, Davey, and Drake may exhibit rotted hulls, but shoot dieback is rare in these cultivars.

To reduce the incidence of hull rot, avoid excess nitrogen fertilization and follow cultural practices that increase uniformity of hull split and decrease drying time of the hulls without sacrificing yield or kernel quality. For example, prune trees to increase the amount of sun exposure that the almonds receive, but not to the point where yield is sacrificed. This will improve uniformity of hull split.

Not irrigating or reducing irrigation at the onset of hull split dramatically reduces the incidence of hull rot.

Harvest your crop as soon as fruit at eye level show 95 to 100% hull split. Early harvest lowers the incidence of twig dieback. Fungi remain in the hulls, and once fruit are removed, twig dieback ceases. Shake your trees twice, if necessary, to remove later-maturing fruit. After harvest, removing any infected fruit remaining on the tree may help to reduce the level of inoculum in the orchard. It will also serve to eliminate overwintering sites for the navel orangeworm.

Spring fungicide treatment to control brown rot blossom and twig blight will not effectively control hull rot in the fall. Fungicide treatments are not effective against this disease, and no materials are registered for control of hull rot.

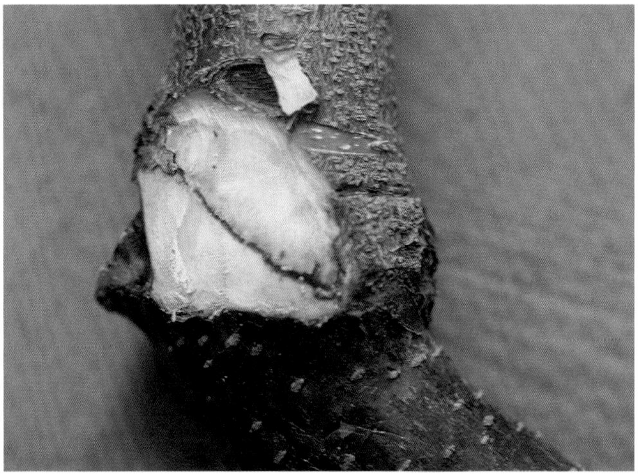

JERRY K. UYEMOTO

Brown, necrotic areas develop at the graft union of almond trees affected by brownline and decline. The surface of the wood at the graft union becomes pitted.

Almond Brownline and Decline
Peach Yellow Leafroll Phytoplasma

Almond brownline and decline is a disease that affects certain almond cultivars grown on Marianna 2624 rootstock. It is thought to be caused by peach yellow leafroll phytoplasma, transmitted primarily by budding from an infected scion source. The phytoplasma is a microbe that multiplies and spreads within the phloem of the infected tree. When the phytoplasma reaches the graft union, cells of the Marianna rootstock die. This results in a layer of brown, necrotic cells in the phloem at the graft union, which prevents spread of the phytoplasma into the rootstock but also interferes with movement of nutrients between scion and rootstock.

Affected trees have leaves that droop and appear wilted. Current-season shoot growth is abnormally shortened or absent. If bark is removed from the graft union, brown necrotic areas can be seen. The brown areas at the graft union may be scattered around the trunk or may form a continuous line. You may need to check several places around the circumference of the trunk to determine whether this symptom is present. Over time, the surface of the wood at the graft union becomes mildly to severely pitted. Trees decline and become unproductive.

Brownline and decline has been observed only on Carmel, Peerless, and Price scions. To avoid the disease, obtain trees from nurseries that use budwood sources known to produce healthy trees.

Almond trees severely affected by union mild etch develop shallow grooves in the wood along the graft union.

Union Mild Etch and Decline

Like brownline and decline, union mild etch and decline affects almond trees grown on Marianna 2624 rootstock. The disease has been observed only on a few cultivars, including Aldrich, Butte, Carmel, Mission, Peerless, and Price. Affected trees have pale foliage that drops prematurely. Shoot growth is not stunted or lacking as it is on trees with brownline and decline. Some trees decline and die while others recover and appear normal, although their productivity is reduced significantly. Severely affected trees develop shallow grooves in the wood along the graft union.

The causal agent of mild etch and decline has not been identified. To reduce the likelihood that the disease will develop in your orchard, use planting stock from nurseries known to use budwood sources that produce healthy trees.

Almond Kernel Shrivel
Peach Yellow Leafroll Phytoplasma

Almond kernel shrivel has occurred in certain locations of the northern San Joaquin Valley on almond trees planted on peach rootstock. Affected trees bloom later than healthy trees, new shoot growth is stunted, and leaves are pale and smaller than normal. Trees develop thin canopies and the kernels of all nuts are shrivelled at harvest. Almond kernel shrivel is caused by the peach yellow leafroll phytoplasma and is known to affect only almond scions grown on peach

Kernels are severely shriveled in all the nuts (right) from trees affected by almond kernel shrivel.

A typical overall symptom of noninfectious bud failure is sparse foliage concentrated on new growth.

rootstock. Peach yellow leafroll phytoplasma can be transmitted by certain insect vectors, and it is presumed that diseased almond trees are infected in this manner. The phytoplasma can also be transmitted in infected budwood. At the present time, the only management recommendations for almond kernel shrivel are to remove diseased trees and use planting material produced using budwood sources known to be free of the disease.

BUD FAILURE DISORDERS

Bud failure disorders can affect primarily vegetative buds (noninfectious bud failure), flower buds (nonproductivity), or both (dormant bud drop). Flower bud failure results from a loss of flower buds or a failure of the flowers to set and may be associated with a virus (as in almond calico) or genetically induced (as in nonproductive syndrome).

Noninfectious Bud Failure

Noninfectious bud failure, also known as crazy top, is a genetic disorder. It has never been transmitted from an infected plant to a healthy one but is perpetuated by vegetative propagation. Seedlings inherit it from their parents, but the exact nature of the causal factor is not completely known.

The disorder is associated with specific cultivars. Before 1940, bud failure was observed sporadically in Nonpareil and Peerless, but the problem became acute when the Jordanolo cultivar was released in 1937. Since that time, noninfectious bud failure has appeared in other new cultivars including Jubilee, Merced, Thompson, Harvey, Carmel, and Price. A few affected Mission trees also have been observed.

The disorder occurs to some extent throughout all almond-growing regions in California, but it is most severe in the southern San Joaquin Valley and western and northern Sacramento Valley, where high summer temperatures accentuate its occurrence. Incidence and severity of bud failure within a given orchard is variable and can involve parts of a tree, a few random trees, groups of trees, or most trees of a certain cultivar.

Symptoms and Damage. The most conspicuous symptom of noninfectious bud failure is failure of vegetative buds to grow in the spring. If you cut open and examine individual vegetative or shoot buds during fall or winter, they often will appear necrotic. Flower buds are not affected by this disorder once they have developed. If symptoms are severe, however, fewer flower buds will develop because vegetative growth is reduced (fruit bearing wood), and flowering time may be delayed. The percentage of fruit set is generally good, considering the fewer number of blossoms produced.

The percentage of vegetative buds on a shoot that fails determines the severity of the symptoms. Buds are often viable at the base of the shoot, nonviable in the middle, and occasionally viable at the tip of the shoot.

The following spring, viable buds grow into shoots, often at right angles to the branch (especially on Nonpareil), and they are unusually vigorous. Look for bud failure symptoms just after growth begins in spring and for the next 6 to 8 weeks (through April). Examine previous years' shoots for necrotic, failed buds (some of these buds may have already abscised and dropped). Later in the season new shoot growth

This shoot illustrates "crazy top," a pattern of growth associated with noninfectious bud failure.

Crazy top caused by noninfectious bud failure is easiest to see during dormancy when the leaves are off the tree.

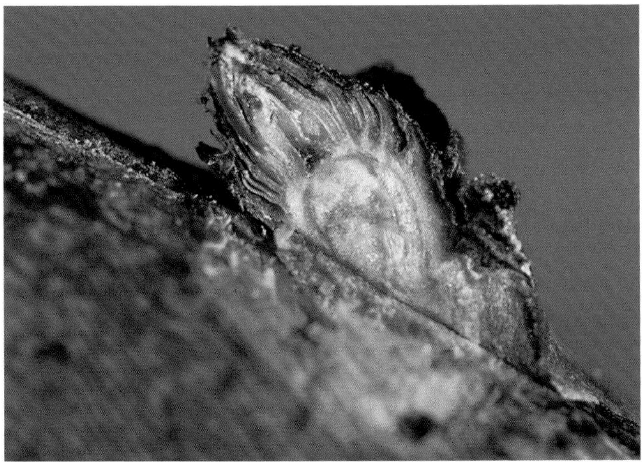

Cross-sections of vegetative buds from trees with noninfectious bud failure frequently reveal brown areas of dead tissue.

The band of rough, cracked bark located to the left of the nut cluster is one of the symptoms associated with severe noninfectious bud failure.

may obscure symptoms and make it difficult to distinguish between vegetative bud drop caused by noninfectious bud failure and bud failure from other causes.

Trees that have shown bud failure symptoms for several years have sparse foliage because fewer shoots are produced, and no leaves occur at the nodes of older twigs. Unusually large numbers of branches are clustered at the base or tip of lateral branches. Often branches have changed their direction of growth several times—some grow back upon themselves. This growth behavior and the right-angled shoots give the branch system a crooked, tangled appearance. Names such as "crazy top" refer to this condition.

"Rough bark" is another symptom associated with severe noninfectious bud failure, but its occurrence is sporadic. This symptom occurs when necrosis from the shoot bud extends to the surrounding stem tissue. It shows up on new shoots during summer or fall, often starting as a discolored area at the base of a leaf. As the shoots get older, these areas continue to develop, expand, and become more prominent. Rough bark frequently occurs on the middle or upper portions of shoots in 5- to 6-inch (12.5- to 15-cm) bands along the branch. Viable buds do not form in these areas and symptoms can be quite dramatic on young trees. Entire shoots may be devoid of viable lateral buds, and few normal buds will develop at the shoot tips.

The effect of noninfectious bud failure on yield depends on the severity of the disorder in a particular tree. Symptoms may increase with age, but yield reductions usually reach a plateau in most cultivars; yields may fluctuate about this plateau, depending on seasonal temperatures, but are generally stabilized.

Seasonal Development. Vegetative bud necrosis and rough bark develop during midsummer and early fall after exposure to periods of high temperatures. Lack of moisture and other stresses also may accentuate the disorder. Variation in the severity of symptoms from year to year is related to the temperature pattern during the previous growing season.

Symptoms in affected trees can appear at any age from the second growing season to more than 10 years later. The most severe cases usually occur within the first 4 or 5 years. Some trees do not develop symptoms, however, until after 15 to 20 years.

The age at which a tree expresses bud failure symptoms and the severity of these symptoms depend on the relative bud failure potential in the source plant from which buds used for propagation were taken, and the environmental conditions to which the tree is exposed. Trees with bud failure usually develop symptoms at an earlier age and to a greater degree in the hotter areas of the southern San Joaquin Valley than in the cooler summer areas of central California and the coastal valleys. Once bud failure symptoms begin to appear in a tree, the disorder will remain. Although developing buds are killed by high temperatures in July and August, they don't drop off until the following spring.

Management Guidelines. The only known cure for noninfectious bud failure is to replace or top-work affected trees. Trees that aren't severely affected by bud failure can be left in the orchard. Their crop potential may be lower than normal trees, but they will not die from the disorder, and the disorder will not spread from tree to tree.

It is usually not economical to replace bud failure trees unless they are under 7 to 10 years old and have greater than 25 to 30% reduction in yield. Young trees that show symptoms in their second to fourth year or older trees that are severely affected should probably be removed and replanted.

Severely affected trees less than 10 years old also can be top-worked to a different cultivar that is free from bud failure or to the same cultivar. If you use the same cultivar, use propagation material from a source that has a record of low bud failure potential. Although top-working requires skill and postgrafting care to be successful, it is often the fastest method to bring a tree back into production.

Selecting a source of suitable scion wood for top-working or nursery propagation is difficult because the potential for bud failure cannot be identified by visual inspection unless the tree actually shows symptoms. It is important, however, to inspect a bud-source orchard for presence of affected trees. If bud failure trees occur, avoid the entire orchard. It is also important to select a source that has a known record of propagations through commercial use or has been progeny-tested and found relatively free of bud failure. Individual source trees should be tested for freedom from bud failure in the progeny. Lacking this information, it is best to select source orchards located in the cooler, central portion of the Central Valley rather than in the northern Sacramento or southern San Joaquin valleys because of the association of bud failure development over time with high temperature exposure.

If you use cultivars that are known to be affected by this disorder, it is difficult to completely avoid having some bud failure in the orchard, regardless of the source of the vegetative material. However, individual source tree testing in hot areas, as discussed above, is currently the best way to test budwood sources and reduce bud failure in the orchard.

Infectious Bud Failure, Almond Calico

Prunus Necrotic Ringspot Virus

Infectious bud failure and almond calico occur throughout the almond-growing regions in California. They are caused by two different strains of prunus necrotic ringspot virus, and symptoms of both bud failure and calico may appear on the same tree. Prunus necrotic ringspot virus has many strains, and there is great variability among them. Variants of strains can occur in the same tree and, by interfering with each other, markedly affect the distribution of symptoms within a particular tree. Thus, some strains may not particularly affect tree productivity; others can greatly reduce productivity. Likewise, different cultivars can react differently to a particular strain. In addition to infecting almond, prunus necrotic ringspot virus also infects other stone fruits. Infectious bud failure has been observed on Drake and Nonpareil trees. Calico is most severe on Peerless, and Nonpareil; calico symptoms are mild or absent on infected Mission and Ne Plus Ultra trees.

Symptoms and Damage. Almond calico symptoms first appear in most cultivars as a mottling of the leaves on infected branches. In subsequent years, leaves produced in the early part of the season develop chlorotic rings or bands that gradually become creamy white, although these symp-

Trees with almond calico have bare shoots with thin foliation and few fruit. Almond cultivars that grow upright and erect resemble the tree in this photo when they are infected.

Calico symptoms in almond cultivars with spreading growth frequently cause the tree to appear "brushy."

Leaves with almond calico frequently have chlorotic bands.

toms may not be obvious when spring weather is warm. Characteristic patterns include lines, rings, and chevrons. Some of these leaves may drop prematurely, but leaf symptoms do not normally cause economic losses. Leaf symptoms may not always occur yearly in infected trees, but they are usually accentuated in years with cool spring temperatures.

Moderate to severe flower bud failure may develop without calico symptoms, or it may develop in trees affected by almond calico. This may not occur until 10 years or more following the first appearance of calico symptoms. Lateral flower buds on shoots produced the previous season fail to bloom, or those that do bloom produce few flowers that set. Likewise, lateral leaf buds and even terminal shoot buds may die. Bare shoots occur because vegetative buds are either not produced in the leaf axil ("blind nodes"), or if they are, the bud dies and falls off before leafing out. If vegetative buds do not grow, the shoot will die back to the uppermost live lateral vegetative bud. If most of the lateral vegetative buds die and terminal buds continue to grow, the shoot becomes long and willowy with leaves and fruit only present on the terminal portion. As trees age, flower and vegetative bud failure symptoms become more prominent and result in an overall reduction in yield.

To distinguish infectious bud failure from genetic bud failure, look for leaf symptoms in early spring such as mottled leaves with chlorotic blotches. Trees with noninfectious or genetic bud failure do not have leaf symptoms unless they are also infected with prunus necrotic ringspot virus. Also, genetic bud failure affects leaf buds and not flower buds. Infectious bud failure affects both leaf and flower buds.

Trees of the Drake cultivar do not exhibit prominent calico leaf symptoms, but they do develop flower and leaf bud failure. Affected Mission trees commonly have no calico leaf symptoms or bud failure, but they may fail to set fruit several years following infection and tend to have leaves that are longer in proportion to width. These symptoms resemble those caused by nonproductive syndrome (see below). Because Mission trees do not exhibit diagnostic leaf and bud failure symptoms, prunus necrotic ringspot virus may be difficult to identify in this cultivar unless indexing is performed to determine if the virus is present.

Seasonal Development. Prunus necrotic ringspot virus is thought to be transmitted in pollen that is produced in flowers on infected branches. Pollen may infect other trees during bloom when bees carry infected pollen to healthy trees. Leaf symptoms may or may not occur during the first year of infection. In the second and third year, however, leaves on affected branches show line patterns and chlorotic mottling. Although this disease does not kill the tree, leaf symptoms recur throughout the tree's life, and flower and vegetative bud failure symptoms may gradually intensify as trees age.

Management Guidelines. Both the scion cultivar and the rootstock of nursery trees can be infected with almond calico virus. This disease can be avoided by obtaining trees that have been propagated from source trees tested free of prunus necrotic ringspot virus planted on rootstocks that come from ringspot-free seed trees. This can be accomplished by buying certified nursery trees, if they are available.

If a few isolated trees in your orchard have almond calico, consider removing them. Their pollen may serve as a source of inoculum for other trees during bloom.

If the disease is widespread in your orchard, you may eventually want to replant the entire orchard when production appears to be affected. This decision should be based on the overall productivity and profitability of the trees. Do not replant a small portion of the orchard if the disease is widespread. When newly planted trees come into bearing, they are likely to be infected by pollen from older infected trees.

Nonproductive Syndrome

Large, vigorous, upright trees with nonproductive or low production syndrome are often referred to as "bull trees" or "rooster trees." The terms "bull" and "rooster" are general ones, however, and have also been used in reference to trees affected with virus-induced bud failure. Trees with nonproductive syndrome have varying production ranging from very low to near normal, distorted fruit with softer-than-normal shells, and long, somewhat narrow leaves. In recent years, this condition has been most commonly observed in the Mission cultivar, where it is referred to as "bull mission," but a similar condition also occurs in Nonpareil, Carmel, and Fritz.

In Mission, the severity of symptoms may vary greatly from one tree to the next, and even from one branch to the next within the same tree. Productivity, as measured by the percentage of fruit set, can range from zero to near normal levels (30 to 50% set). Fruit on affected trees tend to have distorted shapes; they might be longer and narrower than normal, or they can be shorter and plumper. Frequently, nuts on affected Mission trees have soft shells with a prominent wing extending from the suture, instead of the hard shell that is characteristic of that cultivar.

Nonproductivity results from a range of abnormalities in the reproductive system, including defective pistils, ovules, and pollen grains, as well as from flower and fruit drop. Not all of these abnormalities occur on every tree that is affected, but their prominence increases with the increasing severity of the condition. These abnormalities sometimes occur simultaneously with other variations, such as bitter kernels.

In addition to reproductive abnormalities, nonproductive bud sports (single branches that are characteristically different from other branches on the same tree) with crinkled leaves and rough bark occasionally are observed in single trees in orchards.

Symptoms associated with nonproductive syndrome appear to be the result of genetic changes, or mutations, within cells. Mutant cells may occur in the growing points (buds) of trees, but their presence may not be detected until a large proportion of the cells in the branch are affected. If the mutations that cause nonproductive syndrome occurred in the plant supplying buds used for propagation, the resulting offspring trees may have the disorder if a large proportion of cells were affected in the buds used. Affected trees in an orchard often occur in groups of approximately ten trees, which can be associated with nursery bundles of ten. In nurseries, trees from a specific bud source are frequently grouped together. In addition, affected trees are unusually vigorous, which increases the likelihood of being grouped together during size grading.

Because other conditions such as young age, high vigor, some mineral deficiencies, lack of pollination and fruit set, or specific diseases can reduce the percentage of fruit set and modify fruit characteristics, careful diagnosis of nonproductive syndrome is important. Shot hole fungus, for instance, can reduce crops and produce distorted fruit. Some viral diseases, such as almond calico, are also associated with nonproductivity and can resemble this condition to a certain degree. Corroborative indexing, a method of identifying viruses (see "indexing" in the glossary), can be used to distinguish between viral and nonviral conditions.

If trees with nonproductive syndrome appear in an orchard, replace the ones that have significantly reduced yields by either replanting or grafting. Selection of nonaffected source trees for propagation material is essential but

Fruit on Mission trees with nonproductive syndrome tend to have distorted shapes.

A normal fruit on a healthy Mission tree is shown here.

On trees with dormant bud drop, most of the lateral buds may fall without opening while the terminal vegetative bud always develops normally.

ROGER A. DUNCAN

The ground underneath trees affected by dormant bud drop may be littered with fallen buds during bloom.

difficult because lack of visual symptoms on the bud source tree does not prove absence of nonproductive syndrome.

Dormant Bud Drop

Dormant bud drop occurs when dormant buds fail to open and fall to the ground during the normal bloom period. Both flower and vegetative buds are affected, which distinguishes dormant bud drop from other bud failure disorders. Dormant bud drop occurs primarily on coarse-textured soils of the northern San Joaquin Valley, but orchards in other areas are affected occasionally. Dormant bud drop appears to be most severe on Carmel; Wood Colony, Price, Mission, Butte, Fritz, and Nonpareil also can be severely affected. Dormant bud drop often occurs in the same orchards year after year and can result in serious economic losses.

Buds affected by dormant bud drop appear to be normal early in the dormant season, but they fail to swell normally late in the dormant season and remain tightly closed at the beginning of bloom. Affected buds fall from the tree during bloom and have a small spot of brown, necrotic tissue where they were attached to the shoot. When bud drop is severe, almost all buds except terminal buds may fall from some trees. Terminal buds develop normally, often resulting in willowy shoots with few spurs. Dormant bud drop appears to be most severe in springs preceded by wet fall weather.

The cause of dormant bud drop is uncertain, but the disorder is associated with bacterial canker. Bud drop is most severe in young trees (4th-leaf to 9th-leaf) of replanted orchards growing in coarse-textured soils, usually with high populations of ring nematode, which are known to predispose trees to bacterial canker. Affected buds are colonized by the bacterial canker pathogen, *Pseudomonas syringae*, and the numbers of bacteria are generally higher on buds from trees with a history of severe bud drop.

Preliminary results indicate that the severity of dormant bud drop can be reduced by maintaining healthy, vigorous trees, controlling nematodes, applying monthly supplements of nitrogen (35–40 lb N per acre [39–45 kg per ha]) and calcium during the growing season, and applying a foliar copper spray in the fall. Fertilizer supplements appear to reduce bud drop severity even when leaf analysis for nitrogen and calcium indicate that levels of these nutrients are in the "adequate" range. For the latest information on dormant bud drop, check with your UCCE farm advisor.

ENVIRONMENTALLY CAUSED DISEASES

Pathogens and genetic disorders may not be the only causes of disease symptoms on leaves and fruit. Adverse weather, soil conditions, nutritional disorders, and the like produce symptoms as well.

Gumming, "Split Pit," and Corky Growth

Occasionally, almond fruit show protrusions of gum during development. You may find gum pockets, cracked shells, and aborted, dark, gummy, or poorly developed kernels. Gum may also be seen on the surface of the hull. In nuts of Jordanolo and a few other cultivars, a thin layer of corky material may be stuck to the kernel. These conditions are associated with larger-than-normal fruit size and are generally the result of a rapid growth rate during early stages of fruit development in April and May. The innermost layer of the shell hardens and stops growing. It eventually cracks because the outer layers of the shell are still growing. Gum is produced within the fruit as a result of this injury. Such gummy fruit may be difficult to shake from the tree. Callus tissue may develop from cracks and grow next to the kernel to form corky material that dries and sticks to the kernel when mature. These conditions can cause reject kernels.

Fruit tend to grow more rapidly in the early stages of development if certain growth conditions such as a light

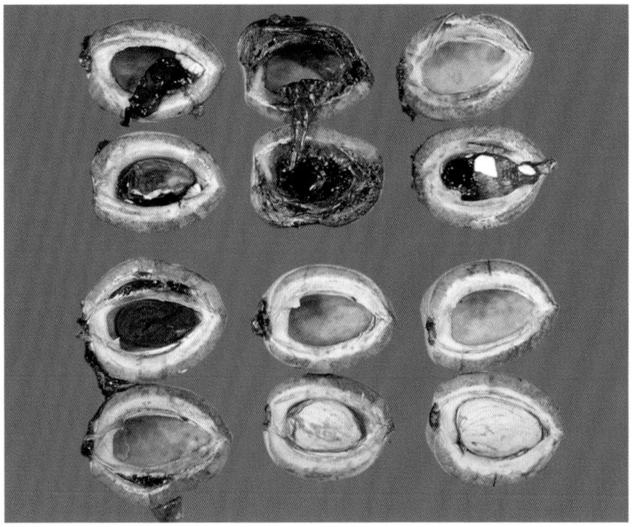
DON C. EDWARDS

Gum pockets, cracked shells, and aborted, gummy, or poorly developed kernels may all result from rapid growth during early stages of fruit development in April and May, especially in Thompson and Jordanolo cultivars.

Excess sodium produces tip burn on leaves of almond trees. Identical leaf symptoms may also be produced by chloride toxicity.

crop exist. Early-season moisture stress followed by irrigation can aggravate this condition. Thompson and Jordanolo are more subject to gumming, split-pit, and corky growth than other almond cultivars. To avoid these problems, correct any orchard condition that contributes to excessively vigorous growth and light crops, especially if you are growing susceptible cultivars like Thompson and Jordanolo.

Nutritional Disorders

An excess or deficiency of minerals in the soil is harmful to the tree because it interferes with its growth and development. Symptoms produced by trees with nutritional disorders may be confused with symptoms caused by other stresses such as diseases, herbicide toxicities, and environmental pollutants. To identify nutritional deficiences or toxicities, evaluate leaf symptoms in conjunction with analysis of soil (toxicities only), or leaf tissue, or both. Leaf analysis is discussed in the chapter "Managing Pests in Almonds." A section on diagnosing mineral deficiencies and toxicities can be found in the *Almond Production Manual*, listed under the suggested reading.

Toxicities. Nutrient toxicities most likely to occur in California almond orchards are excesses of boron, chlorine, and sodium. Toxic levels of these minerals may be present in the soil or may build up because of poor-quality irrigation water. Symptoms caused by those toxicities are most evident from July through September.

Boron. Boron toxicity and damage is frequently encountered on the west side of the San Joaquin and Sacramento valleys as new plantings expand into soils high in boron. Boron toxicity causes shoots to die back at the tip and the bark to become cracked and corky. Gummy deposits form around cracks. Midribs on the undersides of leaves also develop corky areas.

Chloride and Sodium Toxicity. Toxicities caused by excess levels of chloride or sodium occur in limited sites scattered throughout all almond-producing districts. Both produce leaf-tip burn symptoms. Use leaf tissue analysis to distinguish between sodium and chloride toxicities.

Deficiencies. The most common nutrient deficiencies in California almonds are, in order of importance, nitrogen, zinc, potassium, boron, and iron. Copper, magnesium, and manganese deficiencies can also occur, but these are rare.

Nitrogen. Almost all orchards will become nitrogen-deficient if nitrogen is not applied. Nitrogen deficiency in a block of trees is difficult to diagnose by symptoms alone because all the trees in the block may look the same. Have a leaf analysis performed to confirm nitrogen deficiency.

Early in the season, trees may show reduced shoot growth and leaves may be small and pale. Foliage also may be sparse. In autumn the leaves become yellow and drop earlier than normal. Insufficient or excess soil moisture may produce similar symptoms.

Zinc. Zinc deficiencies are common in the sandy soils of the San Joaquin and Sacramento valleys, but they can appear on

Trees deficient in zinc have small, chlorotic leaves that appear in tufts.

Potassium deficiency results in pale leaves with necrotic tips and edges. Frequently, the leaf tips curl upwards, creating the symptom known as "Viking's prow."

trees grown in heavier soils as well. Symptoms appear early in the season, especially if the deficiency is severe. Zinc-deficient trees bloom and leaf out later than normal trees; in some cases they may be as much as a month behind in development. When leaf buds do open, leaves are small, chlorotic, and appear in tufts, giving rise to the term "little-leaf." If the deficiency is not so severe, leaves are only slightly smaller than normal but show chlorotic areas between the lateral veins. In severely deficient cases, terminal dieback sometimes occurs. As the season progresses, subsequent shoot growth tends to mask zinc deficiency symptoms that were present earlier, especially on older trees. Rapidly growing young trees may show symptoms throughout the season. Fruit on zinc-deficient trees are smaller than normal.

KYOTO URIU

Almond trees deficient in boron may exhibit shoot dieback.

Potassium. Potassium deficiences can be found in the northern Sacramento Valley and in sandy areas of the San Joaquin Valley. Leaf symptoms appear in early to mid summer; leaves become pale, resembling nitrogen deficiency, and leaf size and shoot growth are reduced. Symptoms often first appear along the middle of shoots. The tip and upper margins of leaves on vigorously growing shoots become necrotic. Leaf tips curl upward, causing a condition referred to as "Viking's prow." Overall fruit size is reduced.

Boron. Boron deficiency is limited to a few areas in the northern San Joaquin Valley and the north and west sides of the Sacramento Valley, although recent evidence suggests that boron deficiency could be more widespread. In severe deficiencies, the terminal and subterminal leaves on vigorously growing shoots develop tip scorch, and eventually the entire leaf may become necrotic and fall off. Frequently the tips of these shoots die back. In the following year, several buds below the dieback produce shoots, resulting in a brushy appearance. If deficiencies are moderate, brown, gummy areas develop in the green fruit, and gum appears on the surface of the fruit of some cultivars, such as Peerless. These fruit often drop in May or June and can result in loss of the entire crop. With some cultivars, such as Butte, boron deficiency results in a lack of fruit set and an increase in vigor.

Iron deficiencies generally result in leaf chlorosis early in the growing season.

CLYDE L. ELMORE

The almond leaf on the left is healthy; the others have been damaged by paraquat drift.

CLYDE L. ELMORE

Leaf symptoms produced by dichlobenil on these prune leaves are similar to those that occur on almond leaves. The leaf tip chlorosis may spread around the entire leaf margin and cause interveinal chlorosis, depending on herbicide dosage.

Iron. Iron chlorosis is occasionally found in the Paso Robles area and in a few scattered Central Valley locations. Chlorotic leaves are produced early in the season and a fine netting of green veins may develop. Chlorosis may disappear as the season progresses. In severe cases, chlorotic leaves may be uniformly yellow, develop scorch, and drop. Iron chlorosis symptoms are associated with calcareous soils or heavy, poorly drained soils.

Herbicide Symptoms

Tree injury is a risk if herbicides are misused and may also occur under some conditions despite proper herbicide use. Dosage rate, organic matter content, rainfall or irrigation, careless application, soil texture, and depth of soil all bear on potential tree injury by herbicides. Tree injury can range from severe damage to slight effects on growth and leaf color that do not affect yield or quality.

Most crops generally display specific symptoms from a particular herbicide. Familiarity with these symptoms and with the history of herbicide use in and around the field can help you diagnose herbicide injury.

Oxyfluorfen (Goal). Oxyfluorfen can cause damage if warm weather occurs in late winter while the soil is wet. Under these conditions, the chemical can evaporate from the soil surface and cause burning at leaf edges. If the spray drift contacts new foliage, brown spots develop on the leaves.

Paraquat. Paraquat drift damage on almond leaves appears as small, round, yellow spots that may turn brown but do not fall out, as they do in shot hole. Also, no small spot, or fruiting structure, develops in the center of the lesion as with a shot hole infection. If a heavy concentration is sprayed on leaves, it will burn the foliage, but new growth is normal. Tree trunks of newly planted trees can show brown spotting if the bark is still green when sprayed.

Dichlobenil (Casoron, Norosac). Leaf tips of mature leaves damaged by dichlobenil become chlorotic and burnt. Depending on the herbicide dosage, this chlorosis may spread around the entire leaf margin. Interveinal chlorosis and small burned areas may also develop. Symptoms usually appear in early to mid summer and become progressively more severe through the season.

ART H LANGE

Dalapon injury results in marginal necrosis of mature leaves.

Almond trees accidentally sprayed with glyphosate in the fall produce small, narrow, puckered leaves the following spring.

Simazine symptoms on almond foliage include marginal or interveinal chlorosis.

Dalapon. Mature leaves injured by dalapon develop marginal necrosis that gradually moves into the leaf blade. Blotches of dead tissue develop, similar in appearance to sodium or chloride burn injury, and if the leaves are still developing the leaf cups upward.

Glyphosate (Roundup). Two different types of damage symptoms may be caused by glyphosate depending on the time of year it is applied. If glyphosate is applied in the fall and gets on the trees, symptoms appear the following spring on new growth. Affected trees are slow to leaf out in spring and produce many small, puckered, almost needlelike leaves. Tree growth is slow throughout the season until the effect of the herbicide wears off. When glyphosate is applied in the spring and gets on the trees, young terminal leaves on the tree may turn yellow. High leaf concentration of the herbicide will cause these leaves to die. New growth that follows may appear yellow or blotchy yellow, or it may be completely normal.

Simazine (Princep). Simazine symptoms are frequently seen in almonds, especially in trees of the Mission and Merced cultivars grown in sandy soils that are highly alkaline. A slight overdose, sometimes due to failure to turn off the sprayer on turns, can give mature leaves the characteristic ivory yellow marginal or interveinal chlorosis. Chlorosis starts on the leaf margins and moves toward the center, leaving the midvein and some of the lateral leaf veins green. High concentrations may also cause marginal burn. Symptoms usually appear in the early to mid summer and become more severe during periods of high temperatures.

Norflurazon (Solicam). Injury by norflurazon causes veinal chlorosis of leaves and a pinkish, light purple to completely white color on new shoot growth.

2,4-D or MCPA. The phenoxy herbicides usually cause young limbs and leaf petioles to grow in a twisted manner. Symptoms are most frequently observed in early to mid summer. High concentrations of these chemicals may scorch leaves. New growth is also affected, but the symptoms gradually disappear.

Norflurazon injury causes veinal chlorosis in the leaves.

Frost Damage

In midwinter, dormant almond flower buds are relatively insensitive to frost damage down to 20°F (–7°C). Sensitivity to frost injury increases gradually as the buds come out of dormancy. When the tree blooms, sensitivity increases tremendously. From midbloom until fruit are the size of a pea or larger, the crop can be significantly damaged by temperatures only slightly below freezing. Spring frost can kill flowers and young fruit.

The ability of almond flowers and fruit to withstand freezing temperatures may be related to the presence and density of certain species of bacteria commonly found on plant surfaces. These bacteria are believed to serve as nuclei around which ice crystals form. If the bacteria are absent or inactivated, almond blossoms and fruit can resist ice formation down to temperatures as low as 24°F (–4.5°C). The most common ice-nucleating bacterium found on almonds is *Pseudomonas syringae*. This is the same bacterium responsible for causing bacterial blast, a disease associated with freezing temperatures, and bacterial canker.

Wind Injury

Proper pruning and training of young trees helps prevent wind damage. Insect pests and disease can render a tree susceptible to wind damage. Scaffold crotches may be weakened by wood-boring insects, and strong winds will cause the scaffolds to break. Wood-rotting fungi weaken older trees by causing the structural wood to rot. These trees may appear to be healthy but can be blown over by strong winds.

Sunburn Damage

The bark of young tree trunks and exposed limbs on older trees may be sensitive to sunburn. Sunburned bark is cracked, somewhat sunken, and susceptible to attack by wood borers and wood-rotting fungi. Use tree wrap or paint trunks of young trees with whitewash or a white interior water-based paint to protect them against sunburn.

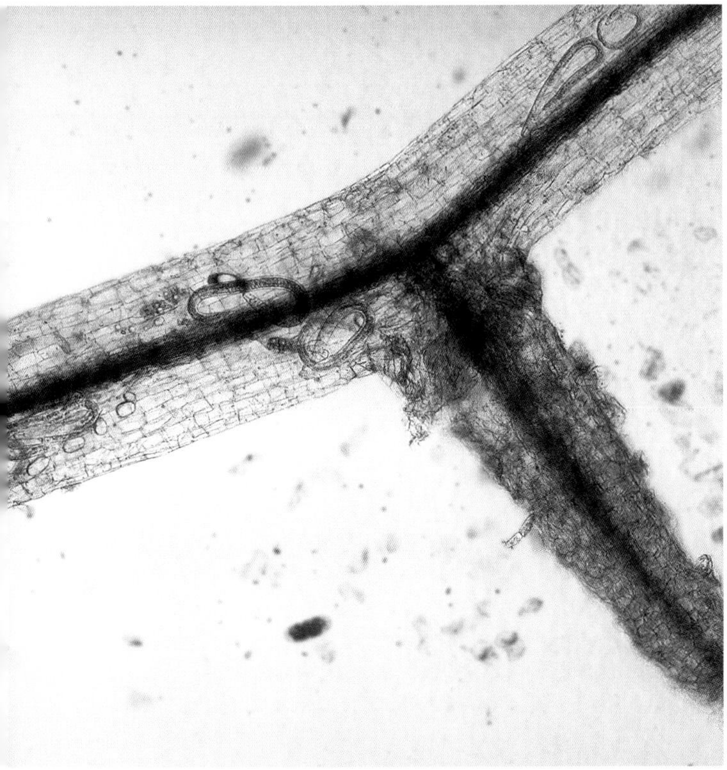
Root lesion nematodes feed inside the roots of host plants.

Nematodes

The nematodes that occur in orchards are tiny unsegmented roundworms that live in the soil or inside plant roots. Some feed on or in the roots of orchard trees, interfering with their uptake of water and nutrients, diverting nutrients that would otherwise aid growth or fruit production, transmitting viruses, or making trees susceptible to certain diseases. The life cycles and feeding behaviors of nematodes depend on the species; damage symptoms are similar for all harmful nematodes, except where other pathogens are involved.

Description and Damage

Plant parasitic nematodes that damage almond trees include root knot nematodes, root lesion nematode, ring nematode, and dagger nematode. They use specialized mouthparts to pierce the cells of the host root and withdraw the cellular contents. Nematode feeding reduces the ability of roots to take up water and nutrients, causing nutrient deficiency and water stress symptoms in affected trees. Nematode feeding also has what is called a "sink effect": it causes photosynthate produced in the foliage to be diverted to the roots. The overall result is a reduction in nut size and number. Aboveground symptoms of nematode injury include lack of vigor, stunted growth, twig dieback (most noticeable on unpruned trees), slight yellowing of foliage, and reduced yield. All of these symptoms can have a number of causes and are not by themselves diagnostic of nematode injury. Laboratory analysis of soil samples is necessary to diagnose a nematode problem.

Damage occurs more frequently to trees growing on sandy soils and is most severe when trees are young or when their root systems are restricted by conditions such as shallow hardpans or stratified soil layers. When high nematode populations develop on the roots of young, nonbearing trees, those trees may never become economically productive.

Root Knot Nematodes
Meloidogyne spp.

At least four root knot nematode species occur in California almond orchards: the northern root knot nematode (*Meloidogyne hapla*), the southern or cotton root knot nematode (*M. incognita*), the Javanese root knot nematode (*M. javanica*), and the peanut root knot nematode (*M. arenaria*). *Meloidogyne incognita* is most often associated with damage to almond; *M. hapla* must reach relatively high population levels to cause damage. The life cycles of these four species are similar and they are managed with the same techniques.

However, the resistance of some rootstocks may vary depending on the species of root knot nematode.

The root knot nematode spends most of its life cycle inside the roots of the host plant (Fig. 49). Females produce eggs in small gelatinous masses on the root surface. The first stage of the juvenile nematode develops inside the egg. After the first molt, second-stage juveniles hatch from the eggs and move through the soil to invade new roots.

Roots are invaded just behind the root tip. Within several weeks of infection, new adult females become established at feeding sites in or near the conductive tissue. Some of the root cells at the feeding site enlarge greatly to form giant cells. Other cells proliferate to form swellings or galls. The swollen adult female remains at the feeding site and produces eggs. An individual may produce as many as 2,000 eggs when conditions are favorable. Adult males remain wormlike in shape after their final molt and leave the root without feeding.

Populations may build up quickly on susceptible rootstocks, and they can occur wherever roots develop because the second-stage juveniles are able to move in any direction through moist soil to find new root sites. In many orchards grown on the susceptible Lovell rootstock, populations reach a peak after 3 or 4 years and then decline, so the trees that survive beyond 6 or 7 years have low populations and little root galling. On the other hand, root knot nematodes do remain a problem in some older orchards on Lovell rootstock. Root knot nematodes are most prevalent and damaging on coarse-textured soils (sand, loamy sand, and sandy loam). They usually do not cause severe problems in finer-textured soils.

Feeding by root knot nematodes within the roots interferes with the roots' uptake of water and nutrients. Young trees can be severely stunted when planted in sites that are already infested with root knot nematodes, even when they are grown on nematode-resistant rootstock. Root knot nematodes can invade resistant Nemaguard rootstock and feed, but they cannot reproduce there. Trees planted in heavily infested soil may grow very poorly the first year or two until the nematodes' inability to reproduce on the resistant rootstock results in a gradual reduction of the population.

Water stress increases the severity of root knot nematode damage. If trees show symptoms that suggest root knot nematode damage, examine the roots for galling, which tends to be concentrated on young feeder roots. Have soil or root samples analyzed for nematodes to confirm your diagnosis.

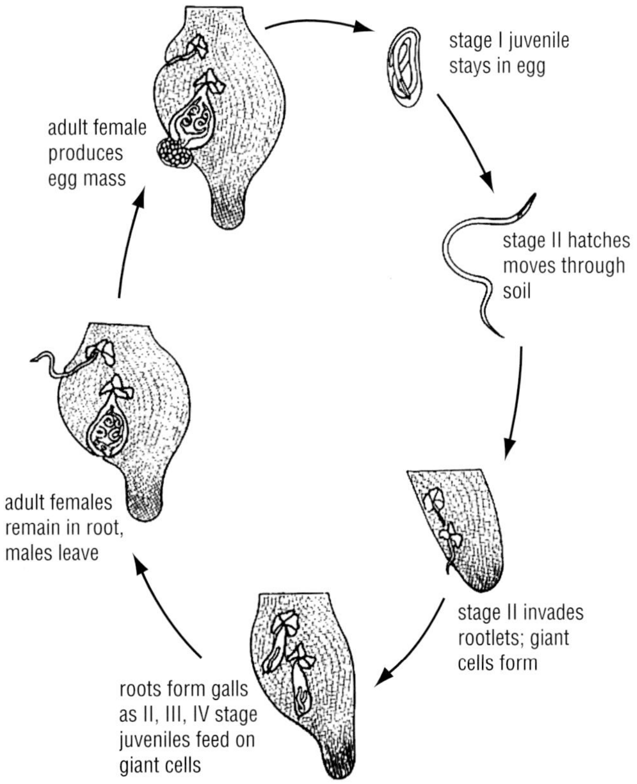

Figure 49. Root knot nematodes spend most of their life cycle in galls on roots. Second-stage juveniles hatch from eggs and invade new sites, usually near root tips. Nematodes induce some root cells to grow into giant cells, where the nematodes feed. As feeding continues, root cells proliferate to form galls around the infection sites.

Root Lesion Nematode

Pratylenchus vulnus

Although several root lesion nematode species occur in California orchard soils, *Pratylenchus vulnus* is the only species known to damage almond trees. Other species of root lesion nematode feed on the roots of weeds and cover crops and may reach high populations in orchard soils, but they do not usually damage almonds unless they occur in abundance within the roots of the trees. It is important to have the root lesion nematode species identified when you have your orchard soil analyzed for nematodes.

Root lesion nematodes feed on the surface of host roots and also penetrate roots to feed inside. Females lay eggs inside roots and in the soil. After hatching, all developmental stages may feed on roots or move through the soil to find other host roots. If roots become unsuitable, those feeding inside the roots will leave to search out new host roots. *Pratylenchus vulnus* feeds on a number of woody perennial hosts, including walnut, fig, plum, prune, rose, cherry, grape, and peach.

Feeding by root lesion nematodes restricts the flow of nutrients within larger roots and may kill smaller roots. Feeder roots are most susceptible, and a lack of feeder roots and major structural roots is one symptom of root lesion nematode infestation. In some cases, dark lesions may form on the surface of infected roots, but this symptom is not a reliable diagnostic characteristic, especially with the rootstocks used for almond. In any case, you should use soil analysis to confirm the presence of root lesion nematodes.

Damage from root lesion nematodes can be severe when trees are replanted in infested sites that have not been treated to reduce nematode populations. Stunting and foliage discoloration, due in part to nutrient deficiencies caused by the nematode infestation, often occur on trees within a certain area of the orchard. Root lesion nematode damage is increased by conditions that restrict root growth, such as shallow hardpans, sandy subsurface soil, herbicide injury, and poor land preparation.

Ring Nematode
Mesocriconema xenoplax

Ring nematode, which has been known in the past by the scientific names *Criconemella xenoplax*, *Macroposthonia xenoplax*, and *Criconemoides xenoplax*, does not penetrate roots as do root knot and root lesion nematodes. It remains on the outside of the root and withdraws nutrients from root cells with its piercing mouthparts. Ring nematode is characterized by the ringlike appearance of the outer body layers, as seen under a microscope.

Feeding by ring nematode can greatly reduce the number of small feeder roots and result in poor tree growth. It is considered an especially important nematode because high populations predispose trees to bacterial canker. If trees are replanted in infested sites that have not been treated to reduce nematode levels, they can be killed within the first year. Ring nematode builds up on the roots of orchard trees and vines growing on coarse-textured soil or clay soil; it usually does not cause problems on finer sandy loam soils.

Dagger Nematode
Xiphinema americanum

Dagger nematode is a relatively large nematode that, like ring nematode, is an external parasite of roots. It is of concern because it transmits strains of tomato ringspot virus. Dagger nematode occurs in a wide variety of soil types in all almond growing areas; it is most common in northern California. Dagger nematode does not live very deep in the soil, so it ought to be more susceptible to control with nematicides; however, some nematicides are effective while others are not.

Management Guidelines

You can minimize the harmful effects of nematodes by selecting planting sites that are free of harmful nematode populations; planting resistant rootstocks; using procedures such as crop rotation, fallowing, and soil fumigation to reduce nematode populations before planting; following careful sanitation practices to avoid spreading harmful nematodes from infested sites; and following practices after planting, such as proper fertilization, that help reduce nematode buildup and increase the ability of orchard trees to withstand nematode damage.

Root knot nematodes may cause severe galling of tree roots.

Soil Sampling. Have your orchard soil analyzed for nematodes by a professional laboratory before you select a management strategy. Soil analysis will tell you which nematodes are present and give you an estimate of their numbers. This information will help you decide if a site is suitable for planting an orchard, if preplant treatments such as fumigation or long-term rotation are needed to reduce nematodes before planting, and if you should use a nematode-resistant rootstock. Preplant soil analysis is always a good idea, but it is especially important for sites where high nematode populations are most likely to occur: where tree crops have been grown, where broadleaf annual crops have recently been grown, where soil is coarse-textured or has a subsurface layer, or where bacterial canker is prevalent. Use soil analysis in established orchards to confirm a diagnosis of nematode injury.

Check with your UCCE farm advisor or local pest control professionals to find a commercial laboratory that will analyze your soil for nematodes. Some laboratories will collect the samples for you, or you may collect them yourself. If you take your own samples, be sure to make arrangements with the laboratory ahead of time to make sure they will be prepared to analyze the samples promptly and to see whether they have specific recommendations about how you should take the samples and package them. It may take about 2 weeks to get the results of a soil analysis. Be sure to keep the laboratory reports as a part of your orchard's permanent history.

Before taking samples, draw a map of the orchard or field to be sampled and divide the site into sampling blocks of 5 acres (2 ha) or less that adequately represent differences in soil type, drainage patterns, cropping history, and crop injury. Make sure each block is an area that can easily be managed as a unit. Take a separate sample from each block. In a field that is not yet planted, collect samples from

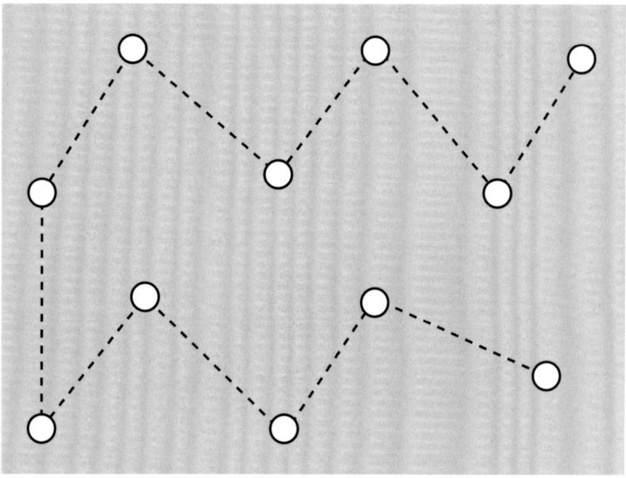

Figure 50. Recommended pattern for collecting soil samples from a sampling block in a fallow field.

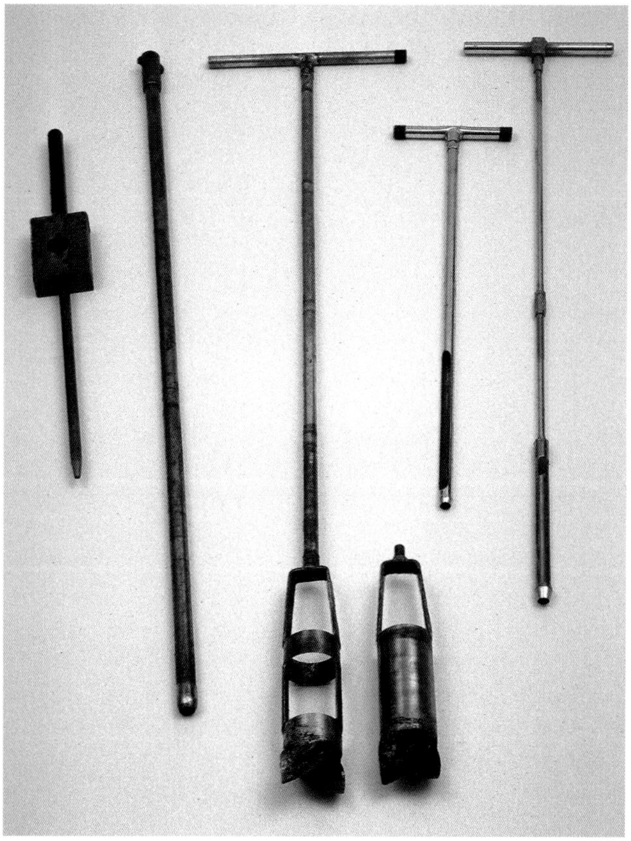

Figure 51. Tools for taking soil samples. The Veihmeyer tube (left) has a slotted hammer for driving the tube into the soil and removing it. Soil augers (center) have a variety of bits for different soil samples. Oakfield soil tubes (right) are usually easiest to use, especially for samples down to 2 feet (60 cm). These tools can be used for monitoring soil moisture and sampling for nematodes.

throughout each block, following the pattern illustrated in Figure 50. In an established orchard, take separate samples from the soil around trees that show symptoms and from around nearby healthy trees for comparison. Sample soil that is frequently wetted at the edge of the canopy and include feeder roots when possible.

In fallow fields or established orchards, you can take soil samples for nematode analysis at any time of year. It is best to sample soil that is moist, preferably within a week after rain or an irrigation. If you are planning to prepare a field that has an existing crop, it is best to sample the field while that crop is still present. Be sure to sample below the root zone of that crop, where nematodes from a previous woody-rooted crop may be concentrated. You can use a shovel or various types of soil tubes or augers to collect soil samples (Fig. 51). For each sample, collect subsamples at various depths from 6 inches to 3 feet (15–90 cm). Take 15 to 20 subsamples from each block or 5 subsamples from around individual trees, mix the samples thoroughly in a clean bucket, and take about 1 quart (about 1 l) of this mix as the sample for the block or location in the orchard that is suspected of having nematode damage.

Put the soil sample into a durable plastic bag or other moistureproof container, seal the container, label it, and place it in the shade or in a cooler while you finish taking samples. Attach a label to the outside of each sample container: include your name, address, and phone number; the crop present; the crop you intend to grow; the location, soil characteristics, and cropping history of the orchard site; and any injury symptoms that were present. Use an ice chest or cardboard box insulated with newspapers to keep the samples cool (50° to 60°F, or 10° to 15.5°C) until you deliver them to the laboratory. Do not freeze the samples. Deliver them to the lab immediately.

If you want the samples analyzed for ring nematodes, request that sugar flotation or centrifugation be used for nematode extraction. Root knot and root lesion nematodes are also detected with these methods. Request that root lesion nematodes be identified to species if they are found; presence of *Pratylenchus vulnus* indicates a need for management action.

Nematode damage thresholds are not well established for almond. As a general rule, if soil analysis identifies root knot nematodes, ring nematode, or damaging species of root lesion nematodes on a site where tree crops have been grown, soil fumigation or a long-term rotation to nonhost crops such as grains is recommended before you plant the site to almonds. If root knot nematodes are present, use resistant rootstocks.

Field Selection and Preparation—The "Replant Problem"

Whenever possible, avoid planting new orchards on land that has previously been planted to woody crops. Almond trees planted on an old orchard or vineyard site are likely to suffer

Table 19. Suggested Schedule for Preparing Orchard Site Following a Tree or Vine Crop.

Season	Action
summer-fall	Remove trees or vines, destroy residues[1], deep-cultivate to remove residual roots and break up cultivation pans or soil layers. Sample for nematodes.
winter-spring	Fallow or plant a winter crop of small grains after soil temperature is below 64°F (18°C).
spring-summer	Level or grade site for irrigation and proper drainage, cultivate, and do other operations necessary to prepare site for planting. Allow soil to dry.
late summer-early fall	Rip the soil. Apply 1 inch (2.5 cm) of water if necessary to break up surface clods. Fumigate in September or October.
winter-spring	Observe waiting period required by fumigant label. Plant young trees on resistant rootstock if root knot nematodes are present.

1. Treatments that kill the root systems of the previous perennial crop shorten the fallow period needed. Treat the tree or vine stumps as soon as possible after harvest and before November 1.

from poor vigor and are often accompanied by symptoms of nutrient deficiency. This phenomenon, called the "replant problem," is caused by nonpathogenic microbes that became established on the old root system as well as by pathogens (nematodes, fungi, and bacteria) and other factors. Nematode populations build up in the soil as deep as the tree roots penetrate. For several years after the old trees are removed, the remaining roots act as food sources for nematodes and provide them some protection from fumigation. After an herbaceous crop, harmful nematode populations will be lower, will not be as deep in the soil, and will be less protected because the roots of these crops decay more quickly.

When you plant an almond orchard on a site that has grown trees or vines before, proper preparation is essential to the successful establishment of the orchard. Table 19 gives an example of how to prepare an orchard site for planting after a tree or vine crop.

Soil Fumigation

Fumigation of the soil reduces nematode populations but does not eradicate them. Following fumigation, nematode populations will increase over time and eventually reach harmful levels. Proper fumigation keeps nematodes below damaging levels for as long as 6 years, long enough for trees to develop extensive root systems that are better able to tolerate nematode damage. If soil sampling indicates that root knot nematodes are the only harmful species present and you are planting on Nemaguard rootstock, you may want to fumigate just the tree rows or, if you are replanting individual trees, just the planting sites. If lesion or ring nematodes are present, fumigate the entire orchard.

The effectiveness of soil fumigation is influenced by soil preparation, soil moisture, soil texture, soil temperature, and the quantity of chemical applied. How soil moisture, texture and temperature affect chemical rates is detailed in *UC Pest Management Guidelines: Almond.* As a general rule, the best time of year to fumigate is from September until mid-November. This is when soil moisture and temperature are optimal for effective fumigation. The actual time to fumigate varies from one region to the next, and more flexibility in timing is possible with sandy soils. If soil conditions are not optimal, nematode control will be less effective and populations will return to damaging levels more quickly. This can reduce the period of time new trees are protected by anywhere from 6 months to 6 years.

Soil Preparation. Table 19 includes a suggested program for preparing an orchard site before fumigation. Thoroughly work the soil to remove clods, which the fumigant cannot penetrate. Whether you are fumigating an entire orchard site or individual tree sites for replanting, remove the old tree stumps and as many of the old roots as possible that are larger than about ½ inch (about 1 cm) in diameter; nematodes within roots can escape the fumigant's effect. Remove stumps and roots before you fumigate individual planting sites when you replace trees in an established orchard. To improve the effectiveness of fumigation of individual tree sites, backhoe to a depth of 5 feet (1.5 m) and cave in the sides of the hole. Before you fumigate, make sure to add extra soil when backfilling to compensate for the soil settling that will occur after you plant.

Soil Moisture. Correct soil moisture is essential for successful fumigation. If the soil is too wet, the fumigant will not disperse properly in the soil. If soil is too dry, gaseous fumigant will disperse too quickly and liquid fumigant will be tightly adsorbed by soil particles, reducing its effectiveness. Determine the soil moisture by taking soil samples from throughout the orchard, making sure to sample all areas that have different textures. You may want to follow a sampling plan such as that illustrated in Figure 50. At each sampling location, take 1-pint (0.5-l) soil samples from depths of 1, 2, 3, 4, and 5 feet (30, 60, 90, 120, and 150 cm). Weigh each sample, dry the samples at 221°F (105°C) for 24 hours, and then weigh again. Be sure to subtract the weight of the container from each sample weight. Calculate the percentage of soil moisture by subtracting the dry sample weight from the wet sample weight, dividing the result by the dry sample weight, and multiplying by 100.

Variations in soil moisture and texture in the soil profile can prevent adequate penetration of the fumigant. For example, if a layer of sandy loam with moisture in excess of 12% is at the 3-foot (90-cm) depth beneath a layer of loam with 12% moisture, the fumigant will not penetrate beyond 3 feet unless the sandy loam layer is permitted to dry out before treatment.

Soil Texture. Soil texture is a way of describing the proportion of different-sized particles that make up a given soil.

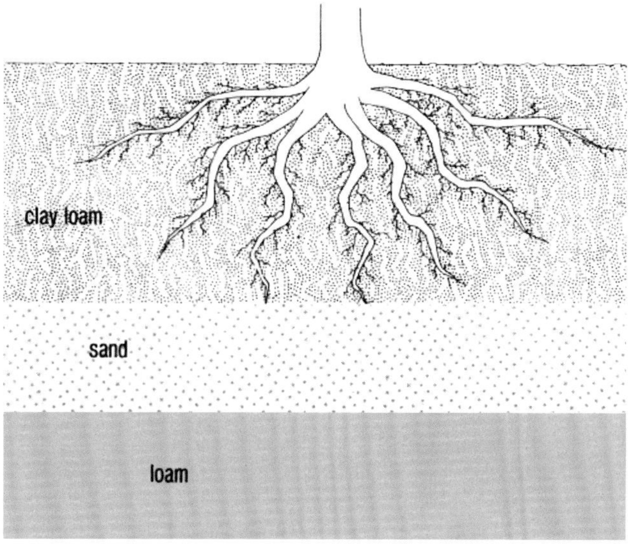

Figure 52. Abrupt boundaries between different soil textures act like pans, even though they may not be dense.

Coarse-textured soils have higher proportions of sand and larger pores, through which fumigant can quickly disperse. Fine-textured soils have high proportions of clay and silt, with much smaller pores, and are able to hold a greater proportion of water. Fumigants move more slowly and are more readily diluted in fine-textured soils. The finer the soil texture, the higher the application rate needed for fumigant to penetrate to the 5-foot (1.5-m) depth.

Most soils are not uniform, but vary in texture throughout the profile. Determine the soil textures present in your soil profile when you are taking samples to determine soil moisture. If you find abrupt boundaries (Fig. 52), they may prevent adequate penetration of the fumigant. Use backhoeing, ripping, or deep plowing to eliminate such boundaries, hard pans, or clay pans before you fumigate.

Soil Temperature. Check the soil temperature 12 inches (30 cm) deep with a thermometer to be sure it is within the ranges specified in *UC IPM Pest Management Guidelines: Almond.* If the soil is too cold, the fumigant will not move properly through the soil and will not penetrate soil aggregates and root fragments. If the soil is too warm, the fumigant will degrade and volatilize too quickly.

Application. Both gas and liquid nematicides are available for preplant fumigation of almonds. All are restricted-use materials, so you need a permit to use them. Liquid formulations may require preirrigation before fumigation. The materials are applied with irrigation water to move them into the soil profile. Gas is injected into the soil and the treated area is immediately covered with a plastic tarp to keep the gas from dissipating. Fumigants are toxic to plant roots; be sure to observe the waiting periods specified on the fumigant's label before planting. Information on the most recent application procedures and restrictions for currently available fumigants can be obtained from county agricultural commissioners' offices.

Postplant Nematicides

The proper application of a postplant nematicide can effectively reduce populations of ring and root lesion nematodes. Although costly, the expense of the application may be justified if the nematodes are causing severe damage. Successful treatment requires the uniform delivery of an adequate amount of nematicide throughout the root zone. The best way to achieve this is to deliver the nematicide via low-volume irrigation in spring or fall. The best results are obtained when you apply the nematicide to preirrigated and drained soil, using a 4-hour irrigation with 3½ hours of nematicide injection. More specific information on materials, rates, and restrictions are given in *UC IPM Pest Management Guidelines: Almond,* listed under "Pesticide Application and Safety" in the suggested reading.

Ring Nematode. Postplant treatment can substantially improve the trees' vigor if they are suffering from ring nematodes. Where ring nematodes are associated with bacterial canker, nematicide treatment in mid-October will greatly reduce or halt bacterial canker the following spring. This is discussed in greater detail in the section on bacterial canker in the chapter "Diseases."

Root Lesion Nematode. Nematicide treatments can increase nut yields if root lesion nematodes are limiting the trees' productivity. However, nematicides do not kill nematodes that are inside the tree roots, so any protection will be short-lived.

Table 20. Resistance of Almond Rootstocks to Nematode Pests.

Rootstock	Root knot nematodes	Root lesion nematodes	Ring nematode
almond	susceptible	susceptible	susceptible
Nemaguard peach	resistant	susceptible	susceptible
Nemaguard peach–almond hybrid	resistant	susceptible	susceptible
Lovell peach	susceptible	susceptible	somewhat tolerant
Marianna 2624 plum	resistant	somewhat tolerant	susceptible

Rootstock Selection

Resistance or tolerance to nematodes is a consideration in selecting rootstocks. Some rootstocks are resistant to root knot nematodes, and a few have some resistance to root lesion or ring nematodes (Table 20). If you are planting in coarse-textured soil or in locations where soil analysis has shown root knot nematodes to be present, choose rootstocks that are resistant to these nematodes.

Be aware that a resistant rootstock will not protect trees that are planted in soil that is heavily infested with root knot nematodes. These nematodes can feed on resistant roots but cannot reproduce. By using these rootstocks, you can prevent low populations from building to damaging levels, but you cannot prevent high populations from severely damaging newly planted trees. Also, cover crops that favor the buildup of root knot nematodes may allow the nematodes to overwhelm the rootstock's resistance. Where ring nematodes are present and bacterial canker is a potential problem, you may want to use Lovell peach for compatible scions, as long as you do not expect root knot nematodes to be a problem.

Fallow and Crop Rotation

Fallowing a site before planting or replanting an orchard can reduce soil-dwelling nematode population densities. However, it may not control the nematodes remaining in live roots from the previous orchard. For a fallow period to be effective, the roots remaining from the previous orchard must somehow be killed. Weeds must be controlled by cultivation or herbicide application, because many weeds are good hosts for harmful nematode species.

Grains are not hosts for the root lesion nematode, *Pratylenchus vulnus*. Some wheat varieties, such as NK 916 and Stacey, may not grow optimally under California conditions, but they are reported to reduce ring nematode populations. Sorghum–sudangrass hybrids such as Germain's SS-222 are also antagonistic to ring nematode, but they are hosts for root knot nematode species and may increase dagger nematode populations if planted too many years in succession. Piper sudangrass is a host for northern root knot nematode, *Meloidogyne hapla*.

If a grain rotation crop is a host for root knot nematode species that attack almonds, you can prevent root knot nematodes from reproducing before the crop is harvested or plowed under in the spring by planting it in the fall after soil temperatures drop below 64°F (18°C). By following this planting schedule when you use small grain rotations, you can reduce root knot nematode populations. It may take as many as four seasons of small grains to reduce nematode population densities the same amount as a soil fumigation. Preplant rotations of sudangrass or a sorghum–sudangrass hybrid in summer and barley or Cahaba white vetch in winter can be effective tools for reducing nematode populations only if the remaining roots of the old orchard have been destroyed.

Sanitation

After taking steps to prepare the orchard site and reduce nematode populations, follow good sanitation practices to avoid introducing nematodes into the orchard. Nematodes can be introduced in roots of infested planting stock, in infested soil on rootstocks or field equipment, or in runoff water from infested fields or orchards. Always use certified nematode-free rootstock. This greatly reduces the likelihood that nematodes will be introduced on the roots of the young trees. You can minimize the spread of nematodes by working noninfested orchards before you move equipment into infested orchards, cleaning roots and soil off equipment after you work infested soil, and avoiding the use of runoff water from infested sites to irrigate noninfested sites.

Cover Crops

A number of the cover crops that may be grown in orchards are hosts for nematode pests of almonds (see Table 21 at the end of the next chapter). It is best to avoid cover crops that are hosts for root knot nematode and also to avoid intercropping with broadleaf crops such as cotton, lettuce, or alfalfa during the first 2 years of establishing an orchard on coarse-textured soils. Root knot nematodes may build up on these crops to levels high enough to damage trees on resistant rootstocks. Where nematodes are a concern, either choose cover crops that are not hosts for the species that damage almonds or choose nematode-resistant varieties of cover crops when they are available.

Cover crops that are broadly useful for managing nematode pests in California orchards are barley, Blando brome, and Cahaba white vetch. Some cover crops will help reduce populations of some nematode species but increase the populations of other species. For maximum effectiveness, rotate the cover crops you use. More information about cover crops can be found in *Covercrops for California Agriculture*, listed in the suggested reading.

Vegetation Management

The management of orchard floor vegetation in almond orchards has several goals:

- controlling weed species
- achieving a nearly bare orchard floor just prior to harvest
- improving infiltration of irrigation water
- encouraging a shift to less-invasive plant species
- reducing other pest problems
- maintaining orchard accessibility to farm equipment during the winter and spring rainy season
- improving infiltration of precipitation to reduce runoff during the rainy season

A number of problems can develop if the orchard floor is not managed properly or if weed populations build up in the orchard. Weeds can compete with orchard trees for water and nutrients; competition is greatest during an orchard's first 3 or 4 years of growth. Competition increases as irrigation water is applied to smaller areas of the orchard floor, becoming progressively stronger for basin flood, furrow, sprinkler, micro-sprinkler, and drip irrigation respectively. Tall weeds compete with trees for sunlight in newly planted orchards, and young orchards may take longer to come into production if badly infested with weeds. Dense weed growth around trees also can interfere with harvest operations. If dense weed growth raises humidity in the orchard, diseases like scab and Alternaria leaf spot may increase in severity. Weeds in row middles can interfere with harvest and contaminate harvested nuts. Weeds and certain cover crops may serve as hosts or food sources for pests such as ants, mites, root knot nematodes, and rodents. Certain weeds or cover crops such as dandelion and clover may attract gophers, which can injure trees directly or cause problems at harvest with their mound building. Weed growth around the bases of trees can provide shelter for voles (meadow mice), increase the likelihood of crown rot, and create a fire hazard when dry. Weeds such as mustards and fiddleneck that bloom at the same time as almonds compete with trees for pollinators. Weedy orchards and orchards with cover crops are more likely to suffer frost damage because the presence of vegetation can reduce the temperature by a few degrees, as compared with moist, vegetation-free soil. However, mowing weeds or cover crops before bloom reduces competition for pollinators and the potential for frost injury. Also, sprinkler irrigation of a low cover crop reduces the potential for frost injury.

Weed Management

A plant's life cycle affects both its importance as a weed and the management strategies that may be needed for its control. The plants that grow in almond orchards can be grouped as annuals, biennials, or perennials, based on their life cycles. Annuals complete their life cycle of germination, growth, flowering, and seed production within 1 year. Winter annuals germinate in the fall, grow through the winter, flower in late winter and early spring, produce seed in the spring, and die by early summer. Summer annuals germinate in late winter, spring, or early summer and produce seed in summer or fall; they die in fall or early winter. Biennials complete their life cycle in 2 years, usually producing leaves close to the ground during the first year and flowering and producing seed in the second year. Perennials live for 3 or more years, usually dying back in the winter and regrowing from underground vegetative structures. Perennials do not normally produce seed during their first year of growth. Some biennials, such as bristly oxtongue, little mallow (cheeseweed), and sweetclovers, behave as short-lived perennials in the mild climates of most California growing areas.

Winter annuals are the least troublesome weeds because they grow when competition for water usually is not an issue. They may be controlled by mowing, flaming, spring cultivation, or herbicide application. Many winter annuals can serve as desirable cover crops in almond orchards; maintaining resident vegetation or cover crops in the middles improves accessibility of the orchard under wet conditions. Summer annuals tend to cause more serious problems because they compete for the more limited water supply in the summer, may interfere with distribution of irrigation water, and, if uncontrolled, cause problems at harvest. Perennials usually are the hardest to control because they form extensive rhizomes, stolons, tubers, or tap roots that regrow after mowing or cultivation. Often their roots are competing for water and nutrients at the same depth as the trees' roots. Their management requires repeated cultivations and herbicide treatment to destroy underground structures. It is best to deal with established perennials, especially field bindweed, before an orchard is planted.

Management Methods

Orchard floor management fundamentally depends on the irrigation system used. The frequency of weed management activities increases for orchards with sprinkler, micro-sprinkler, or drip systems, where irrigation water is placed in the tree row. This is because weed germination and growth is enhanced by the moisture, and herbicides break down more quickly in moist soil. Weed management activities are needed less often in orchards where trees are on berms and furrow or basin flood irrigation is used, because the soil surface in the tree row remains dry. Several methods are available for managing vegetation on the orchard floor; successful management programs combine a number of these methods. Mowing, herbicide application, cultivation, soil solarization, flame cultivation, mulches, and cover crops can be useful tools to manage weeds. Orchard floor management typically is divided into two physical areas: the tree row and the tree middle. Greater attention is focused on maintaining the tree row free of weeds. Orchard middles may contain vegetation during the winter, spring, and early summer, but vegetation should be eliminated prior to harvest.

Cultivation. Two cultivations perpendicular to each other may be used in newly established orchards, especially those slated for organic production. Cultivation may also be used in the spring to turn under a cover. If you turn the ground cover under and smooth the bare soil before bloom, you will end up with a warmer orchard environment and reduced risk of frost injury. Cultivation may be done in orchard middles during the season, but the surface must be firm prior to harvest. Cultivation can be combined with the application of systemic herbicides to effectively control established infestations of perennial weeds, because cultivation cuts the perennial structures (stolons, rhizomes, or rootstocks), which breaks dormancy of buds and allows the herbicide to kill a larger number of buds on the perennial structures. Cultivation should be timed to dislodge weeds when they are small, preferably under 2 inches (5 cm) tall. For best control of nutsedges, cultivate before they reach the 5-leaf stage to prevent the formation of new tubers. Drawbacks to cultivation include damage to soil structure, which affects water infiltration, and spread of perennials within the orchard if herbicides are not used. Cultivation cuts tree roots in the upper 6 to 8 inches (15–20 cm) of soil, reducing the volume of feeder roots, which could make a critical difference in how well the trees can use water in orchards on shallow soils. Excessive cultivation can lead to increased dust, erosion, and injury to tree roots or trunks.

Mowing. Mowing or flailing can be used for total orchard floor vegetation management or, more commonly, to control vegetation only in the middles between tree rows. Mowers are available that can be used in the tree row to mow weeds around the trees. Orchard middles often are mowed 8 to 13 times each season. Mowing keeps most weedy species in check but allows the soil to remain covered and allows plants to maintain root systems that improve infiltration. Mowing the entire orchard floor may force insects such as lygus bug to move out of the orchard into a crop such as cotton where they cause damage. Mowing alternate row middles on a weekly basis encourages these insects to remain in the orchard on the unmowed vegetation. Mowing can promote a shift to species that tolerate mowing, particularly common purslane, which can remain green and interfere with harvest. These weedy species are more likely to dominate when mowing resident vegetation rather than a planted and managed cover crop. Healthy cover crops are more resistant to invasions of new weeds than resident vegetation is.

Mulches. One alternative to mowing all season involves early-season mowing of a cover crop followed by rolling a ring roller over the cover crop to injure the stems. The injured cover crop can continue to produce seed, but it uses less water and forms a mulch that shades out weeds in the middles. A cover crop that produces a large amount of biomass can also provide a mulch over the tree row, using a specially designed mower for row middles that blows the mown vegetation into the tree rows where the resulting mulch suppresses weed growth. Cereal cover crops, especially forage oats, work particularly well for this "mow and throw" technique because the biomass does not degrade as quickly as does that of broadleaf vegetation. Vetch may be combined with the cereal for added nitrogen. Plant the cover crop in October and mow in late March (the exact timing for best results depends on the location). Be sure to control vegetation in the tree rows by mowing or applying an herbicide before you blow on the mulch. For effective suppression of weed growth, the mulch needs to be several inches thick. Mulch around the bases of trees reduces soil temperatures, slowing the growth of new plantings, and may increase problems with voles or crown diseases. If you use mulching for strip weed control, you may want to have the mulch moved away from the crown by hand and then use hand hoeing as needed for basal weed control. Any mulch used in almond orchards must not exceed the amount that will decompose prior to harvest. Mulches present at harvest make the harvest more difficult.

Soil Solarization. Soil solarization involves covering moist, prepared soil with special plastic and allowing the sun to heat the moist soil to temperatures that are lethal to weed seeds and other plant propagules. The technique's main application in orchards is for preplant weed control. However, a shortcoming of preplant solarization is that viable new weed seeds may be brought to the surface during planting. While some control can be achieved in newly planted orchards, older orchards have too much shade for the technique to be effective. If clear plastic for solarization is placed around tree trunks, trees will be damaged.

Black plastic placed over the soil surface during planting inhibits weed growth and greatly reduces the amount of irrigation water required during the early establishment of new trees. However, this technique does not control weed propagules as well as preplant solarization with clear plastic. Black plastic should be kept away from the trunks of young trees or injury will result.

The susceptibility of common orchard weeds to preplant solarization is given in Figure 53. You can find more information on soil solarization in *Soil Solarization: A Nonpesticidal Method for Controlling Diseases, Nematodes, and Weeds*, listed in the suggested reading.

Flaming. Specially designed propane flamers can be used to control weeds in the tree row for strip weed control and for basal weed control around larger trees. They can also be used for total orchard floor weed control as an alternative to cultivation. Using a flamer, you can achieve successful weed control if trips through the orchard are timed to kill weeds while they are less than 2 inches (5 cm) tall. The correct travel speed can be determined by checking mortality after the flamer moves over weeds. Weeds are being killed if a gentle pressure to their leaves creates a water-soaked appearance, which indicates that cell membranes have been ruptured. Broadleaf weeds are more sensitive to flaming than are grasses, which have their growing point at the ground level, where it is somewhat protected; some grasses such as johnsongrass are very tolerant of flaming. Flaming selects for grasses on the orchard floor. To control larger weeds, you need to move the flamer more slowly through the orchard. A higher-speed treatment injures weeds, retarding their growth and having the same effect as chemical mowing. If you use flaming for strip weed control in young orchards, control weeds before they are taller than 1 inch (2.5 cm) to avoid injuring the trunks or burning trunk guards with the flame. Be sure that leaves and other trash are cleared away from tree trunks; this material may hold a fire that can girdle the tree trunk. Grass weeds may require additional controls such as spot-treatment with herbicides or hoeing.

The advantages of flaming include broad-spectrum weed control, low cost, and lack of chemical residue; it can also be used when the orchard floor is too wet for cultivation. Disadvantages include a lack of residual control, poor control of some grasses, critical timing requirement to ensure adequate control, hazards associated with handling pressurized flammable gas, and the potential for fire. Repeated flaming will lead to perennials and grasses dominating the orchard floor vegetation unless other controls are used. Flaming will not control perennials unless done repeatedly for several seasons.

Herbicides. Herbicides used in almonds are classified as preemergence herbicides if they are applied before weeds emerge, or postemergence herbicides if they are applied after weeds emerge. Herbicides available for use in almond orchards are listed in Figure 53. Maintaining a weed-free strip in the tree row can be accomplished using winter application of a preemergence herbicide. Often, some weeds have already emerged, which requires addition of a nonselective postemergence herbicide in the herbicide mix. Most preemergence herbicides continue to be effective for 2 to 5 months after application. Delaying an application can move the control period into late spring or early summer. Delayed applications also have the advantage of lowering the risk of off-site movement of herbicides by reducing the likelihood that heavy rainstorms will occur after the herbicide is applied. The weed-free strip can also be maintained using postemergence herbicides. Using only postemergence herbicides requires more frequent access to the orchard for applications, often in November/December, February/March, May/June, and July. Quite often the February/March application or the May/June application can be avoided when temperatures remain cool because weed growth is slower.

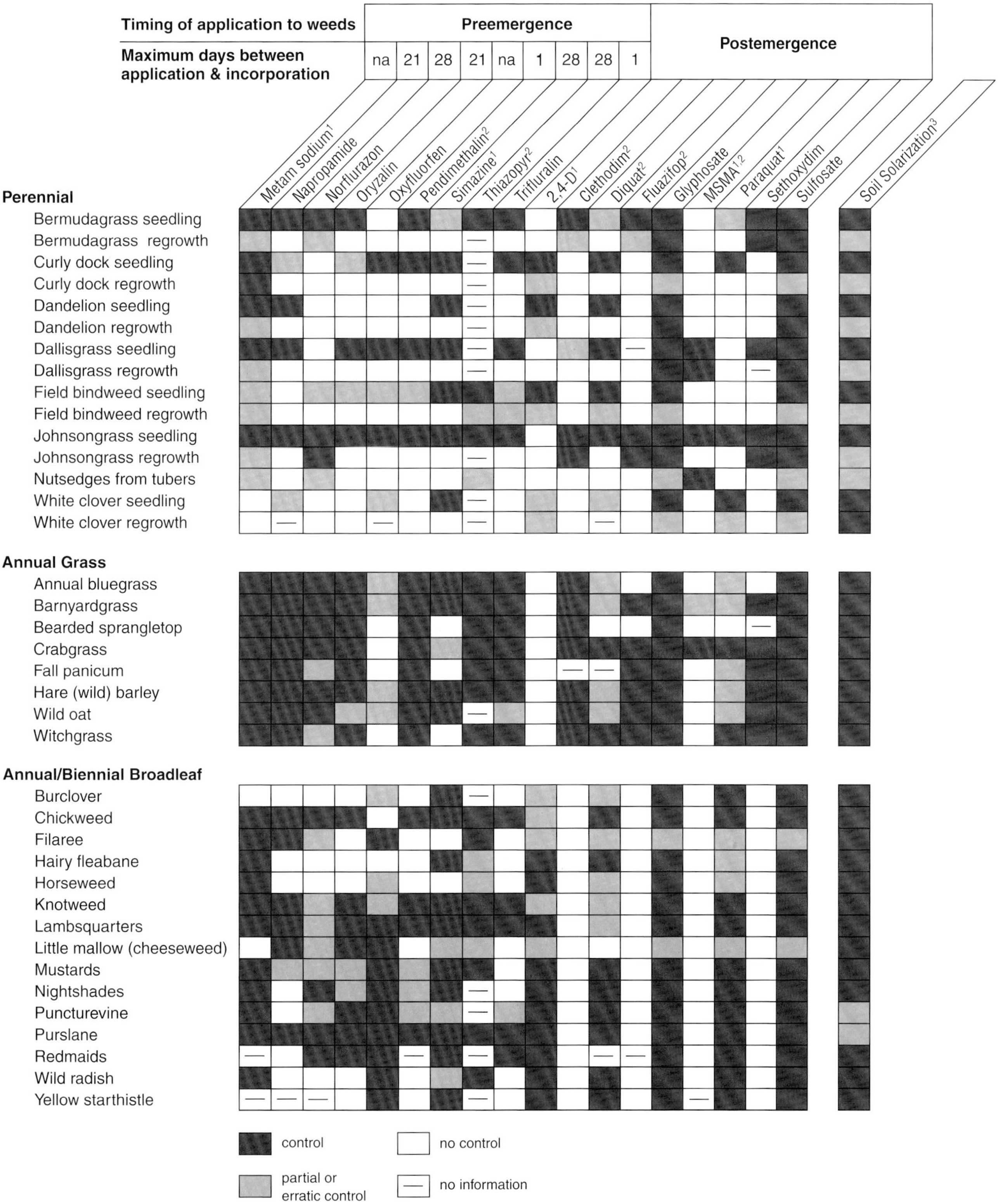

1. Restricted-use pesticide. Permit required for purchase or use.
2. Nonbearing orchards only.
3. Solarization controls perennial structures in upper few inches of soil.

REGISTRATION STATUS OF THESE COMPOUNDS MAY CHANGE. CHECK WITH LOCAL AUTHORITIES BEFORE USING.

Figure 53. Susceptibility of weed species to herbicides available in almonds and to soil solarization.

When using herbicides, choose materials and rates according to the weed species you need to control, your soil type, your irrigation method, and the age of your orchard. Combinations of materials or sequential treatments with different materials are often needed, since no single herbicide available for use in almonds will control all weed species. Figure 53 lists the susceptibility of common weed species to the herbicides available for use in almond orchards. For current registration status of herbicides, check with your UCCE farm advisor or county agricultural commissioner. ***ALWAYS READ AND FOLLOW LABEL DIRECTIONS CAREFULLY.*** Be sure spray equipment is calibrated carefully and functioning properly. Proper calibration and use of spray equipment are discussed in detail in *The Safe and Effective Use of Pesticides*, listed in the suggested reading.

Preemergence Herbicides. Preemergence herbicides kill susceptible plants as they germinate and begin to grow. To be effective, the chemical must be moved to where weed seeds germinate, typically in the surface 2 inches (5 cm) of soil. Movement of herbicide into this surface soil can be accomplished by rain, with a light irrigation (about ½ inch, or about 1 cm), or mechanically. Some preemergence herbicides must be moved into the soil immediately, while others may remain on the soil surface for a short time before incorporation without losing their effectiveness. Label directions specify how quickly a particular herbicide must be incorporated. The allowable amount of time between application and incorporation is listed for preemergence herbicides in Figure 53.

Most preemergence herbicides used in almonds are effective only against germinating weed seeds. Oxyfluorfen is effective on some weed species as a postemergence herbicide when applied to the seedlings' foliage. Preemergence herbicides may remain active in the soil for several weeks up to a year, depending on the chemical, application rate, soil conditions, and amount of rainfall or frequency of irrigation. In areas of the orchard that remain wet, such as around low-volume drip emitters or during periods of prolonged wet weather, herbicide activity dissipates more quickly as a result of leaching and accelerated breakdown caused by microbial activity and hydrolysis. In higher-rainfall locations or where unusually heavy rainfall is expected, you can split applications to prolong control. The use of split applications on coarse-textured soil prolongs the period of activity and reduces the risk of injury, because less herbicide is leached into the active root zone of the crop.

In some cases, you can achieve adequate weed control with lower application rates than those specified on the label. If a weed spectrum in the orchard is particularly sensitive to an herbicide, a lower rate may be used. If you use herbicides for strip or basal weed control around trees planted on berms in areas with annual rainfall of less than 11 inches (28 cm), lower rates may be effective on the berms. The treated soil of the berms remains drier during rains, and if it is above the level of surface irrigation water it has a lower rate of leaching and breakdown. Also, because the berms are drier, fewer weeds are likely to germinate.

Plan to discontinue the use of preemergence herbicides 1 or 2 years before you replant an orchard. This will eliminate the possibility that you will injure the following crop or the replant trees with herbicides, some of which may persist in the soil for a year or more. When replacing trees in the orchard, be sure to fill around the roots of the new tree with soil from 6 inches (15 cm) or deeper to avoid exposing the roots to herbicide residues (Fig. 54).

Postemergence Herbicides. Postemergence herbicides are called foliar-applied herbicides because they are sprayed on the foliage of weeds that have emerged. They are classified as *contact herbicides* if they kill only the plant parts that are sprayed. Contact herbicides are most effective on weed seedlings and young weeds. *Translocated* or *systemic herbicides* are transported via the plant's vascular system from contacted foliage to other parts of the plant, including roots and

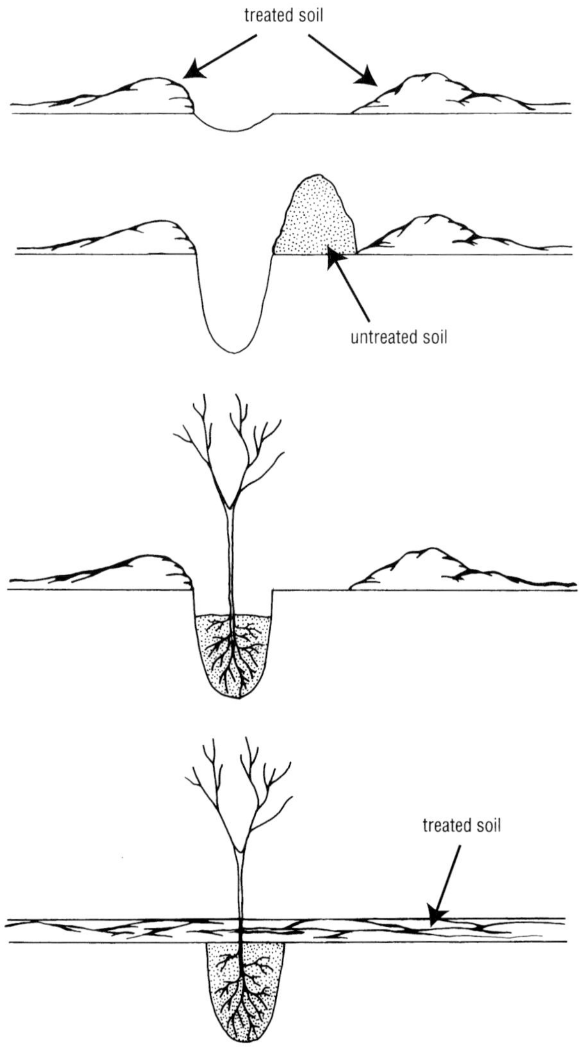

Figure 54. When replanting in soils treated with herbicide, avoid placing treated soil around the roots of the young trees.

rhizomes. They are more effective on actively growing weeds and are the best choice for controlling perennials.

Use postemergence herbicides when monitoring indicates they are needed. Apply them in conjunction with preemergence herbicide if weeds are present when the preemergence treatment is made. Make additional applications later in the season for weeds that are not controlled by preemergence herbicides and for spot-treatment of perennials.

Sprayers are now available that apply postemergence herbicides only where they detect the presence of weed foliage. Often called "smart sprayers," they reduce the amount of herbicide applied by 40 to 80% when weed populations are low or weeds are small. They can be used also for strip weed control or for total orchard floor control.

Management during Orchard Establishment

Begin your weed management program before you plant the orchard. Do a weed survey in the spring to learn or record what species are present before you cultivate. Surveys made in late winter, summer, and fall the year before you prepare the orchard for planting will give you the full spectrum of weed species that are present so you can plan the most cost-effective management strategies.

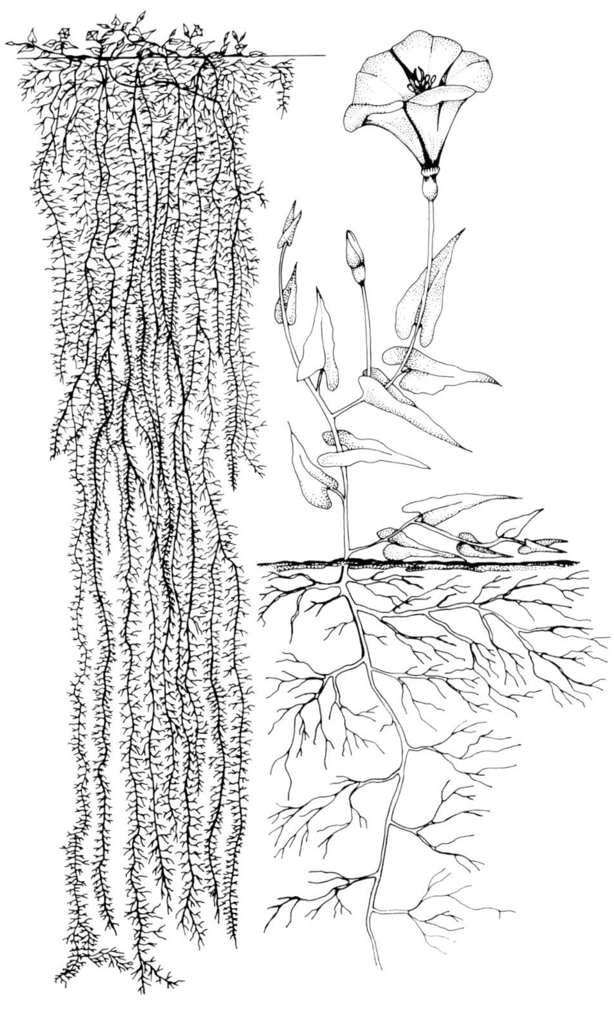

Figure 55. The storied rhizome system of field bindweed. The ability of new plants to form from the deeply penetrating root system makes established infestations of field bindweed extremely difficult to control. (Redrawn from B. F. Klitz, Journal of the American Society of Agronomy, 22:216–234, 1930.)

Perennials are easier to control before trees are planted. Established johnsongrass, bermudagrass, and dallisgrass can be destroyed by repeated cultivations during the summer before planting. This exposes rhizomes and stolons to drying, which will eventually kill the perennial structures that have been established. However, this does not eliminate the reservoir of seeds, which will continue to germinate for a period of years. Preplant solarization kills seeds in the upper few inches of soil, but later cultivations will bring viable seed to the surface. Control emerging seedlings with herbicides or cultivations to prevent them from forming perennial structures. In early fall, treat regrowth of perennials with glyphosate when the plants are beginning to flower, and cultivate again after 10 to 20 days to expose the root systems to further drying. Watch for regrowth the following spring and use glyphosate to spot-treat any plants that appear. If field bindweed is present, the above program will help reduce infestations, but it is virtually impossible to eradicate this weed's deep perennial root system (Fig. 55). Oryzalin and trifluralin applications help control bindweed seedlings and reduce the infestation. Competition from cover crops, especially those containing oat or vetch, also helps reduce bindweed infestations.

You can apply preemergence herbicides before planting in the spring to the entire orchard or to strips 4 to 6 feet (1.2–1.8 m) wide where trees are to be planted. Some preemergence herbicides should not be applied to newly planted orchards, so be sure to follow label restrictions. When planting trees into ground that has been treated with herbicide, set aside at least 6 inches (15 cm) of surface soil and be sure to place deeper soil around the roots of the young trees. This will prevent herbicide injury to the young roots. After the roots are covered, replace the surface layer so that the treated soil remains on top, where weed seeds will try to germinate (see Fig. 54). Some preemergence herbicides can be applied after planting and incorporated with a shallow cultivation or irrigation.

Weed fabrics can be used to prevent weed growth around the tree base. Using weed fabrics instead of tillage adjacent to the tree avoids the potential for damage to the young trunk and roots. Water will move through the fabric but weeds do not emerge through it. Use two rectangles to cover the area on each side of the tree. The rectangles can then be moved away from the tree as needed to cover the areas wetted by drip emitters or micro-sprinklers.

Winter annuals may be controlled with cultivation after the trees are planted or with preemergence herbicide treatments in late fall. Choose materials based on the weed species noted in your previous year's late-winter survey. Repeat the fall preemergence treatment on the tree row every year if you are using a strip weed control program. Use postemergence herbicides when monitoring indicates they are needed; be sure they are not sprayed onto green wood, cracked bark, or developing foliage. In orchards with fine-textured soils or where rainfall is high, you can split the preemergence treatment into two applications to prolong the effective period of weed control. Make the first application in fall and the second in late winter. A postemergence herbicide can be combined with the first or second treatment to control weeds that have already emerged.

Management in Established Orchards

Several options are available for weed management in established orchards. The weed control program you choose depends on the spectrum of weeds present, tree spacing, soil type, the irrigation method you will use, the need for erosion control, the potential for frost injury, economic considerations, and your own personal preference. You may wish to rely, singly or in combination, on cultivation, mowing, or herbicides. Herbicides may be used to control weeds around the bases of trees (basal weed control), to control weeds in the tree rows (strip weed control), or to manage the entire orchard floor. When cultivating or mowing to control weeds in the row middles, you also need to control weeds around tree bases with strip weed control (if you do not cultivate across tree rows) or basal weed control (if you work the row middles in both directions). As an alternative to herbicides, you may use hand-weeding, mulching, or flame cultivation for basal or strip weed control.

Regardless of the management strategies you adopt, you must take steps to avoid introducing problem weeds and to eliminate conditions that favor weed development. Clean your field equipment after working infested ground to keep from spreading weed seeds and perennial weed structures. When working several orchards, be sure to work those infested with problem weeds last or thoroughly clean the equipment between orchards. Control weeds in and around irrigation ditches and consider installing screens in canals used as sources of surface water to reduce the spread of seeds and perennial structures such as rhizomes and root fragments. Provide proper drainage in the orchard to eliminate areas of standing water, which favor the development of dallisgrass, sprangletop, and curly dock. If you let the top 2 to 4 inches (5–10 cm) of soil dry out between irrigations, you will discourage the establishment of weed seedlings and increase the effectiveness of herbicides, which dissipate more rapidly in saturated soil.

Cultivation. Clean cultivation can keep orchards free of weeds most of the year. Growers often use cultivation for the first few years in newly planted orchards to control weed seedlings and reduce established weeds. A combination of cultivation and herbicide applications is the most efficient way to reduce or eliminate infestations of perennials such as johnsongrass. Cultivation may be the most practical method if you are using furrow or flood irrigations, which require periodic activity to remove and replace furrows or checks. Cultivate when you remove furrows or checks to destroy weed seedlings that germinate after an irrigation.

A typical cultivation program involves discing or harrowing in late winter or early spring to turn the winter cover crop under, followed by additional cultivations from late spring through summer as needed to control weed growth. In nonirrigated orchards, one additional cultivation in early summer may be sufficient; irrigated orchards may require four cultivations or more. One way to prepare weed-free orchards for harvest is to disc and roll or float the orchard floor and use an herbicide just before or in the last irrigation.

If you allow resident vegetation or a seeded cover crop to grow in the winter, it will help prevent soil erosion and improve soil structure and drainage, helping to reduce problems with root and crown diseases. However, the presence of orchard floor vegetation may increase the risk of frost injury. If frost is a concern, cultivate before the bloom period or apply sprinkler irrigation to a low cover crop. In summer, keep cultivations as shallow as possible to conserve moisture and reduce potential damage to the trees' feeder roots.

Cultivation of the entire orchard floor requires discing or harrowing in both directions, which may cause problems. The heavy equipment involved can compact soil, decreasing water penetration and increasing erosion on sloping land. Equipment can injure the trunk or crown, increasing the trees' susceptibility to certain diseases such as Ceratocystis canker and crown gall. Use hand-weeding or herbicides around tree trunks to avoid injuring the trees. Cultivation destroys feeder roots in the top 6 inches (15 cm) of soil, rendering the nutrients and water present in the cultivated layer unavailable to the tree. Keep cultivations as shallow as possible. Cultivation can spread perennial weeds by carrying rhizomes or perennial roots to new locations in the orchard. A postemergence application to mature weeds 7 days before a cultivation will reduce the spread of viable perennial structures. Be sure to clean all cultivation equipment before you move it from an infested site to an uninfested site. Cultivation of dry soil creates dusty conditions that can increase mite problems and contribute to particulate air pollution.

Total Mowing. A total mowing program uses mowing instead of cultivation to manage the ground cover. Winter ground cover may be a volunteer cover provided by annual plants or a fall-seeded cover crop. If the volunteer ground cover is not adequate, it can be supplemented with an annual fall-seeded cover crop such as wheat, cereal rye, or barley at 50 to 80 pounds per acre (56–90 kg/ha). Also, a number of self-seeding cover crop mixes of annual plants are available commercially.

Begin mowing each season before frost is a concern. After that initial mowing, mow as necessary to keep problem weeds from seeding. As a general rule, you should mow resident ground cover before it is 6 to 8 inches (15–20 cm) high so as to allow desirable winter annuals such as redmaids, chickweed, and annual bluegrass to reseed, while preventing seed production by less-desirable species. As with cultivation, use hand weeding or herbicides to control vegetation around tree trunks to avoid mechanical injury to the trees. Following the last irrigation before harvest, flail the ground cover as close to the ground as possible without disturbing the soil surface. The objective is to have the orchard free of plants by the time harvest begins. If the orchard is mowed frequently enough, the clippings will have sufficient time to decompose and will not interfere with harvest.

A program of total mowing obviates some of the problems of cultivation. Equipment is lighter, so soil compaction is less, and the roots of the cover crop maintain good water penetration and reduce soil erosion. Increased accessibility during wet weather allows the ground application of fungicides during bloom for more effective control of diseases such as brown rot. During dry weather, dusty conditions are reduced. Tree roots can grow closer to the surface and make better use of water and nutrients. The equipment used for mowing is less expensive and easier to operate.

Disadvantages include a cooler orchard during frost season. You can minimize this condition by keeping the ground cover mowed closely during late winter and early spring. The cooling effect can be advantageous during summer, but the ground cover will consume more water and nutrients. Perennials and grasses tend to increase under a total mowing program, because they have well-developed roots systems and outcompete the seedlings of annual weeds. The perennials most likely to become problems are bermudagrass, field bindweed, silverleaf nightshade, and dandelion. Nutsedge may become a problem on coarser soils of the San Joaquin Valley. Contour checks for flood irrigation cannot be used with total mowing.

Strip Weed Control. Strip weed control involves the use of herbicides, flaming, or mulching to maintain a weed-free strip 2 to 6 feet (0.6–1.8 m) wide in the tree row. Hand-hoeing in the tree row may be a viable option for small orchards. Strip weed control can be used with both young and established orchards, with flood or furrow irrigation that uses permanent berms down the tree rows, and with drip or sprinkler irrigation. It reduces equipment travel by eliminating the need for cross-discing or mowing in two directions and prevents tree injury by eliminating the need to cultivate close to the trunks. Weed control in the tree rows reduces some competition for nutrients and water and keeps the tree crowns drier. Strip weed control has the same advantages for soil structure as total mowing and, when herbicides are used, requires a smaller quantity of chemicals than does total reliance on herbicide.

With strip weed control, perennials can become established quickly in the tree rows because there is no competition from annual weeds. Watch for their appearance both in tree rows and in row middles. Spot-treat with systemic herbicides to keep perennials from becoming established. For a detailed description of strip weed control, see *Nontillage and Strip Weed Control in Almond Orchards* listed under "Weeds" in the suggested reading.

Basal Weed Control. Basal weed control involves the use of herbicides, hand-hoeing, mulching, or flaming to control weeds in a 4- to 8-square-foot (0.4- to 0.8-sq m) area around the base of each tree. This system uses less herbicide than strip weed control but requires more mowing or cultivation. The cross-discing or mowing that is necessary to control weeds between trees prevents the use of permanent berms or low-volume irrigation systems. Basal weed control eliminates the possibility of trunk injury from mowing or tillage equipment. In young orchards, the safest preemergence herbicides to use around tree bases are napropamide, oryzalin, oxyfluorfen, pendimethalin, and trifluralin.

Total Reliance on Herbicides. Herbicides can be used to keep the entire orchard floor free of weeds, or they can be used in a combination of basal or strip weed control and chemical mowing. Total orchard floor weed control with herbicides requires the fewest equipment trips of any control strategy, controls perennials more easily, and greatly reduces competition for water where the annual rainfall is 11 inches (28 cm) or less. Initial costs are higher than with other methods.

Total reliance on herbicides has certain disadvantages. It requires more careful monitoring of the weed population and often requires the application of combinations of herbicides or sequential treatments to achieve total control. No single chemical will control effectively all of the weeds that infest California orchards. Soil compaction will be a problem on some soils, so periodic shallow cultivations may be needed. The orchard will be less accessible during wet weather than it would be if vegetation were maintained during winter months. Sloping land will be subject to increased erosion of the topsoil. Perennials may become established quickly in the absence of competition, especially nutsedges and bermudagrass, which tolerate most postemergence herbicides. Repeated applications of some herbicides may lead to the buildup of species tolerant to that herbicide. For example, repeated use of simazine increases the buildup of tolerant annual grasses such as crabgrass and witchgrass that are not controlled by this herbicide. In addition, resistance to herbicides in weeds that were formerly controlled is a growing problem. Annual ryegrass has become resistant to glyphosate in some orchards. Strategies for avoiding resistance are found in *Herbicide Resistance: Definition and Management Strategies* listed under "Weeds" in the suggested reading.

Chemical Mowing. An alternative to mechanical mowing involves using a low rate of an herbicide such as glyphosate, sulfosate, or oxyfluorfen to inhibit the growth of plants without killing them, thus maintaining a ground cover in the orchard. Chemical mowing requires fewer trips through the orchard than mechanical mowing; two or three trips per season may be sufficient, depending on the plants present. By maintaining a ground cover, you decrease erosion and increase accessibility as compared to total orchard floor weed control. Chemical mowing has advantages in orchards where irrigation water is limited or unavailable and where mechanical mowing is difficult before bloomtime. If you use this technique, remove weeds around trunks by hand or with herbicides. Chemical mowing can be combined with strip weed control to reduce weed competition around trees and minimize drift hazards from the glyphosate used for mowing.

The application timing and rates used for chemical mowing depend on the vegetation present, its stage of growth, and growing conditions. Chemical mowing works best on annual weeds. Young winter annuals can be controlled with very low rates applied in January or February. Summer annuals and older winter annuals are harder to control. Plants are harder to control in spring or summer, and when stressed by periods of drought they require even higher rates of herbicides. If you don't start a chemical mowing program until March or April, it is best to mow mechanically first. Perennials or annuals not controlled by chemical mowing may quickly take over because of the reduced competition. Spot-treat these weeds with standard rates of a systemic herbicide as they emerge or they will become the dominant species and will be extremely difficult to manage.

Continued use of low rates of an herbicide for chemical mowing may result in a shift in the weed population; careful monitoring is essential. Chemical mowing may select for herbicide-resistant plants. Use alternate strategies within season or between seasons: for example, use mechanical mowing in combination with chemical mowing within the same season.

Weed Monitoring

Identification of the weed species that are present and their emergence dates is essential for making decisions about which herbicides to use and whether postemergence treatments are necessary. Knowledge of what weeds are present is also important for planning control programs that involve solarization, cultivation, flaming, or mowing, whether or not you plan to use herbicides. Monitor your orchard for weeds at least three times each year, in fall, winter, and late spring or early summer.

Rains in September and early October stimulate germination. These months are typically warm, so growth of winter annuals is accelerated. If rains do not begin until after temperatures have cooled, winter annuals will grow more slowly. Early rains are of concern because weeds will be larger, requiring higher rates of postemergence herbicides in orchards where herbicides are used, or earlier hand-weeding or flaming to control weeds at a young stage.

Fall Monitoring. After the first rains in the fall, look for winter annual weeds in tree rows to check the effectiveness of any preemergence herbicide applications, and check the ground cover in row middles for perennial seedlings (see Fig. 56). Use appropriate postemergence herbicides, flaming, or hand-weeding if needed.

Winter Monitoring. An examination of untreated ground in February will show you the full range of winter annual species present. Keep a record of what you find to help you plan weed control activities for the orchard (see Fig. 57). February monitoring following a fall preemergence herbicide treatment will tell you which weeds were not controlled. Record these and use the results to plan the next year's weed control program. If a large number of weeds are present, consider treating with a postemergence herbicide.

If preemergence herbicide application was delayed, monitoring will determine if a postemergence herbicide should also be applied. Cold winters delay weed emergence, reducing the need for a postemergence herbicide. A warm winter means that winter annuals will be larger than normal, requiring earlier control or higher rates of postemergence herbicide.

Spring Monitoring. Monitor in late spring or early summer, after summer annuals have germinated (see Fig. 56). Spring monitoring the year before you plant the orchard will indicate which species are present; after the orchard is in place, it will tell you which species have not been controlled by preemergence herbicides. Monitoring at this time also tells you what perennial weeds are present.

Where you use cultivation for weed control, monitor at least 2 weeks before you plan to cultivate as a check for the presence of perennial weeds. If present, treat perennials with a translocated herbicide to kill underground structures so the weeds will not be spread by cultivation. This treatment is most effective 1 to 2 weeks before cultivation. Monitor again a few weeks after the cultivation to check for regrowth of perennials, and treat again if necessary.

Keep records of your monitoring results. By knowing which species are present, you will be able to make correct decisions on cultural and chemical controls. Information collected over a period of years tells you how weed populations may be changing and how effective your control operations have been. When monitoring for weeds, pay special attention to perennials. It is a good idea to draw a diagram of the orchard and mark where perennials are found. This way you can quickly return to see how well your control actions are working. Two examples of weed monitoring forms are illustrated in Figures 56 and 57.

Identifying Major Weed Species

In addition to being classified as annual, biennial, or perennial based on their growth habit, weeds can be classified as broadleaf, grass, or sedge based on botanical characteristics. Figure 58 illustrates some of the structures used to identify weeds. Broadleaf seedlings have two seed leaves (cotyledons), whereas grass and sedge seedlings have one seed leaf. The seed leaves of each broadleaf weed species have a characteristic shape, texture, and color, which makes the seedlings relatively easy to identify. Seed leaves and usually the first true leaves are different from the leaves that form

ORCHARD LOCATION ________________ HERBICIDE(S) ________________

DATE ________________ APPLICATION ________________

COMMENTS ________________

	LATE FALL		LATE SPRING	
	Treated	Untreated	Treated	Untreated
ANNUAL GRASS				
annual bluegrass				
barnyardgrass				
crabgrass				
fall panicum				
hare (wild) barley				
sprangletop				
wild oat				
witchgrass				
________				
________				
________				
ANNUAL BROADLEAF				
cheeseweed (mallow)				
clovers				
fiddleneck				
filaree				
groundsel				
hairy fleabane				
horseweed				
knotweed				
lambsquarters				
mustards				
pigweeds				
prickly lettuce				
puncturevine				
purslane				
starthistle				
wild radish				
________				
________				
________				
PERENNIAL				
bermudagrass				
curly dock				
dallisgrass				
dandelion				
field bindweed				
johnsongrass				
nutsedge				
________				
________				
________				

Figure 56. Example of form used in fall and spring to monitor for weed species present in the orchard. Checking in late fall, after rains or postharvest irrigation stimulates germination, will tell you what winter annuals are present. Checking in late spring or early summer will tell you what summer annuals and perennials are present. Use a scale from 1 to 5 to indicate level of infestation: 1 = very few weeds, 2 = light infestation, 3 = moderate infestation, 4 = heavy infestation, 5 = very heavy infestation.

ORCHARD LOCATION ______________________ HERBICIDE(S) ______________________

DATE ______________________ APPLICATION ______________________

COMMENTS ______________________

	SUMMER		MIDWINTER	
	Treated	Untreated	Treated	Untreated
ANNUAL GRASS				
annual bluegrass	____	____	____	____
barnyardgrass	____	____	____	____
crabgrass	____	____	____	____
fall panicum	____	____	____	____
hare (wild) barley	____	____	____	____
sprangletop	____	____	____	____
wild oat	____	____	____	____
witchgrass	____	____	____	____
____	____	____	____	____
____	____	____	____	____
____	____	____	____	____
ANNUAL BROADLEAF				
cheeseweed (mallow)	____	____	____	____
clovers	____	____	____	____
fiddleneck	____	____	____	____
filaree	____	____	____	____
groundsel	____	____	____	____
hairy fleabane	____	____	____	____
horseweed	____	____	____	____
knotweed	____	____	____	____
lambsquarters	____	____	____	____
mustards	____	____	____	____
pigweeds	____	____	____	____
prickly lettuce	____	____	____	____
puncturevine	____	____	____	____
purslane	____	____	____	____
starthistle	____	____	____	____
wild radish	____	____	____	____
____	____	____	____	____
____	____	____	____	____
____	____	____	____	____
PERENNIAL				
bermudagrass	____	____	____	____
curly dock	____	____	____	____
dallisgrass	____	____	____	____
dandelion	____	____	____	____
field bindweed	____	____	____	____
johnsongrass	____	____	____	____
nutsedge	____	____	____	____
____	____	____	____	____
____	____	____	____	____
____	____	____	____	____

Figure 57. Example of form used in summer and winter to monitor for weed species present in the orchard. Checking in summer will tell you what weeds are present and not controlled. Checking in midwinter will tell you what was not controlled by preemergence treatments. Use a scale from 1 to 5 to indicate level of infestation: 1 = very few weeds, 2 = light infestation, 3 = moderate infestation, 4 = heavy infestation, 5 = very heavy infestation.

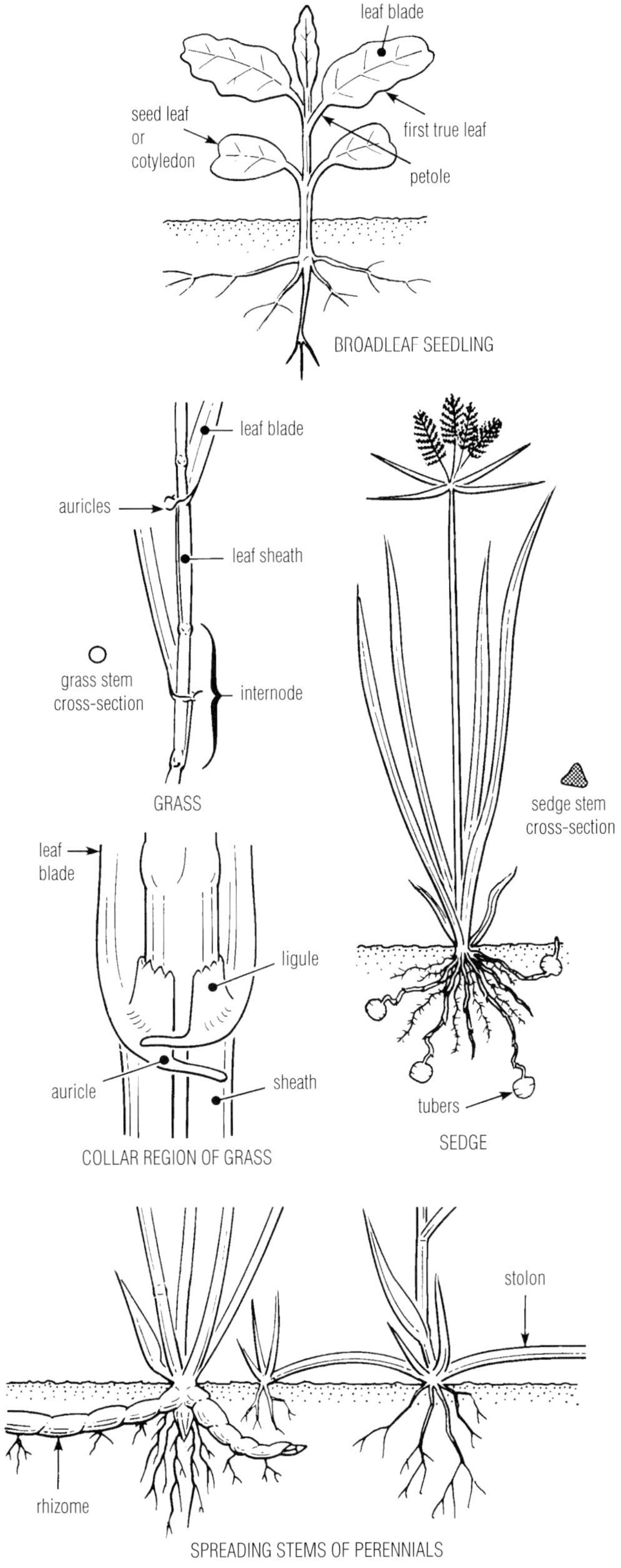

Figure 58. Vegetative parts commonly used to identify weeds.

later. Seedlings of grasses are more difficult to identify because the seed leaves of different species are similar, and they differ little from true leaves that form later. Characteristics of the collar region, where the leaf blade joins the leaf sheath, are used to distinguish grasses. Perennial grasses may be distinguished by the appearance of rhizomes or stolons. Sedges are similar in appearance to grasses; most species found in orchards have three-sided stems that are triangular in cross-section. The most important perennial sedges are yellow and purple nutsedge, which form characteristic tubers or nutlets on their rhizomes.

Photographs and detailed descriptions of all the weeds mentioned in this chapter can be found in *The Grower's Weed Identification Handbook,* listed under "Weeds" in the suggested reading.

Perennial Grasses and Sedges

Johnsongrass *Sorghum halepense*

Johnsongrass occurs commonly in annual crops, orchards, and other locations where soil is moist. It is one of the most troublesome of the perennial grasses. Mature plants form leafy clumps that may reach heights of 6 or 7 feet (2 m). Johnsongrass reproduces from underground stems (rhizomes) and from seed. Johnsongrass seed are often spread in irrigation water and can remain dormant for many years. Repeated cultivation of dry soil the summer before planting can effectively control established infestations. Care must be taken not to move viable rhizomes to uninfested locations. If johnsongrass invades an established orchard, it is best controlled with herbicides. Control seedlings before they begin to form rhizomes, which may occur 3 to 6 weeks after germination. If rhizomes become established, repeated applications of translocated herbicide will be needed to control them. In newly planted orchards, sethoxydim can be used to reduce johnsongrass. In established orchards, glyphosate should be used.

A. Johnsongrass seedlings have broad, light green leaves with smooth leaf sheaths that may have a maroon tinge. The midvein appears as a broad, white line at the base of the first leaf.

B. The johnsongrass ligule is a dense fringe of hairs. The presence of a ligule helps distinguish young johnsongrass plants from barnyardgrass.

C. Johnsongrass rhizomes are thick, fleshy, and segmented. Each segment can form roots and shoots.

D. The mature johnsongrass plant grows in stout, spreading, and very leafy patches, which may be up to 6 to 7 feet (2 m) tall. The leaves have a prominent whitish midvein, which snaps readily when folded over. The flower head is large, open, well branched, and often reddish tinged.

A.

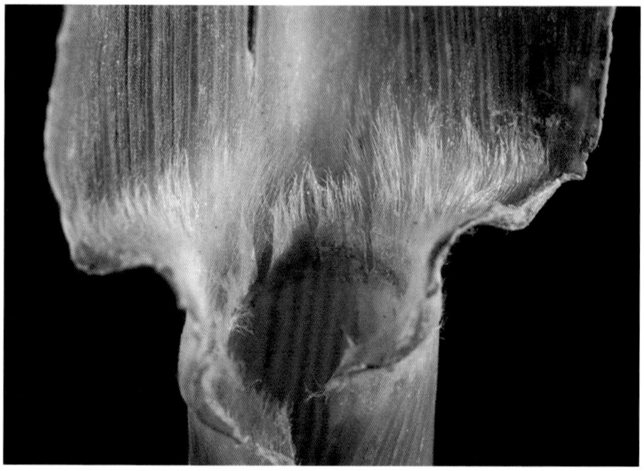
B.

C.

D.

Dallisgrass ***Paspalum dilatatum***

Dallisgrass occasionally becomes a problem in orchards, usually in low, wet areas. It is a bunchgrass that reproduces from seed and very short rhizomes. Dallisgrass is undesirable in ground covers because it may inhibit the growth of other plants and is able to quickly invade tree rows. Dallisgrass is highly competitive with trees for water and nutrients. It produces large quantities of seed and tends to become dominant in mowed groundcover because mowing stimulates seed production. Seed are transported easily in water or by machinery. Because it grows in a clump, dallisgrass can be a problem at harvest by interfering with nut pickup. Like other perennial grasses, dallisgrass can be controlled before planting the orchard by repeated cultivation of dry soil. However, seedlings will still need to be controlled, which can be accomplished with preemergence herbicides or with cultivations. If dallisgrass becomes established in your orchard, you can control it with repeated applications of translocated herbicide.

E. The ligule of dallisgrass is firm and membranous with a few spreading hairs at the margins and no auricles.

F. Mature dallisgrass plants form loose bunches that may reach 1 to 4 feet (30–120 cm) in height. Flower heads often droop, and they consist of three to six spikes that arise from different points along the flower stem. Dallisgrass plants often grow close to the ground in mowed orchards. Rhizomes (not shown) are very short, with shortened internodes that give the appearance of concentric rings.

Bermudagrass *Cynodon dactylon*

Although commonly used as a turfgrass, bermudagrass is a troublesome weed in orchards. It reproduces from rhizomes, stolons, and seed, and it becomes a problem in mowed orchards because mowing reduces competition from other weeds and allows bermudagrass to spread. However, exposure to direct sunlight in a dry soil kills stolons and rhizomes. Before planting the orchard, control bermudagrass by repeatedly cultivating dry soil. In established orchards, use herbicides to control plants that become established. Even if mature plants are controlled, their seed may be viable for two years. You can use cultivations to control seedlings. Solarization will control seed, stolons, and rhizomes in the upper few inches of the soil.

G. Bermudagrass forms dense mats of spreading, branching stolons that have short leaves and erect stems 4 to 18 inches (10–45 cm) tall. There is a papery sheath around the stolon at the base of each leaf. Three to seven slender flower spikes arise from a single point at the tip of the stem.

Nutsedge *Cyperus* spp.

Nutsedge, also called nutgrass, can be distinguished from grasses by its three-sided stem, which is triangular in cross-section. Two species are found in California: yellow nutsedge, *Cyperus esculentus*, is the most common; purple nutsedge, *C. rotundus*, tends to be found in wetter locations south of Fresno County. Nutsedges can reproduce from seed, but for the most part they reproduce from nutlets, or tubers, that form on their rhizomes. The tubers are spread easily by cultivation. Nutsedges may become troublesome in an orchard where herbicides are used for all weed control, since most chemicals do not control nutsedges and they spread quickly in the absence of competition. To prevent the formation of tubers, you must kill the young plants before they reach the 5-leaf stage. Some preemergence and postemergence systemic herbicides will provide partial control. Deep tillage of dry soil reduces purple nutsedge populations. The tubers of purple nutsedge are more susceptible to drying than are those of yellow nutsedge.

H. Mature nutsedge plants are 1 to 2 feet (30–60 cm) tall with a flower head at the tip of each stem. Flowering stems are triangular in cross-section and the stiff leaves

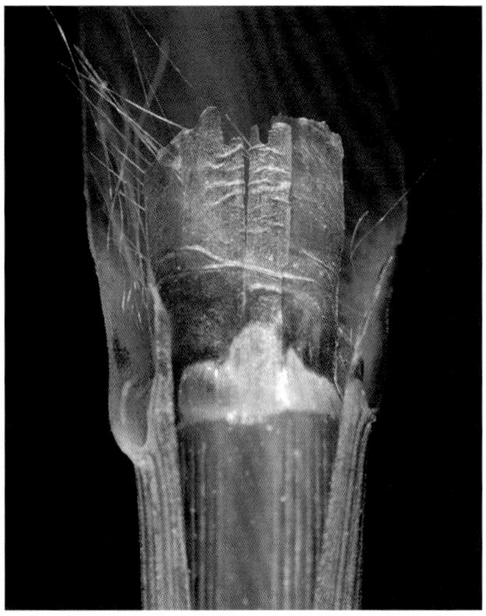

E.

F.

G.

H.

I.

are V-shaped in cross-section. Yellow nutsedge has three long, leaflike bracts at the base of each flower head. These bracts are short in purple nutsedge.

I. Tubers, or nutlets, form on the extensive root system of the nutsedge plant. Tubers can sprout to form new plants and are the main means for dissemination of nutsedge. The tubers of yellow nutsedge (shown here) are produced singly at the end of rhizomes, whereas the tubers of purple nutsedge are produced in chains along rhizomes. Yellow nutsedge tubers have a pleasant nutlike flavor, while purple nutsedge tubers are bitter.

Perennial Broadleaves

Curly Dock *Rumex crispus*

Curly dock grows in wet areas and usually becomes a problem where drainage is poor or where orchards are overirrigated. It is a member of the buckwheat family, which is characterized by jointed stems and membranous sheaths surrounding the stem at the base of each leaf. Curly dock regrows from a fleshy taproot after mowing or discing. Deeper cultivation will control curly dock. Preemergence herbicides will control seedlings but will not control the deep taproot of an established plant. Some foliar-applied translocated herbicides will control established plants.

J. Curly dock plants regrow in the spring from a rosette of leaves. Plants remain low-growing if mowed; if unmowed they can reach heights of 2 to 5 feet (0.6–1.5 m). Long, loosely branched flower heads form at the ends of stems. The flower heads turn reddish brown when mature. Stems die back in the fall except in warmer areas, where the plant may overwinter.

J.

Field Bindweed *Convolvulus arvensis*

Field bindweed, also called perennial morningglory, is one of the most troublesome broadleaf perennial weeds in many crops. In almonds, it competes with trees for moisture and nutrients during summer months and can be a serious problem in young orchards. Bindweed plants can be prostrate or can climb vinelike up other vegetation. The showy white or reddish flowers are funnel shaped and borne singly along the stems. Bindweed is difficult to control after the seedling stage because it forms a deep perennial rootstock and produces many seed. Established infestations are nearly impossible to eradicate because the seed can remain dormant for up to 60 years. Cultivation can spread bindweed infestations by moving fragments of viable rootstock. Repeated deep cultivations of dry soil may help reduce infestations, but care must be taken not to transport viable rootstock fragments on field equipment. A program of treating bindweed plants with glyphosate, cultivating, and treating regrowth when flowers begin to form will reduce infestations substantially if carried out over a period of years.

K. Seed leaves of field bindweed are nearly square, with a deep notch at the tip. True leaves are arrowhead-shaped, with petioles that are grooved on the upper surface. Young plants regrowing from rootstock lack seed leaves.

L. The mature bindweed plant grows prostrate, with slender stems, about 3 feet (90 cm) or longer; the plant can also grow upward while entwining other plants. Flowers are funnel shaped, white to light red, and are borne singly on slender stalks in leaf axils.

K.

L.

Dandelion *Taraxacum officinale*

Dandelion is a commonly occurring perennial that is most troublesome in mowed orchards with fine-textured soils or orchards where chemical mowing is used. It is a host for tomato ringspot virus, which causes yellow bud mosaic. The mature dandelion plant consists of a rosette of leaves that sends up hollow stalks 3 to 12 inches (7.5–30 cm) long, topped with a single, bright yellow, disclike flower head. Dandelion reproduces from its familiar windblown seed and regrows from a strong, deep taproot. It is difficult to control with herbicides and tolerates close mowing. Cultivation can spread fragments of the taproot, which can regrow.

White Clover *Trifolium repens*

White clover can be a desirable cover crop, especially in orchards where a perennial cover is maintained in tree middles. However, the plants are aggressive, use a lot of water, and may invade tree rows, where they become difficult to control. White clover plants are low-growing, and the stems root at the nodes. The trifoliate leaves have a distinctive white crescent on each leaflet.

Biennial Broadleaves

Little Mallow (Cheeseweed) *Malva parviflora*

Little mallow, also called cheeseweed, grows as a biennial or short-lived perennial in the milder climates of the Central Valley and coastal valleys. Plants may be prostrate or may grow as tall as 5 feet (1.5 m), with flowers that are purple or pink and white. The fruit is shaped like a tiny wheel of cheese, giving the weed the common name "cheeseweed." Seed remains viable for many years and can germinate and emerge from deep in the soil. This, plus the fact that little mallow tolerates many herbicides, means that it may become predominant in orchard cover crops. Mature plants form dense bushes with woody stems that interfere with orchard activities. Under a mowing program, little mallow becomes a low-growing and thick-stemmed plant that does not degrade well and causes problems at harvest. Use hand-weeding or herbicides to control little mallow plants while they are small; otherwise they quickly grow a deep, tough tap root, and become difficult to eliminate. Discing will control seedlings and young plants. Napropamide controls germinating seedlings. Oxyfluorfen can be used during the dormant season, in late spring, and during the summer.

M.

M. Seed leaves of little mallow are triangular or heart-shaped. True leaves are round or kidney-shaped with scalloped edges. There is a red spot at the base of each leaf where it attaches to the petiole.

Bristly Oxtongue *Picris echioides*

Bristly oxtongue, like little mallow, grows as a biennial or a short-lived perennial in almond production areas. The spiny leaves and stems can be unpleasant to work around and may interfere with orchard activities. Bristly oxtongue plants produce large quantities of seed but can be controlled by discing. Plan cultivations to uproot the plants before they set seed.

N.

N. The rough, hairy leaves of bristly oxtongue have a warty appearance. Clusters of yellow flowers form near the tips of stems.

Winter Annual Grasses

Annual Bluegrass *Poa annua*

Annual bluegrass is an excellent ground cover in orchards. It is low-growing, and because it dies out in early summer it does not compete with trees for irrigation water. It can be distinguished from other grasses by its typical leaf tip and blade.

O.

O. The mature annual bluegrass plant grows as dense, low-spreading tufts, 3 to 12 inches (8–30 cm) tall, and often roots at the lower nodes. The most distinguishing feature of the mature plant and the seedling is the blunt leaf tip shaped like the bow of a boat. The leaf blade is often crinkled at midsection.

Wild Oat *Avena fatua*

Wild oat is a tall plant that can become a problem around tree trunks, where it creates moist conditions that favor crown diseases and provides shelter for voles. This weed is difficult to control because it may emerge several times during the year, and the large seed can germinate deep in the soil, beyond the effective zone for most preemergence herbicides. Effective postemergence herbicides are available, but several treatments may be necessary to control young plants. Wild oat seedlings may be distinguished from most other grasses by their large seed coat, which usually remains attached for a long time.

P.

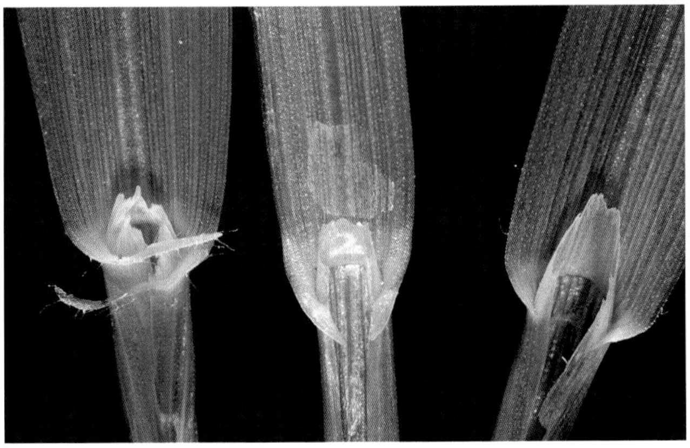

Q.

R.

P. Mature wild oat plants are 1 to 4 feet (30–120 cm) tall, with flowers nearly identical to those of domestic oats.

Q. The collar region of wild oat (right) has a tall, pointed ligule with toothed margins and no auricles. Also shown are wheat (middle) and barley (left).

Hare Barley (Wild Barley)
Hordeum murinum* ssp. *leporinum

Hare barley is less bothersome than wild oat because the mature plants are shorter and they dry up earlier in the summer. Seedlings are similar to those of wild oat but lack the large attached seed coat. Mature hare barley plants are 6 to 24 inches (15–60 cm) tall. The long awns or bristles of the individual flowers give the flower head a bushy appearance. Long, clawlike auricles are present at the collar (see photo Q), but may not be prominent on the young plant. When mature, the flower head disintegrates into the familiar "foxtails."

Other winter annual grasses that can become weed problems in almond orchards include ripgut brome (*Bromus diandrus*), soft brome or soft chess (*Bromus mollis*), and Italian ryegrass or annual ryegrass (*Lolium multiflorum*). These are described and illustrated in the *Grower's Weed Identification Handbook* listed under "Weeds" in the suggested reading.

Winter Annual Broadleaves

Mustards

Several species of mustard may occur in California orchards. Those found most commonly are black mustard (*Brassica nigra*), wild mustard (*Sinapis arvensis*), and birdsrape mustard (*B. rapa*), also known as common yellow mustard and wild turnip. Mustards are frequently used in ground covers or as green manure crops because their tap roots help loosen heavier soils and their flowers attract a number of beneficial insects. However, they are not a good ground cover for almond orchards because they bloom at the same time as the almond trees, and so compete for pollinators. Because of their large size, mustard plants can interfere with cultivation, and plants around the bases of trees create conditions that favor crown diseases and provide refuge for voles. They compete for water, and this is of special concern in lower-rainfall areas (annual average less than 11 inches, or 28 cm). In milder coastal locations, mustards may grow year-round.

R. All mustard seedlings have broad seed leaves with a deep notch at the tip. The first true leaves are bright green above, pale green below, deeply lobed, and often hairy.

S. Mustards are characterized by large lower leaves, usually more or less deeply cut into irregular lobes, and distinctive seed pods. The mature plant is erect and about 2 to 6 feet (0.6–1.8 m) tall. Leaves grow alternately on the stem. Many yellow flowers rise at the tip of the stalk. They have four petals, which spread in the form of a cross. The flower stalk extends during the blooming season, and maturing seed pods have long beaks.

Redmaids, *Calandrinia ciliata,* can be seen growing in the foreground of this photo.

Wild Radish ***Raphanus raphanistrum***

Wild radish is similar to mustards in its growth habit and appearance. Plants reach heights of 2 to 5 feet (0.6–1.5 m), and flowers range in color from white to purple or sometimes yellow. The seed pods and seed are much larger than those of mustards. As with mustards, wild radish is not a good ground cover for almond orchards because it blooms at the same time as the almond trees.

S.

Redmaids (Desert Rockpurslane) ***Calandrinia ciliata***

Redmaids are common in Central Valley orchards. Mature plants may be from 6 to 12 inches (15–30 cm) tall with narrow, succulent leaves. The showy, reddish purple flowers appear in early spring (see photo S). Redmaids makes a good ground cover for almond orchards because it is low-growing, dries up by early summer, and is easy to control in the tree rows.

Common Chickweed ***Stellaria media***

Chickweed is a low-growing, spreading plant that makes a good orchard ground cover when combined with other winter annuals such as bluegrass. Chickweed has thin, succulent stems that trail along the ground and root where nodes contact moist soil. Chickweed dries up with the onset of hot weather in Central Valley orchards, although it can continue growing through the summer in shady, cool locations where moisture is adequate. Under high-nitrogen conditions, chickweed can form dense mats 12 to 14 inches (30–35 cm) tall. It can become a problem if it invades tree rows and forms dense, wet mats around the bases of trees.

T.

T. Chickweed stems are trailing, weak, and slender, with a line of hairs down the side and sometimes rooting at nodes that are in contact with the soil. The mature leaves are ovate, ¼ to 1 inch (6–25 mm) long, and opposite on the stem. Chickweed flowers are small but showy, with five deeply cut white petals. They are borne singly on long slender stalks arising at the base of the leafstalk.

Burclover ***Medicago*** **spp.**

Two or three different species of burclover (bur medic) occur in California. Medics are used commonly in cover crops, where they fix atmospheric nitrogen. California burclover, *Medicago polymorpha,* is most likely to occur as resident vegetation in orchards. Seedlings may emerge in spring in milder coastal locations.

U.

U. California burclover stems are up to 2 feet (60 cm) long and tend to trail along the ground, but they may grow upright. The trifoliate leaves resemble those of clover and usually have red-tinged midveins. Small, bright yellow flowers form in clusters at the end of stems. The seed pod is a bur that contains several yellowish or tan kidney-shaped seed.

V.

W.

X.

Filaree *Erodium* spp.

Several species of filaree are common winter annuals or biennials throughout California. They are desirable ground cover in orchards because they are low-growing and do not compete with trees. They dry up in the summer and decompose by harvest.

- **V.** Filaree plants are erect or spreading, and the stems may be anywhere from 3 inches to 2 feet (7.5–60 cm) long. The leaves are lobed or finely divided, depending on the species. In the early stages, leaves form a rosette close to the ground.
- **W.** Filaree flowers may be rose, lavender, purple, or violet. The characteristically long, pointed fruit are borne in clusters. At maturity, the fruit separates into five parts, each consisting of a long, spirally twisted beak attached to a seed.

Summer Annual Grasses

Barnyardgrass *Echinochloa crus-galli*

Barnyardgrass grows in dense clumps or patches that may be tall or may spread along the ground. It forms dense mats when mowed. Several varieties that differ in growth habit and floral appearance occur in California. Because barnyardgrass plants produce huge quantities of seed, the weed can be difficult to control without the use of herbicides.

- **X.** Barnyardgrass is the only common summer annual grass that has no ligule or auricles. This characteristic helps distinguish young barnyardgrass plants from young johnsongrass plants.

Y.

Z.

A.

Y. The mature barnyardgrass plant is stout, grows upright, and varies in height from 6 inches to 6 feet (15–180 cm). It often roots at the lower nodes, forming dense tufts. The leaf sheath is flattened and often reddish tinged at the soil line. Lower spikes of the flower head are spaced apart; top ones are crowded together.

Bearded Sprangletop ***Leptochloa fascicularis***

Bearded sprangletop occurs in Central Valley orchards, usually on alkaline soils. It is most abundant on fine-textured soils and in wet locations where herbicides are leached or degraded more quickly. Good drainage in the orchard helps discourage this weed.

Z. Bearded sprangletop has no auricle, but it has a long, thin, delicate ligule.

A. Mature bearded sprangletop plants form large, upright tufts 12 to 40 inches (30–100 cm) tall. Flowers are elongated, upright, and highly branched, and are straw-colored when mature.

Large Crabgrass ***Digitaria sanguinalis***

Large crabgrass, also called hairy crabgrass, occurs commonly in orchards, especially in the San Joaquin Valley. The low-growing plant roots deeply at the nodes, becoming difficult to remove once established. Large crabgrass has a papery ligule but no auricles; there are small tufts of hairs where the leaf blade meets the sheath. Large crabgrass is easily controlled with napropamide, norflurazon, or dinitroanilines such as trifluralin, oryzalin, and pendimethalin.

B.

B. Flower heads of large crabgrass resemble those of bermudagrass, but flowering branches usually arise separately from the stalk, whereas bermudagrass flower branches all arise from the same point.

Fall Panicum ***Panicum dichotomiflorum***

Fall panicum, or smooth witchgrass, occurs commonly in cultivated orchards. It is found frequently in orchards where simazine is the only preemergence herbicide used. The thick, flattened stems of fall panicum may be erect, but more commonly they are spreading. Leaves are rolled in the budshoot. No auricles are present. The ligule is a dense fringe of hairs. Fall panicum grows in a clump and can be a problem at harvest if it remains in the orchard. In addition, its seed heads can clog up harvest equipment.

Witchgrass ***Panicum capillare***

Witchgrass, also called tumbleweed grass, ticklegrass, and witches' hair, occurs commonly in orchards grown on sandy soils. It has a tough, wiry stem that frequently causes difficulty at harvest by interfering with nut pickup and contributing to the accumulation of trash in the harvested crop.

C. Mature witchgrass plants are bushy and branched at the base, and they have a fuzzy appearance. The large, highly branched flower heads break off easily when mature and blow around in the wind. The stem, leaf sheath, and leaf are covered with long, coarse hairs. The ligule is a fringe of hairs.

C.

Summer Annual Broadleaves

Common Knotweed ***Polygonum arenastrum***

Common knotweed, also known as prostrate knotweed, grows well in most orchards and may build up under a total mowing program. It becomes troublesome at harvest because its tough, wiry stems can get entangled in sweeping machinery. If it invades tree rows, knotweed competes for moisture during summer months.

D. Seed leaves of prostrate knotweed are long, narrow, rounded at the tip, and have whitish streaks or blotches. The true leaves are much broader and emerge from a membranous sheath that encircles the stem. Nodes usually are swollen.

E. Mature knotweed plants may be prostrate or erect. Stems are thin, tough, and extensively branched. The tough stems may become entangled in cultivation equipment.

D.

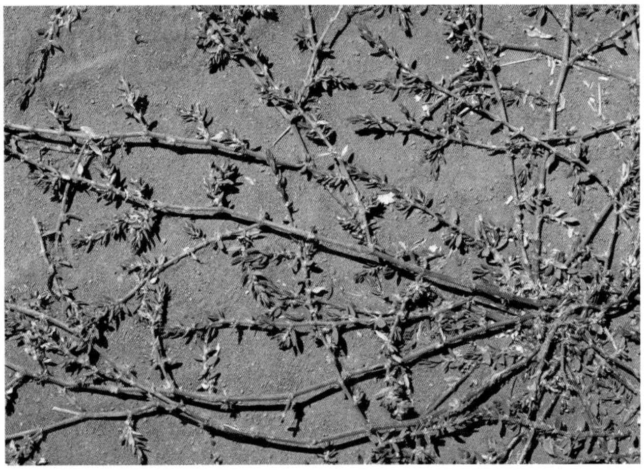

E.

Spotted Spurge ***Chamaesyce maculata***

Spotted spurge, also known as prostrate spurge, is another weed species that may build up under a mowing program because of its low-growing growth habit. Flowers are tiny and inconspicuous, and the plant has a milky, sticky sap. Spurge seed are highly attractive to ants. Spurge infestations may encourage the buildup of ant populations that become damaging to almond nuts, and the presence of spurge seed may interfere with baiting programs.

F.

G.

F. Seed leaves of prostrate spurge are oval, with a mealy texture on the upper surface and reddish tinge below. True leaves form oppositely along stems.

Puncturevine ***Tribulus terrestris***

Puncturevine plants are prostrate in open areas but grow erect in dense vegetation. Single yellow flowers arise from leaf axils. At maturity, the fruit breaks apart into five nutlets, each of which has two hard, sharp spines. Puncturevine is drought-resistant, and its seed burs are easily spread on shoes or vehicle tires. Tractors frequently deposit the seed on the periphery of the ground cover strip, where they germinate. The plant then grows over into the strip and creates a problem at harvest. Mature plants are harder to control than the seedlings, but the seed can germinate beyond the effective depth of some preemergence herbicides. A stem and seed weevil that attacks puncturevine controls populations of the weed more effectively in undisturbed areas than in cultivated situations.

G. Puncturevine seed leaves are thick, elongated, and brittle. They are grayish below and green above, with a groove along the prominent midvein. True leaves have 8 to 12 leaflets.

H. The mature puncturevine plant grows prostrate on open ground but almost erect in dense vegetation. The yellow flowers are borne singly in leaf axils. The fruit consists of a cluster of five spiny nutlets or burs; it breaks apart at maturity, and each bur has two spines.

H.

Common Purslane ***Portulaca oleracea***

Common purslane occurs commonly in orchards on coarse-textured soils and grows rapidly under both wet and dry conditions. The thick, succulent growth can create a moist environment favorable to crown diseases if infestations develop around the bases of trees. Horse purslane, *Trianthema portulacastrum*, is similar to common purslane but has broader leaves and purple flowers. Hand-weeding or spot-treatment with a postemergence herbicide may be needed to control purslane that emerges after preemergence herbicides have broken down; otherwise, it may become a problem at harvest.

I.

I. Mature common purslane forms a mat of highly branched, reddish stems up to 3 feet (90 cm) long that may be prostrate or may stand up to 1 foot (30 cm) tall. Leaves are succulent and often reddish tinged. Small, yellow, cup-shaped flowers are borne singly in stem axils or at the tip of stems. They usually open only on sunny mornings.

Horseweed ***Conyza canadensis***

Horseweed, also known as mare's tail, is a common weed in California orchards, cultivated fields, and disturbed areas. Because it is tolerant to most herbicides available for use in almonds, it may increase over time. If left undisturbed in the tree row, the tall woody stems will hinder harvest operations. If it is kept mowed throughout the summer, however, it should not cause problems at harvest.

J. Undisturbed horseweed plants grow up to 10 feet (3 m) tall, branching near the top. Small flower heads with yellow centers form at the ends of the branches. Horseweed plants are shorter and more highly branched if mowed, and they may be confused with hairy fleabane. Horseweed leaves are dark green, while leaves of hairy fleabane are grayish green. Also, horseweed flowers are yellow and more showy than the whitish flowers of hairy fleabane.

K. The seed leaves and first true leaves of both horseweed and hairy fleabane are oval, narrowing to a stalk at the base. Later leaves are narrower. These two related species are difficult to distinguish in the seedling stage.

J.

K.

Hairy Fleabane ***Conyza bonariensis***

Hairy fleabane, a close relative of horseweed, is often found growing in the same locations. It is more tolerant of available preemergence herbicides and can produce seed after several mowings. The hard stems do not degrade easily and can cause harvest problems. They can also interfere with orchard activities such as the movement of irrigation pipes or drip lines.

L. Mature hairy fleabane plants are 1½ to 3 feet (45–90 cm) tall and are highly branched with very narrow leaves. Numerous flower stalks form on the upper parts of branches. The small, dull white flowers are not showy.

L.

Cover Crops

An orchard cover crop can consist of the resident vegetation, one or more seeded cover crops, or a blend of resident and seeded vegetation. The cover crops that will work best in your orchard depend on the irrigation system, the age of the orchard, the soil conditions, the location, and the weather. If you use flood or sprinkler irrigation, you will have more cover crop options than if you use drip or microsprinkler systems. For information on which cover crops are best suited to your own situation, consult with cover crop specialists and your local UCCE farm advisor.

Cover crops have a number of advantages. They can attract beneficial insects and allow natural enemies of a number of almond pests to build up, but care must be taken to avoid problems with damaging species of mites and plant bugs that may move into trees. Cover crops contribute to a reduction in ant damage to almond nutmeats. Legume cover crops contribute to the orchard's seasonal nitrogen requirement if worked into the soil. Deep-rooted plants such as grasses improve water penetration and reduce soil erosion. Water consumption by grass cover crops helps dry the soil in late winter; this is particularly beneficial in orchards with heavy soils where root and crown diseases such as Phytophthora root and crown rot are likely to be problems. Cover crops should always be kept away from the trunks of trees, or they may increase problems with Phytophthora root and crown rot and provide habitat for voles. Some cover crops may reduce populations of harmful nematode species, but others such as clovers may increase harmful species. By providing a firmer surface, cover crops can make orchard activities such as ground spraying possible under wet conditions. Winter cover crops substantially reduce the runoff of insecticide residues from dormant-season sprays. Cover crops help reduce dusty conditions that favor mite pests. Competition from desirable cover crop species helps keep weeds from building up.

Cover crops can also have a negative impact on pest management in almonds. Legume cover crops attract pocket gophers and certain plant bugs that may damage almonds. Cover crops that allow root knot nematode populations to build up may negate the benefits of nematode-resistant rootstocks, which prevent the multiplication of root knot nematodes but are susceptible to infestation. If not properly managed, cover crops can lead to increases in relative humidity, longer periods of wetness, and lower temperatures, which may be conducive to development of a number of diseases. Dense cover crops, especially grasses, favor the buildup of voles, which can cause severe damage to orchard trees. Perennial legume cover crops can be difficult to control once they are established, and they can become competitive with trees.

Native or resident vegetation is the least expensive cover crop, and it can be as useful as a seeded cover crop for most purposes. Properly managed, a cover crop of desirable resident vegetation increases the accessibility of the orchard to heavy equipment during wet weather, minimizes the need for additional water, improves water penetration, and provides habitat for beneficial insects.

Seeded cover crops most commonly used in almond orchards include vetch, oat, Blando brome, winter annual clovers, and bur medic. Cover crops that include vetch work best in orchards that are flood-irrigated and cultivated. Cahaba white vetch is resistant to nematode pests; vetch reseeds itself if managed properly. Blando brome is an excellent cover crop for almond orchards, but it is expensive to plant and requires strict management to maintain a stand. Mowing must wait until about the first of June for the brome to reseed. When mowed, the brome cover crop provides a heavy mulch that reduces water use, inhibits growth of summer weeds, and decomposes fairly well before harvest. Clovers, especially perennial clovers, can create some management problems. They encourage the buildup of gophers, may be difficult to keep out of the tree row, and use large amounts of water. Subterranean clover is probably the best choice for an almond orchard because it competes well with weeds, reseeds well, and can be mowed in time to break down before harvest.

Table 21 lists a number of cover crops, their horticultural characteristics, and their effects on pest problems in almonds. More information about cover crops can be found in *BIOS for Almonds* and *Covercrops in California Agriculture*, listed in the suggested reading.

Table 21. Cover Crops and Their Effects on Pest Management in Almonds.

Cover Crop	Horticultural benefits and requirements	Effects on arthropod pests and natural enemies	Effects on diseases and nematode pests	Effects on weeds and vertebrate pests
		ANNUAL LEGUMES		
bell bean *Vicia faba*	N contribution[1]: 50–200 lb tolerates high and low pH tolerates some high mowing for frost control roots penetrate 2–3 feet depending on moisture plant early to mid-fall flowers 40–60 days after planting	populations of predaceous wasps, aphid predators may attract plant bugs	host for root knot nematodes[2] SRN[3]	not competitive with weeds; best mixed with annual grass or vetch
berseem clover *Trifolium alexandrinum*	N contribution[1]: 50–400 lb tolerates high pH and salinity tolerates close mowing; frequent mowing recommended best adapted to mild winters; can be grown as a summer annual in colder areas roots penetrate upper 2 feet plant early fall; will reseed flowers May-June	high populations of bigeyed bugs may attract plant bugs	host for root knot nematodes[2] SRN[3]	highly competitive with weeds, especially when mowed frequently attractive to rabbits
crimson clover *Trifolium incarnatum*	N contribution[1]: 70–140 lb tolerates mowing to 3–5 inches tolerates wide range of climatic and soil conditions plant early fall; will reseed if moisture adequate in spring flowers April-May	high populations of minute pirate bugs and aphid natural enemies	host for root knot nematodes[2] SRN[3]	competitive with weeds
rose clover *Trifolium hirtum*	N contribution[1]: 50–100 lb tolerates mowing to 2–4 inches tolerates acid soil, poor fertility does not tolerate flooding taproot penetrates over 6 feet plant early fall; will reseed flowers March-May	high populations of minute pirate bugs	host for root knot nematodes[2] SRN[3]	poor competitor unless mowed frequently weed control improved by mowing to 2–4 inches will control summer weeds after seed set if mowed
subterranean clover *Trifolium subterraneum*	N contribution[1]: 50–200 lb tolerates mowing to 2–4 inches plant early fall; will reseed mowing improves establishment and seed production cultivars range in maturity from April to May	buildup of bigeyed bugs populations of spider mites may build up	host for all major root knot nematodes[2] SRN[3]	competes well with weeds if mowed frequently mowing needed during establishment
sweet clovers *Melilotus* spp.	N contribution[1]: 70–165 lb does not tolerate shade strong taproot may help penetrate clay pans biennials, flower May-September seldom used as cover crop in California	populations of predaceous wasps and some general predators may attract plant bugs	host for root knot nematodes[2] SRN[3]	poor weed control unless combined with other cover crops foliage toxic to rodents and livestock
annual medics *Medicago* spp.	N contribution[1]: 50–100 lb frequent mowing to 3–5 inches to increase weed competition and seed set works well in mixes with grasses or other legumes flowers February through May, depending on species.	high populations of spider mites	host for root knot nematodes[2] SRN[3]	competes with weeds when mowed to 3–5 inches

Table 21. Continued

Cover Crop	Horticultural benefits and requirements	Effects on arthropod pests and natural enemies	Effects on diseases and nematode pests	Effects on weeds and vertebrate pests
		ANNUAL LEGUMES (continued)		
Cahaba white vetch *Vicia sativa* × *V. cordata*	N contribution[1]: about 100 lb tolerates low mowing in March and April for frost protection taproot penetrates 3–5 feet plant in fall, flowers April-July; will reseed may require spring irrigation to set seed	attracts a variety of beneficials high populations of spider mites	non-host for most harmful nematode species host for northern root knot nematode host for ring nematode	highly competitive with weeds habitat for voles, rabbits food source and habitat for gophers
hairy vetch *Vicia villosa*	N contribution[1]: 60–100 lb tolerates mowing to 1–2 inches before flowering taproot penetrates 1–3 feet more drought tolerant than other vetches range of varieties adapted to different areas plant mid-fall; will reseed		host for root knot nematodes[1] SRN[3]	highly competitive with weeds provides habitat for voles, rabbits food source and habitat for gophers
purple vetch *Vicia benghalensis*	N contribution[1]: 50–300 lb tolerates moderately close mowing in winter persists well on heavy soils root system penetrates 3 feet plant in early fall, flowers April-May; will reseed may require spring irrigation to set seed	high populations of general predators high populations of spider mites	rank growth increases orchard humidity and potential for foliar diseases host for root knot nematodes[2] SRN[3]	highly competitive with weeds provides habitat for voles, rabbits food source and habitat for gophers
woollypod vetch *Vicia dasycarpa*	N contribution[1]: 50–200 lb tolerates mowing to 5 inches tolerates a range of soil types flowers March-May; will reseed if not mowed for frost protection	aphid predators, minute pirate bugs	host for root knot nematodes[2] SRN[2]	highly competitive with weeds provides habitat for voles, rabbits food source and habitat for gophers
		PERENNIAL LEGUMES		
strawberry clover *Trifolium fragiferum*	N contribution[1]: 100–300 lb low-growing, can be mowed frequently tolerates heat, full sun, low moisture, alkaline soil root system penetrates 3 feet plant in fall or spring, flowers May-June works well in mix with white clover and grass	relatively low populations of beneficials may attract plant bugs	host for northern root knot nematode, poor host for other species host for ring nematode	highly competitive with weeds favors high populations of pocket gophers
white clover *Trifolium repens*	N contribution[1]: 100–200 lb some cultivars are low-growing and more tolerant of frequent mowing tolerates shade root system mainly in top 2 feet plant in fall, flowers April-December	prone to high spider mite populations	host for root knot nematodes[2] SRN[3]	host for tomato ringspot virus competitive with weeds if mowed or grazed spring planting recommended if winter weeds abundant favors high populations of pocket gophers
		ANNUAL GRASSES		
barley *Hordeum vulgare*	accumulates ca 40 lb N/acre[4] cultivars available for all growing areas, fairly drought tolerant mowing delays and prolongs flowering strong root system penetrates over 6 feet crop residue improves water infiltration rate plant in fall or winter, flowers April-July; does not reseed well	aphid natural enemies build up on grain aphids that are not harmful to almonds may reduce root lesion nematode populations	host for root knot nematodes[2] reduces *P. vulnus* populations host for ring nematode	highly competitive with weeds when seeded at high rate mow or till before seed set to avoid attracting rodents

Table 21. Continued

Cover Crop	Horticultural benefits and requirements	Effects on arthropod pests and natural enemies	Effects on diseases and nematode pests	Effects on weeds and vertebrate pests
		ANNUAL GRASSES (continued)		
Blando brome (soft chess) *Bromus mollis*	adapted to all areas below 3,000 feet tolerates mowing to 2 inches for frost protection; stop by early April for seed set improves soil tilth when mowed or cultivated plant in early fall; will reseed	aphid natural enemies may build up on grain aphids that are not harmful to almonds	relatively poor host for root knot nematodes[2] host for ring nematode	highly competitive with weeds
annual fescue *Festuca megalura* *F. myuros*	adapted to wide range of soils and climate tolerates frequent mowing until flowering fibrous root system improves water infiltration and soil structure plant mid-fall, flowers March-June; will reseed if not mowed after flowering begins	harbors few beneficials		highly competitive with weeds
oats *Avena sativa*	accumulates about 12 lb N/acre[4] less tolerant of drought and cold than other cereals fibrous root system improves soil drainage plant fall to mid-winter, flowers April-May	natural enemies of aphids build up on grain aphids that are not harmful to almonds	host for root knot nematodes[2]	less competitive than barley; better as a companion crop with a legume such as vetch or bell bean
cereal rye *Secale cerealis*	accumulates ca 15 lb N/acre[4] good for reducing nitrate leaching tolerates wide range of soil and climatic conditions fibrous root system improves soil drainage, best cereal crop for this purpose tolerates close mowing in winter mow or till while stems still sweet to minimize tie-up of N plant late fall, flowers April-May	natural enemies of aphids build up on grain aphids that are not harmful to almonds	host for root knot nematodes[2]	highly competitive with weeds works well as a companion crop with legumes
annual ryegrass *Lolium multiflorum* *L. rigidum*	accumulates ca 25 lb N/acre[4] excellent for reducing nitrate leaching does well on heavy soils fibrous root system good for improving soil drainage tolerates close mowing in winter plant in fall, flowers May-June; will reseed	harbors few beneficials		reduces *Verticillium* levels in soil when grown without broadleaf companions highly competitive with weeds works well as a companion crop with legumes
sudangrass *Sorghum sudanense*	summer annual useful for improving soil structure can be mowed or flailed several times during growing season	grain aphids and associated beneficials host for southern green stink bug	host for northern root knot nematode, *M. hapla* reduces *P. vulnus* reduces ring nematode	may reduce levels of soil-borne pathogens when worked in as green manure good suppression of summer annual weeds

1. Actual amount of nitrogen (N) made available to orchard depends on growing conditions, how much of the cover crop biomass is worked into the soil, and when the cover crop is worked in.
2. Planting cover crops in fall after soil temperature is below 64°F (18°C) or turning under the cover crop before soil temperature reaches 60°F (15.5°C) in the spring will prevent increase of root knot nematode populations.
3. SRN = suspected host for ring nematode.
4. Grasses reduce soil nitrogen levels, accumulating the N in their biomass. Most of the N is released back into the soil when the grass cover crop decomposes.

Much of the information in this table was obtained from the cover crops database that is available under "Cover Crop Resources" at the University of California Sustainable Agriculture Research and Education Project's World Wide Web site: http://www.sarep.ucdavis.edu/

Suggested Reading

Sources for UC Publications Listed

UC DANR: Communication Services, University of California Division of Agriculture and Natural Resources, 6701 San Pablo Avenue, 2nd floor, Oakland, CA 94608. Free catalog on request. World Wide Web site, http://anrcatalog.ucdavis.edu/

UC IPM: University of California Statewide Integrated Pest Management Project, University of California, One Shields Ave., Davis, CA 95616-8621. World Wide Web site, http://www.ipm.ucdavis.edu/

General

Almond Production Manual. 1996. UC DANR Publication 3364.

Biological Control in the Western United States. 1995. UC DANR Publication 3361.

BIOS for Almonds. Community Alliance with Family Farmers Foundation, P.O. Box 363, Davis, CA 95617, and Almond Board of California.

Common-Sense Pest Control. 1991. W. Olkowski, S. Daar, and H. Olkowski. Taunton Press, Newtown, CT.

Cover Cropping in Vineyards: A Grower's Handbook. 1998. UC DANR Publication 3338.

Covercrops for California Agriculture. 1989. UC DANR Publication 21471.

Honey Bees in Almond Pollination. 1977. UC DANR Publication 2465.

IPM in Practice: Principles and Methods of Integrated Pest Management. 2001. UC DANR Publication 3418.

Pruning Fruit and Nut Trees. 1980. UC DANR Publication 21171.

Soil Solarization: A Nonpesticidal Method for Controlling Diseases, Nematodes, and Weeds. 1997. UC DANR Publication 21377.

Temperate Zone Pomology: Physiology and Culture. 3rd ed. 1993. M. N. Westwood. Timber Press, Portland, OR.

UC Fruit and Nut Information Center. World Wide Web site, http://fruitsandnuts.ucdavis.edu/

Soil, Water, Weather, and Nutrients

CIMIS: California Irrigation Management Information System. Department of Water Resources, P.O. Box 942836, Sacramento, CA, 94236-0001. (916) 653-9847. World Wide Web site, http://www.dpla.water.ca.gov/cimis.html/

DDU. Degree-Day Utility Version 2.3. 1994. Program and documentation for MS-DOS computers. Available from UC IPM.

Degree-days. Documentation and models for a range of pests available at UC IPM World Wide Web site, http://www.ipm.ucdavis.edu/

Determining Daily Reference Evapotranspiration (ETo). 1992. UC DANR Leaflet 21426.

Drip Irrigation Management. 1981. UC DANR Leaflet 21259.

An Easy Way to Calculate Degree-Days. 1986. UC DANR Publication 7174.

Irrigation Scheduling: A Guide for Efficient On-Farm Water Management. 1989. UC DANR Publication 21454.

Managing and Modifying Problem Soils. 1974. UC DANR Publication 2791.

Managing Compacted and Layered Soils. 1976. UC DANR Publication 2635.

Micro-Irrigation of Trees and Vines. 1999. UC DANR Publication 3378.

Organic Soil Amendments and Fertilizers. 1992. UC DANR Leaflet 21505.

Soil and Plant Tissue Testing in California. 1983. UC DANR Bulletin 1879.

Soil Temperatures in California. 1983. UC DANR Publication 1908.

Surface Irrigation. 1995. UC DANR Publication 3379.

Water-holding Characteristics of California Soils. 1989. UC DANR Leaflet 21463.

Western Fertilizer Handbook: Horticulture Edition. 1990. Interstate Printers and Publishers, Danville, IL.

Pesticide Application and Safety

Fumigation of Inhull Almonds on the Farm. 1980. Available from county Cooperative Extension offices or the Almond Board of California, P.O. Box 3130, Modesto, CA 95353.

The Illustrated Guide to Pesticide Safety, Worker's Packet. 1999. UC DANR Publication 21488.

The Illustrated Guide to Pesticide Safety, Instructor's Packet. 1999. UC DANR Publication 21489.

Integrated Pest Management for Stone Fruits. 1999. UC DANR Publication 3389.

Managing Insects and Mites with Spray Oils. 1991. UC DANR Publication 3347.

The Safe and Effective Use of Pesticides. 2nd ed. 2000. UC DANR Publication 3324.

UC IPM Pest Management Guidelines: Almond. Revised continuously. UC DANR Publication 3339. Also available from University of California Cooperative Extension offices and at the UC IPM World Wide Web site.

Insects and Mites

Biological Control and Insect Pest Management. 1979. UC DANR Publication 1911.

California Insects. 1979. J. A. Powell and C. L. Hogue. University of California Press, Berkeley.

Destructive and Useful Insects. 4th ed. 1962. C. L. Metcalf, W. P. Flint, and R. L. Metcalf. McGraw-Hill, New York.

Insect Identification Handbook. 1984. UC DANR Publication 4099.

Insect Pests of Farm, Garden, and Orchard. 7th ed. 1979. R. H. Davidson and W. F. Lyon. Wiley, New York.

Insects, Mites, and Other Invertebrates and Their Control in California. 1994. UC DANR Publication 4044.

An Introduction to the Study of Insects. 6th ed. 1989. D. J. Borror, C. A. Triplehorn, and N. F. Johnson. Saunders College Publishing, Philadelphia.

Natural Enemies Handbook: The Illustrated Guide to Biological Pest Control. 1998. UC DANR Publication 3386.

Diseases and Nematodes

Compendium of Temperate Zone Nut Crop Diseases. 2001. B. L. Teviotdale et al., eds. American Phytopathological Society, St. Paul.

Diseases of Temperate Zone Tree Fruit and Nut Crops. 1991. UC DANR Publication 3345.

General Recommendations for Nematode Sampling. 1981. UC DANR Publication 21234.

Phytonematology Study Guide. 1985. UC DANR Publication 4045.

Plant Pathology. 4th ed. 1997. G. N. Agrios. Academic Press, San Diego.

Weeds

Applied Weed Science. 1985. M. A. Ross and C. A. Lembi. Burgess, Minneapolis.

The Grower's Weed Identification Handbook. 1996. UC DANR Publication 4030.

Growers Weed Management Guide. 1989. H. M. Kempen. Thomson Publications, Fresno, CA.

Herbicide Resistance: Definition and Management Strategies. 2000. UC DANR Publication 8012[1].

How to Identify Plants. 1957. H. D. Harrington and L. W. Durrell. Sage Press, Denver.

Nontillage and Strip Weed Control in Almond Orchards. 1984. UC DANR Publication 2770.

Selective Chemical Weed Control. 1986. UC DANR Publication 1919.

Weeds of the West. 5th ed. 1996. T. D. Whitson, ed. Western Society of Weed Science, Newark, CA. Also available as UC DANR Publication 3350.

Weed Science. 3rd ed. 1996. W. P. Anderson. West Publishing, Minneapolis–St. Paul.

Vertebrates

The Audubon Society Field Guide to North American Birds: Western Region. 1977. M. D. F. Udvardy. A. A. Knopf, New York.

A Field Guide to Western Birds. 3rd ed. 1990. R. T. Peterson. Houghton Mifflin, Boston.

Vertebrate Pest Control Handbook. 4th ed. 1994. J. P. Clark, ed. California Department of Food and Agriculture, Division of Plant Industry.

Wildlife Pest Control around Gardens and Homes. 1984. UC DANR Publication 21385.

Sources of Beneficial Organisms

Directory of Least-Toxic Pest Control Products. Published annually in *The IPM Practitioner*. Biointegral Resource Center, Berkeley, CA.

Suppliers of Beneficial Organisms in North America. 1992. C. D. Hunter. California Department of Pesticide Regulation, Environmental Monitoring and Pest Management, Sacramento, CA.

A Worldwide Guide to Beneficial Animals Used for Pest Control Purposes. 1992. W. T. Thomson. Thomson Publications, Fresno, CA.

[1] Available at the UC DANR World Wide Web site and from UC Cooperative Extension county offices.

Glossary

abdomen. the posterior of the three main body divisions of an insect.

abiotic disorder. a disease caused by factors other than pathogens.

aestivation (estivation). a state of inactivity during the summer months.

air carrier sprayer. a sprayer that uses a blast of air to distribute a pesticide-water mixture into the canopy of the crop being treated (air blast sprayer).

allelopathy. chemical inhibition of one plant by another caused by the secretion of toxic substances.

allowable depletion. the proportion of available water that can be used before irrigation is needed (see Fig. 12).

annual. a plant that normally completes its life cycle of seed germination, vegetative growth, reproduction, and death in a single year.

anther. the pollen-producing organ of a flower (see Fig. 4).

anticoagulant. a substance that prevents blood clotting, resulting in internal hemorrhaging; may be used as a rodenticide.

auricles. the earlike projections at the base of leaves of some grasses used to identify species (see Fig. 58).

available moisture. the amount of water held in the soil that can be extracted by plants (see Fig. 12).

awn. a substantial hair or bristle that terminates a plant part.

axil. the upper angle between a leaf and the stem from which it is growing.

biofix. an identifiable event in the life cycle of a pest that signals when to begin degree-day accumulation or take a management action.

biotic disease. a disease caused by a pathogen, such as a bacterium, fungus, phytoplasma, or virus.

biotype. a strain of a species that has certain biological characters separating it from other individuals of that species.

blank. nut with no kernel—consists of only the collapsed pellicle (skin).

blight. a disease characterized by general and rapid killing of leaves, flowers, and branches.

bulb. an underground storage organ, composed chiefly of enlarged, fleshy leaf bases.

calcareous soils. soils containing high levels of calcium carbonate.

calibrate. to standardize or correct the measuring devices on instruments; to adjust nozzles on a spray rig properly.

calyx. the sepals of a flower, which enclose the unopened flower bud; often referred to as the "jacket" or "shuck" in almonds.

cambium. thin layer of undifferentiated, actively growing tissue between phloem and xylem (see Fig. 2)

canker. a dead, discolored, and often sunken area (lesion) on a root, trunk, stem, or branch.

caterpillar. the larva of a butterfly, moth, sawfly, or scorpionfly.

chorion. the outer membrane of an insect egg.

chlorophyll. the green pigment of plant cells, necessary for photosynthesis.

chlorosis. yellowing or bleaching of normally green plant tissue usually caused by the loss of chlorophyll.

cocoon. a sheath, usually mostly of silk, formed by an insect larva as a chamber for pupation.

collar region. in grasses, the region where the leaf blade and sheath meet; it is used in identifying species (see Fig. 58).

conidium (plural, conidia). an asexual fungal spore formed by fragmentation or budding at the tip of a specialized hypha.

cotyledons. the first leaves of the embryo formed within a seed and present on seedlings immediately after germination; seed leaves (see Fig. 58).

crawler. the active first instar of a scale insect.

crochets. tiny hooks on the prolegs of caterpillars.

cross-pollination. the transfer of pollen from the flowers of one plant to the flowers of another.

cross-resistance. in pest management, resistance of a pest population to a pesticide to which it *has not* been exposed that accompanies the development of resistance to a pesticide to which it *has* been exposed.

crown. the point at or just below the soil surface where the main stem (trunk) and roots join.

cultivar. a variety or strain developed and grown under cultivation.

degree-day (DD). a measurement unit that combines temperature and time used in calculating growth rates.

dehiscence. opening naturally and regularly along lines of weakness; in fruits, opening along sutures to release seed.

delayed-dormant. the treatment period in tree crops, beginning when buds begin to swell and continuing until the beginning of green tip.

developmental threshold. the lowest temperature at which growth occurs in a given species.

diapause. a period of physiologically controlled dormancy in insects.

disease. any disturbance of a plant that interferes with its normal structure, function, or economic value.

dormancy. a state of inactivity or prolonged rest.

economic threshold. a level of pest population or damage at which the cost of a control action equals the crop value gained from that control action.

ectoparasite. a parasite that lives on the outside of its host.

endoparasite. a parasite that lives inside its host.

endosperm. the tissue containing stored food in a seed that surrounds the embryo and is eventually digested by the embryo as it grows (see Fig. 5).

embryo. the small plantlet within the seed; in almond, the embryo develops into the kernel (see Fig. 5).

epidermis. the outermost layer of cells on the bodies of animals or on plant surfaces.

estivation. see *aestivation*.

evapotranspiration. the loss of soil moisture by the combination of soil surface evaporation and transpiration by plants.

feeder roots. the youngest roots with root hairs, most important in absorption of water and minerals.

field capacity. the moisture level in soil following saturation and runoff (see Fig. 12).

frass. a mixture of feces and food fragments produced by an insect in feeding.

fumigation. treatment with a pesticide active ingredient that is in gaseous form under treatment conditions.

girdle. to kill or damage a ring of bark tissue around a stem or root; such damage interrupts the transport of water and nutrients.

hibernaculum (plural, hibernacula). a shelter occupied during the winter by an insect, notably by the peach twig borer.

hibernation. passing the winter in a torpid or dormant state in which the body temperature and metabolic rate drop to very low levels.

honeydew. an excretion from insects, such as aphids, mealybugs, whiteflies, and soft scales, consisting of modified plant sap.

hypha (plural, hyphae). a tubular filament that is the structural unit of a fungus.

inflorescence. a flower cluster.

immune. incapable of being infected by a given pathogen.

indexing. testing a plant for a virus infection, usually by grafting tissue from it onto an indicator plant.

infection. the entry of a pathogen into a host and establishment of the pathogen as a parasite in the host.

inner bark. in older trees, the living part of the bark, comprised of phloem (see Fig. 2).

inoculum. any part or stage of a pathogen, such as a spore or virus particle, that can infect a host.

instar. an insect between successive molts; the first instar is between hatching and the first molt.

internode. the area of a stem between nodes.

invertebrate. an animal having no internal skeleton.

juvenile. in nematology, the immature form of a nematode that hatches from an egg and molts several times before becoming an adult.

larva (plural, larvae). the immature form of an insect that hatches from an egg, feeds, and then enters a pupal stage.

lesion. a well-defined area of diseased tissue, such as a canker or leaf spot.

ligule. in many grasses, a short membranous projection on the inner side of the leaf blade at the junction where the leaf blade and leaf sheath meet (see Fig. 58).

metamorphosis. a change in form during development.

microorganism. an organism of microscopic or small size.

microsclerotia (singular, microsclerotium). very small sclerotia such as those produced by the Verticillium wilt fungus.

molt. in insects and other arthropods, the shedding of skin before entering another stage of growth.

mummy. an unharvested nut or fruit remaining on the tree; the crusty skin of an aphid whose inside has been consumed by a parasite.

mutation. the abrupt appearance of a new, heritable characteristic as the result of a change in the genetic material of one individual cell.

mycelium (plural, mycelia). the vegetative body of a fungus, consisting of a mass of slender filaments called hyphae.

natural enemies. predators, parasites, or pathogens that are considered beneficial because they attack and kill organisms that we normally consider to be pests.

necrosis. the death of tissue accompanied by dark brown discoloration, usually occurring in a well-defined part of a plant such as the portion of a leaf between leaf veins, or in the xylem or phloem in a stem or tuber.

node. the slightly enlarged part of a stem where buds are formed and where leaves, stems, and flowers originate.

nucellus. in plants, the watery tissue composing the chief part of the young ovule in the flower and inside the seed during early development. It furnishes nutrients to the young embryo and is digested by the developing endosperm and embryo (see Fig. 5).

nymph. the immature stage of insects such as plant bugs and aphids that hatch from eggs and gradually acquire adult form through a series of molts without passing through a pupal stage.

outer bark. in older trees, the exterior part of the bark that is dead and sometimes cracked (see Fig. 2).

oviposit. to lay or deposit eggs.

pappus. the modified calyx of flowers in the aster, or sunflower, family; usually takes the form of bristles, scales, or awns.

parasite. an organism that lives in or on the body of another organism (the host); in this manual the term is also used to refer to insect parasitoids, which spend their immature stages on or within the body of a single host.

pathogen. a disease-causing organism.

peduncle. the stem of an individual flower or fruit.

pellicle. the covering (skin) that encloses the kernel; it is white during development but becomes brown at maturity.

perennials. plants that may live three or more seasons and flower at least twice.

pest resurgence. the increase of a pest population following a pesticide treatment to levels higher than before the treatment, as a result of the pesticide having killed natural enemies of the pest.

petiole. the stalk connecting the leaf to a stem.

pheromone. a chemical produced by an animal to communicate with other members of its species. Sex pheromones that attract the opposite sex for mating are used in monitoring certain insects.

pheromone confusion. pheromone mating disruption.

pheromone mating disruption. the disruption of the mating of certain lepidopterous (moth) pests by dispensing synthetic chemicals that mimic the pheromones produced by females to attract males.

phloem. the food-conducting tissue of a plant's vascular system (see Fig. 2).

phytoplasma. prokaryotic microorganism lacking a cell wall that proliferates in the phloem of host plants and may cause disease.

photosynthesis. the process whereby plants use light energy to form sugars and other compounds needed to support growth and development.

phytotoxic. causing injury to plants.

pistil. the female part of a flower, usually consisting of ovules, ovary, style, and stigma (see Fig. 4).

predator. an animal that attacks and feeds on other animals (the prey), usually eating most or all of the prey organism and consuming many prey during its lifetime.

preemergence herbicide. herbicides applied before target weeds emerge.

primary inoculum. the initial source of a pathogen that starts disease development in a given location.

propagule. any part of a plant from which a new plant can grow, including seeds, bulbs, rootstocks, etc.

protectant fungicide. a fungicide that protects a plant from infection by a pathogen.

prothorax. the anterior of the three thoracic segments of an insect.

pupa (plural, pupae). the nonfeeding, inactive stage between larva and adult in insects with complete metamorphosis.

pustule. small, blisterlike elevation of epidermis from which spores emerge.

resistant. able to tolerate conditions harmful to other individuals of the same species.

rhizome. a horizontal underground stem, especially one that roots at the nodes to produce new plants.

rootstock. an underground stem or rhizome; the lower portion of a graft that makes up the crown and root system.

rosette. a cluster of leaves arranged in a compact circular pattern, often at a shoot tip or on a shortened stem.

scion. the portion above a graft that becomes the trunk, branch, and tree top; the cultivar or variety used for that part of a graft.

sclerotium (plural, sclerotia). a compact mass of hyphae, sometimes including host tissue, capable of surviving unfavorable environmental conditions.

secondary infection. infection by microorganisms that enter the host through an injury caused previously by another pathogen.

secondary pest outbreak. the sudden increase in a pest population that is normally at low or nondamaging levels, caused by the destruction of natural enemies by treatment with a nonselective pesticide to control a primary pest.

sedges. a group of grasslike, herbaceous plants that, unlike grasses, have unjointed stems. Stems are usually solid and often triangular in cross-section.

seed leaf. the first leaf (grasses) or first two leaves (broadleaf plants) on a seedling; cotyledon (see Fig. 58).

seedling rootstock. a rootstock propagated from seed.

senescent. growing old; aging.

seta (plural, setae). a bristle.

sheath. the part of a grass leaf that encloses the stem below the collar region (see Fig. 58).

soil profile. a vertical section of the soil through all its horizontal layers, extending into the parent material.

spiracle. an external opening of the system of ducts, or tracheae, that serves as a respiratory system in insects.

spore. a reproductive body produced by certain fungi and other organisms capable of growing into a new individual under proper conditions.

stamen. a flower structure made up of the pollen-bearing anther and a stalk or filament (see Fig. 4).

sticktight. nut with hull firmly adhering to shell.

stigma. the receptive portion of the female flower part to which pollen adheres (see Fig. 4).

stolon. a stem that grows horizontally along the surface of the ground, often rooting at the nodes and forming new plants (see Fig. 58).

stoma (plural, stomata). a natural opening in a leaf surface that serves for gas exchange and water evaporation and has the ability to open and close in response to environmental conditions.

sucker. shoot arising from the trunk or rootstock.

suture. visible seam on hull.

systemic fungicide. a fungicide that can be translocated to some extent within a treated plant.

systemic herbicide. an herbicide that is able to move throughout a plant after being applied to leaf surfaces (*translocated herbicide*).

taproot. a large primary root that grows vertically downward, giving off small lateral roots.

tensiometer. an instrument that measures how tightly water is held by the soil; used for estimating water content of the soil.

thorax. the second of three major divisions in the body of an insect, and the one bearing the legs and wings.

tolerant. able to withstand the effects of a condition without suffering serious injury or death.

translocated herbicide. a systemic herbicide.

transpiration. the evaporation of water from plant tissue, usually through stomata.

treatment threshold. a level of pest population or damage, usually measured by a specified monitoring method, at which a pesticide application is recommended.

true leaf. any leaf produced after the cotyledons (see Fig. 58).

tuber. a much enlarged, fleshy underground stem.

turgid. having the cells fully distended with water.

vascular system. the system of plant tissues that conducts water, mineral nutrients, and products of photosynthesis through the plant, consisting of the xylem and phloem.

vector. an organism able to transport and transmit a pathogen to a host.

vegetative growth. growth of stems, roots, and leaves, but not flowers and fruits.

viroid. a portion of infectious nucleic acid without the protein coat of a virus.

virulent. capable of causing a severe disease; strongly pathogenic.

virus. a small infectious agent, consisting only of nucleic acid and a protein coat, that can reproduce only within the living cells of a host.

wing. extension of the nut shell at the suture line; varies in size according to cultivar.

xylem. plant tissue that conducts water and nutrients from the roots up through the plant (see Fig. 2).

zonate. marked with zones or bands; belted; striped.

zoospore. a motile spore.

Appendix
Almonds

Bloom stages of almonds.

Dormant Bud

First Swell

Green Tip

First Pink

Popcorn

Full Bloom

Petal Fall

Jacket Stage

Jacket Split

Index

Page numbers in **boldface** type indicate major discussions. Page numbers in *italic* type indicate illustrations.